High School Equivalency Test Preparation:

# MATHEMATICS

## Student Edition

**PAXEN**

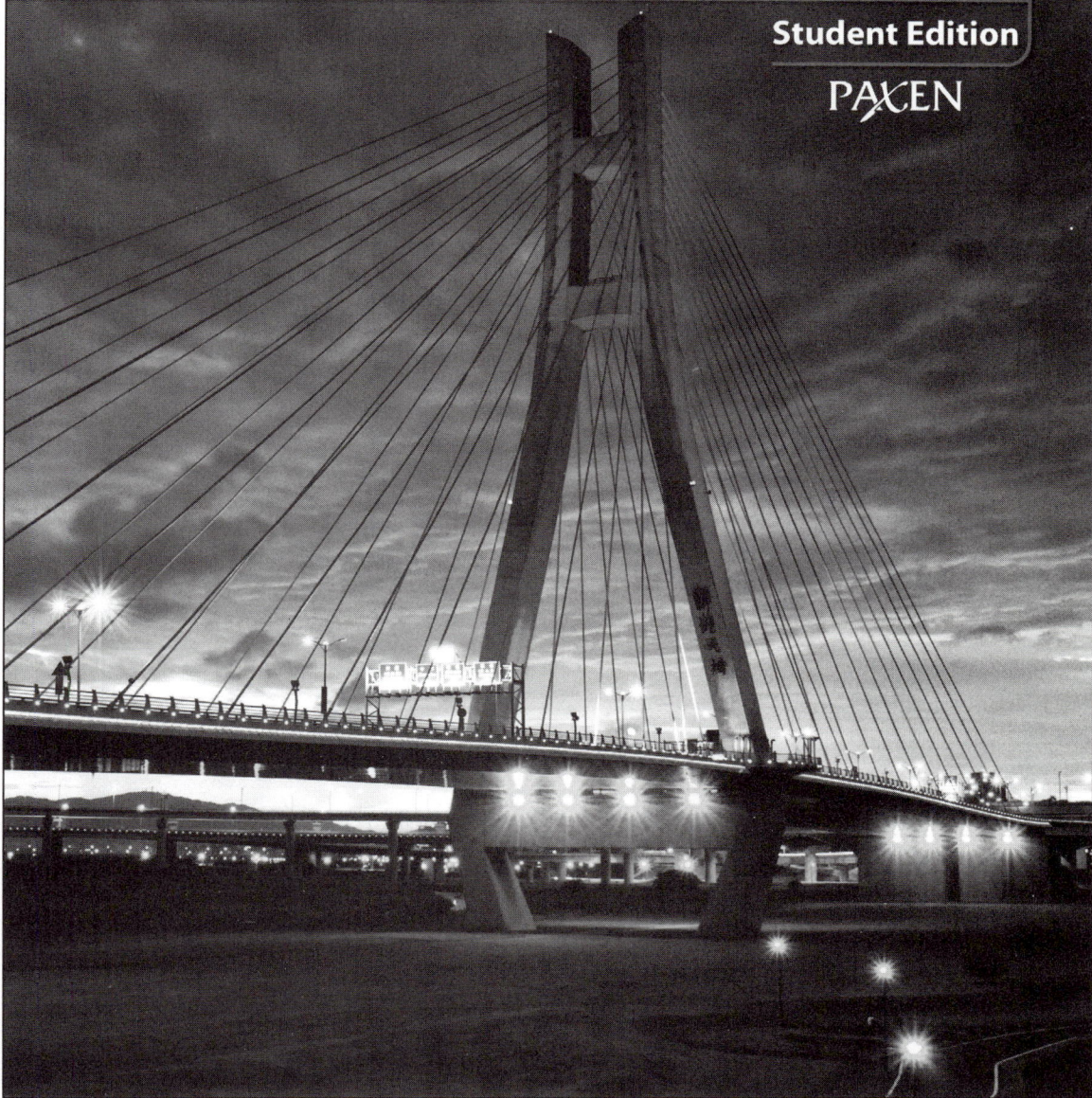

**PAXEN**

Melbourne, Florida
www.paxen.com

# Acknowledgments

For each of the selections and images listed below, grateful acknowledgment is made for permission to excerpt and/or reprint original or copyrighted material, as follows:

## Images

(cover, i) Daniel Aguilera/Getty Images **v** iStockphoto. **vi** iStockphoto.

ISBN: 978-1-934350-58-4

2 3 4 5 6 7    0304    20 19 18 17 16 15                                    Printed in the U.S.A.

4500572004

# High School Equivalency Test Preparation

**Mathematics Student Book**

# Table of Contents

# About High School Equivalency Tests

**S**imply by turning to this page, you've made a decision that will change your life for the better. Each year, thousands of people just like you decide to pursue a high school equivalency certificate. Like you, they left school for one reason or another. And now, just like them, you've decided to continue your education by studying for and taking the high school equivalency tests.

However, these tests are no easy task. The tests, five in all, are spread across the subject areas of Language Arts/Reading, Language Arts/Writing, Mathematics, Science, and Social Studies. Preparation for the tests can involve extensive study and review. The payoff, however, is significant: more and better career options, higher earnings, and the sense of achievement that comes with a high school equivalency certificate. Employers and colleges and universities accept the certificate as they would a high school diploma. On average, certificate recipients earn $10,000 more per year than do employees without a high school diploma or an equivalency certificate.

High school equivalency tests are designed to mirror a high school curriculum. Although you will not need to know all of the information typically taught in high school, you will need to answer a variety of questions in specific subject areas. In Language Arts/Writing, you will need to write an essay.

In all cases, you will need to effectively read and follow directions, correctly interpret questions, and critically examine answer options. The table below details the five subject areas. Since different states have different requirements for the number of tests you may take in a single day, you will need to check with your local adult education center for requirements in your state or territory.

| SUBJECT AREA TEST | CONTENT AREAS |
| --- | --- |
| Language Arts/Reading | Literary Texts<br>Informational Texts |
| Language Arts/Writing (Editing) | Organization of Ideas<br>Language Facility<br>Writing Conventions |
| Language Arts/Writing (Essay) | Development and Organization of Ideas<br>Language Facility<br>Writing Conventions |
| Mathematics | Numbers and Operations on Numbers<br>Data Analysis/Probability/Statistics<br>Measurement/Geometry<br>Algebraic Concepts |
| Science | Life Science<br>Earth/Space Science<br>Physical Science |
| Social Studies | History<br>Civics/Government<br>Economics<br>Geography |

Three of the subject-area tests—Language Arts/Reading, Science, and Social Studies—will require you to answer questions by interpreting passages. The Science and Social Studies Tests also require you to interpret tables, charts, graphs, diagrams, timelines, political cartoons, and other visuals. In Language Arts/Reading, you also will need to answer questions based on workplace and consumer texts. The Mathematics Test will require you to use basic computation, analysis, and reasoning skills to solve a variety of word problems, many of them involving graphics. On the Mathematics Test, questions will be multiple-choice with four or five answer options. An example follows:

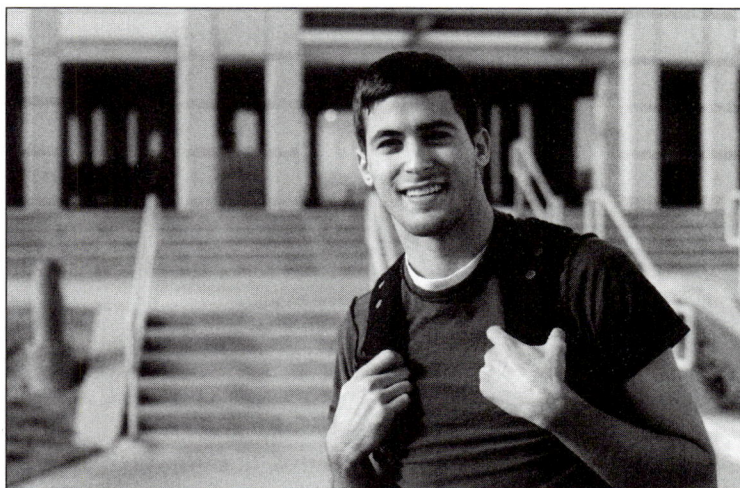

On the other four tests, you will have four answer options for each multiple-choice question.

As the table on page iv indicates, the Language Arts/Writing Test contains two parts, one for editing and the other for essay. In the editing portion of Language Arts/Writing, you will be asked to identify and correct common errors in various passages and texts while also deciding on the most effective organization of a text. In the essay portion, you will write an essay that analyzes texts or provides an explanation or an opinion on a single topic of general knowledge.

So now that you understand the task at hand—and the benefits of a high school equivalency certificate—you must prepare for the tests. In the pages that follow, you will find a recipe of sorts that, if followed, will help guide you toward successful completion of your certificate. So turn the page. The next chapter of your life begins right now.

# About *High School Equivalency Test Preparation*

**A**long with choosing to pursue your high school equivalency certificate, you've made another smart decision by selecting this program as your main study and preparation tool. Simply by purchasing *High School Equivalency Test Preparation,* you've joined an elite club with thousands of members, all with a common goal—earning their high school equivalency certificates. In this case, membership most definitely has its privileges.

For more than 70 years, high school equivalency tests have offered a second chance to people who need it most. To date, more than 17 million Americans like you have studied for and earned high school equivalency certificates and, in so doing, jump-started their lives and careers. Benefits abound for certificate holders. Recent studies have shown that people with certificates earn more money, enjoy better health, and exhibit greater interest in and understanding of the world around them than do those without.

In addition, many certificate recipients plan to further their educations, which will provide them with more and better options. As if to underscore the point, the U.S. government's Division of Occupational Employment Projections estimates that through 2022, about 3.1 million new jobs will require a bachelor's degree for entry.

Your pathway to the future—a *brighter* future—begins now, on this page, with *High School Equivalency Test Preparation.* Unlike other programs, which take months to teach through a content-based approach, *High School Equivalency Test Preparation* gets to the heart of the tests—and quickly—by emphasizing *concepts.* At their core, the majority of the tests are reading-comprehension exams. Test-takers must be able to read and interpret excerpts, passages, and various visuals—tables, charts, graphs, timelines, and so on—and then answer questions based upon them.

*High School Equivalency Test Preparation* shows you the way. By emphasizing key reading and thinking concepts, *High School Equivalency Test Preparation* equips learners like you with the skills and strategies you'll need to correctly interpret and answer questions on the tests. Five-page lessons in each student book provide focused and efficient instruction, while callout boxes, sample exercises, and test-taking and other thinking strategies aid in understanding complex concepts.

Unlike other high school equivalency test preparation materials, which were designed *for* the classroom, these materials were designed *from* the classroom, using proven educational theory and cutting-edge classroom philosophy. For learners who have long had the deck stacked against them, the odds are finally in their favor. And yours.

## HIGH SCHOOL EQUIVALENCY TESTS— FAST FACTS

- About 800,000 people take high school equivalency exams each year.
- Workers with a high school equivalency certificate earn an average of $10,000 a year more than people without a high school diploma or its equivalent.
- Over 3,000,000 students drop out of high school each year.
- Over 85% of Americans have a high school diploma or its equivalent.
- High school dropouts are not eligible for 90% of U.S. jobs.

# About *High School Equivalency Test Preparation: Mathematics*

For those who think mathematics high school equivalency tests are a breeze, think again. The mathematics test is a rigorous exam that will assess your ability to answer a range of questions about various topics. You will answer questions organized across four main content areas: Numbers and Operations on Numbers; Data Analysis/Probability/Statistics; Measurement/Geometry; and Algebraic Concepts. Material in *High School Equivalency Test Preparation: Mathematics* has been organized with these content areas in mind.

*High School Equivalency Test Preparation: Mathematics* helps learners like you build and develop core mathematics skills. A combination of targeted strategies, informational callouts and sample questions, assorted test-taking tips, and ample assessment help to clearly focus study efforts in needed areas, all with an eye toward success on high school equivalency tests.

You may use a calculator to aid you in answering questions throughout the mathematics test. In addition, formulas, such as those shown on the following pages will be supplied.

The **Learn the Skill** section defines and provides additional information about the skill to be studied.

**Callouts** provide strategies and information that you may use to understand and interpret various passages or graphics.

**Test-Taking Tips** offer broad or specific support for answering multiple-choice questions.

---

**LESSON 1**

UNIT 1

## Place Value

### 1 Learn the Skill

**Whole numbers** are written with the digits 0 through 9. A **decimal** is another way to write a fraction. It uses the base-ten place value system. The value of a digit in a number depends on its place. Decimals include place values such as tenths, hundredths, and thousandths. You can compare and order numbers using place value. In some problems, it is helpful to round numbers to a certain place value.

### 2 Practice the Skill

To solve problems on high school equivalency mathematics tests, you must understand place value and how to read, write, compare, order, and round numbers. Read the examples and strategies below. Use this information to answer question 1.

**A** Whole numbers are to the left of the decimal point and the decimals are to the right. Each place in a number is worth 10 times as much as the place to its right and one-tenth as much as the place to its left.

**B** When you compare numbers, the number with the most digits in the whole-number part is greater. For numbers with the same number of places, compare digits with the same place value. If the number of decimal places is different, insert zeros to fill the empty places. Use these symbols when comparing numbers:
= means equals
> means is greater than
< means is less than

| Thousands | | | Units | | | Decimals | | | |
|---|---|---|---|---|---|---|---|---|---|
| hundreds | tens | ones | hundreds | tens | ones | tenths | hundredths | thousandths | ten thousandths |
| | | | | 2 | 0 | . | 3 | 9 | 8 | 1 |

*Multiply by 10 for each cell you move left*

*Divide by 10 for each cell you move right*

43,041 < 43,062      14.359 < 14.370
0.285 > 0.231      12.46 > 11.59

The number 3,062.7 is read and written in words as **three thousand, sixty-two and seven tenths**. When rounded to the hundreds place, 3,062.7 is **3,100**.

☑ **TEST-TAKING TIPS**

Circle the digit you want to round. If the digit to its right is greater than or equal to 5, add 1 to the circled digit. If it is less than 5, do not change it. Change the digits to the right of the rounded digit to zeros.

1  On her income tax form, Val rounds her income to the nearest dollar. If her income is $30,237.59, what will she write on her income tax form?
A  $30,237.00
B  $30,237.50
C  $30,237.60
D  $30,238.00
E  $30,240.00

2                                                                 Lesson 1 | Place Value

---

# Mathematics Formulas

## Measurement/Geometry

### Perimeter

| | | |
|---|---|---|
| Square | Perimeter = 4 × side | $P = 4s$ |
| Rectangle | Perimeter = 2 × length + 2 × width | $P = 2l + 2w$ |
| Triangle | Perimeter = side$_1$ + side$_2$ + side$_3$ | $P = s_1 + s_2 + s_3$ |

### Circumference

| | | |
|---|---|---|
| Circle | Circumference = π × diameter | $C = \pi d$ |
| | Circumference = 2 × π × radius | $C = 2\pi r$ |

### Area

| | | |
|---|---|---|
| Square | Area = side$^2$ | $A = s^2$ |
| Rectangle | Area = length × width | $A = lw$ |
| Parallelogram | Area = base × height | $A = bh$ |
| Triangle | Area = $\frac{1}{2}$ × base × height | $A = \frac{1}{2}bh$ |
| Trapezoid | Area = $\frac{1}{2}$ × (base$_1$ + base$_2$) × height | $A = \frac{1}{2}(b_1 + b_2)h$ |
| Circle | Area = π × radius$^2$ | $A = \pi r^2$ |

### Surface Area

| | | |
|---|---|---|
| Cube | Surface area = 6 × area of base | $SA = 6B = 6s^2$ |
| Prism<br>• Rectangular prism | Surface area = area of 2 bases + area of faces<br>(*P* is perimeter of base) | $SA = 2B + Ph$<br>• $SA = 2lw + 2lh + 2wh$ |
| Pyramid | Surface area = area of base + area of faces<br>(*P* is perimeter of base, *s* is slant height) | $SA = B + \frac{1}{2}Ps$ |
| Cylinder | Surface area = area of 2 circular bases +<br>lateral area (*LA* is circumference × height) | $SA = 2\pi r^2 + 2\pi rh$ |
| Cone | Surface area = area of base + area of curved<br>surface (*s* is slant height) | $SA = \pi r^2 + \pi rs$ |
| Sphere | Surface area = 4 × π × radius$^2$ | $SA = 4\pi r^2$ |

### Volume

| | | |
|---|---|---|
| Cube | Volume = edge$^3$ | $V = e^3$ |
| Prism<br>• Rectangular prism | Volume = area of base × height<br>• Volume = length × width × height | $V = Bh$<br>• $V = lwh$ |
| Pyramid | Volume = $\frac{1}{3}$ × area of base × height | $V = \frac{1}{3}Bh$ |
| Cylinder | Volume = π × radius$^2$ × height | $V = \pi r^2 h$ |
| Cone | Volume = $\frac{1}{3}$ × π × radius$^2$ × height | $V = \frac{1}{3}\pi r^2 h$ |
| Sphere | Volume = $\frac{4}{3}$ × π × radius$^3$ | $V = \frac{4}{3}\pi r^3$ |

### Pythagorean theorem

| | | |
|---|---|---|
| Right triangles only | *a* and *b* are the lengths of legs, and *c* is the<br>length of the hypotenuse | $a^2 + b^2 = c^2$ |

## Numbers and Operations on Numbers

| | | |
|---|---|---|
| Distributive Property | Multiply the terms inside parentheses by the factor outside the parentheses. | $a(b + c) = ab + ac$ |
| Percent | Base × rate = part | $br = p$ |
| Simple interest | Interest = Principal × rate × time | $I = Prt$ |
| Distance | Distance = rate × time | $d = rt$ |
| Total cost | Total cost = (number of units) × (price per unit) | $c = np$ |

### Exponents

| | | |
|---|---|---|
| Addition | Add only expressions with the same base and exponent. | $5a^m + 3a^m = 8a^m$ |
| Subtraction | Subtract only expressions with the same base and exponent. | $5a^m - 3a^m = 2a^m$ |
| Multiplication | Add exponents when multiplying expressions with the same base. | $a^m \bullet a^n = a^{m+n}$ |
| Division | Subtract exponents when dividing expressions with the same base. | $\dfrac{a^m}{a^n} = a^{m-n}$ |

## Data Analysis/Probability/Statistics

### Measures of Central Tendency and Variation

| | | |
|---|---|---|
| Mean | The sum of the data values divided by the number of data values. | $\dfrac{x_1 + x_2 + \ldots + x_n}{n}$ |
| Median | The middle value of an odd number of ordered scores, and halfway between the two middle values of an even number of ordered scores. | |
| Mode | The most common value in a data set. | |
| Range | The difference between the greatest value and the least value in a data set. | $g - l$ |

### Other Formulas

| | | |
|---|---|---|
| Fundamental Counting Principle | One choice has $m$ options; another choice has $n$ options. | number of outcomes = $m \bullet n$ |
| Permutation without repetition | For $n$ options: $n!$ (read "n factorial") | $n! = n(n - 1)(n - 2) \ldots \bullet 2 \bullet 1$ |
| Permutation with repetition | There are $n$ options for $x$ choices. | $n^x$ |
| Combination | A subgroup of $x$ from a group of $n$ | $\dfrac{n! \text{ to } x \text{ places}}{x!}$ |
| Probability | The probability of an event $E$ | $P(E) = \dfrac{\text{favorable outcomes}}{\text{possible outcomes}}$ |

### Algebraic Concepts

| | |
|---|---|
| Distance between two points on a coordinate plane | $d = \sqrt{(x_2 - x_1)^2 + (y_2 - y_1)^2}$ |
| Slope of a line | $m = \dfrac{y_2 - y_1}{x_2 - x_1}$ |
| Slope-intercept form of a linear equation | $y = mx + b$ |
| Point-slope form of a linear equation | $y - y_1 = m(x - x_1)$ |
| Quadratic equation | $ax^2 + bx + c = 0$ |
| Quadratic function | $y = ax^2 + bx + c$ |

# Test-Taking Tips

High School Equivalency tests include 240 questions across the five subject-area exams of Language Arts/Reading, Language Arts/ Writing, Mathematics, Science, and Social Studies. In each test, you will need to apply some amount of subject-area knowledge. However, because all of the questions are multiple-choice items largely based on text or visuals (such as tables, charts, or graphs) or on understanding word problems, the emphasis in *High School Equivalency Test Preparation* is on helping learners like you build and develop core reading and thinking skills. As part of the overall strategy, various test-taking tips are included below and throughout the book to help you improve your performance on the tests. For example:

◆ *Always thoroughly read the directions so that you know exactly what to do.* For example, on the mathematics test, direction lines explicitly state which questions are to be answered using information in a table or other visual. Pay attention to these directions in order to make sure you are correctly matching up these test elements.

◆ *Read each question carefully so that you fully understand what it is asking.* Some questions, for example, may present more information than you need to correctly answer them. Other questions may not provide enough information in order for you to formulate an answer.

◆ *Manage your time with each question.* Because the tests are timed exams, you'll want to spend enough time with each question, but not *too* much time. You can save time by first reading each question and its answer options before examining an accompanying graphic. Once you understand what the question is asking, review the visual for the appropriate information.

◆ *Note any unfamiliar words in questions.* First, attempt to reread the question by omitting the unfamiliar word(s). Next, try to substitute another word in its place.

◆ *Answer all questions, regardless of whether you know the answer or are guessing at it.* There is no benefit in leaving questions unanswered. Keep in mind the time that you have for each test and manage it accordingly. For time purposes, you may decide to initially skip questions. However, note each with a light mark beside the question and try to return to and answer it before the end of the test.

◆ *Narrow answer options by rereading each question and the text or graphic that goes with it.* Although all five answers are *possible,* keep in mind that only one of them is *correct.* You may be able to eliminate one or two answers immediately; others may take more time and involve the use of either logic or assumptions. In some cases, you may need to make your best guess between two options. If so, keep in mind that test makers often avoid answer patterns; that is, if you know the previous answer is B and are unsure of the answer to the next question but have narrowed it to options B and D, you may want to choose D.

◆ *Read all answer choices.* Even though the first or second answer choice may appear to be correct, be sure to thoroughly read all five answer choices. Then go with your instinct when answering questions. For example, if your first instinct is to mark A in response to a question, it's best to stick with that answer unless you later determine that answer to be incorrect. Usually, the first answer you choose is the correct one.

◆ *Correctly complete your answer sheet by marking one lettered space on the answer sheet beside the number that corresponds to it.* Mark only one answer for each item; multiple answers will be scored as incorrect. If time permits, double-check your answer sheet after completing the test to ensure that you have made as many marks— no more, no less—as there are questions.

**Y**ou've already made two very smart decisions in trying to earn your high school equivalency certificate and in purchasing *High School Equivalency Test Preparation* to help you to do so. The following are additional strategies to help you optimize your success on the tests.

## 3 weeks out ...

◆ Set a study schedule. Choose times in which you are most alert, and places, such as a library, that provide the best study environment.

◆ Thoroughly review all material in *High School Equivalency Test Preparation.*

◆ Make sure that you have the necessary tools for the job: sharpened pencils, pens, paper, and, for mathematics, a calculator.

◆ Keep notebooks for each of the subject areas that you are studying. Folders with pockets are useful for storing loose papers.

◆ When taking notes, restate thoughts or ideas in your own words rather than copying them directly from a book. You can phrase these notes as complete sentences, as questions (with answers), or as fragments, provided you understand them.

## 1 week out ...

◆ Take the pretests, noting any troublesome subject areas. Focus your remaining study around those subject areas.

◆ Prepare the items you will need for test day: admission ticket (if necessary), acceptable form of identification, some sharpened No. 2 pencils (with erasers), a watch, eyeglasses (if necessary), a sweater or jacket, and a high-protein snack to eat during breaks.

◆ Map out the course to the test center, and visit it a day or two before your scheduled exam. If you drive, find a place to park at the center.

◆ Get a good night's sleep the night before the tests. Studies have shown that learners with sufficient rest perform better in testing situations.

## The day of ...

◆ Eat a hearty breakfast high in protein. As with the rest of your body, your brain needs ample energy to perform well.

◆ Arrive 30 minutes early to the testing center. This will allow sufficient time in the event of a change to a different testing classroom.

◆ Pack a sizeable lunch, especially if you plan to be at the testing center most of the day.

◆ Focus and relax. You've come this far, spending weeks preparing and studying for the tests. It's your time to shine.

# Unit 1

## Unit Overview

You are surrounded by numbers. Whether paying bills, negotiating a car loan, budgeting for rent or groceries, depositing a check, or withdrawing money, you use basic math skills such as addition, subtraction, multiplication, and division to perform a variety of everyday tasks.

In the same way, numbers and operations on numbers play an important part on mathematics high school equivalency tests, making up about 25 percent of questions. In Unit 1, you will study whole numbers, word problems, fractions, ratios, decimals, percents, squares and cubes, and exponents and scientific notation, all of which will help you prepare for the mathematics test.

## Table of Contents

# Numbers and Operations on Numbers

## Key Terms

**base (number):** the whole amount in a percent problem; the number that is multiplied by itself when associated with an exponent

**common denominator:** the same bottom number shared by two or more fractions; for example, $\frac{1}{3}$ and $\frac{2}{3}$

**cross product:** method by which, in a proportion, the numerator of one fraction is multiplied by the denominator of the other fraction, producing equal products

**cube root:** that number which, when cubed, equals the given number

**cubing:** multiplying a number by itself three times, such as $(2 \times 2 \times 2)$; represented as $2^3$

**decimal:** another way to write a fraction based on the base-ten place value system; for example, $\frac{1}{4}$ would become .25

**denominator:** the bottom number in a fraction that tells the number of equal parts in a whole

**difference:** the answer to a subtraction problem

**dividend:** the initial quantity in a division problem (the number being divided)

**divisor:** the number by which the initial quantity, or dividend, is divided in a division problem

**exponent:** a superscript indicator that tells how many times a base number is used in a multiplication; for example, $4^3 = 4 \times 4 \times 4$

**fraction:** part of a whole or part of a group; indicated by separating two numbers with a fraction bar

**improper fraction:** a fraction in which the numerator is larger than denominator; should be written as a mixed number, for example $\frac{5}{4}$ converted to $1\frac{1}{4}$

**integers:** another name for whole numbers, including positive whole numbers (1, 2, 3, . . .), their opposites (−1, −2, −3, . . .), and zero (0)

**mixed number:** a fraction that includes a whole number, such as $1\frac{2}{5}$

**numerator:** the top number in a fraction that tells the number of equal parts being considered

**order of operations:** the rule for which action is performed first in a series of steps to solve a mathematical expression; parentheses, exponents and roots, multiplication and division, addition and subtraction

**part:** a piece of the whole or base in a percent problem

**percent:** a way to show part of a whole, comparing the partial amount to 100; for example, 50% is 50 parts out of a whole 100

**power:** tells the number of times a base number is used in a multiplication, expressed by an exponent

**product:** the answer to a multiplication problem

**proper fraction:** a fraction, such as $\frac{3}{4}$, in which denominator is larger than the numerator

**proportion:** an equation with equal ratios on each side

**quotient:** the answer to a division problem

**rate:** tells how the base and the whole are related in a percent problem

**ratio:** a comparison of two numbers; written as a fraction, by using the word *to*, or with a colon (:)

**scientific notation:** method of writing very large or very small numbers in a compact form through the use of exponents and powers of 10

**square root:** a number that, when multiplied by itself, equals a given number

**squaring:** multiplying a number by itself one time; for example, $(2 \times 2)$ which is represented as $2^2$

**sum:** the answer to an addition problem

**unit rate:** a ratio with a denominator of 1

**whole numbers:** positive numbers written with the digits 0 through 9 and without fraction or decimal parts

# Place Value

## ① Learn the Skill

**Whole numbers** are written with the digits 0 through 9. A **decimal** is another way to write a fraction. It uses the base-ten place value system. The value of a digit in a number depends on its place. Decimals include place values such as tenths, hundredths, and thousandths. You can compare and order numbers using place value. In some problems, it is helpful to round numbers to a certain place value.

## ② Practice the Skill

To solve problems on high school equivalency mathematics tests, you must understand place value and how to read, write, compare, order, and round numbers. Read the examples and strategies below. Use this information to answer question 1.

**A** Whole numbers are to the left of the decimal point and the decimals are to the right. Each place in a number is worth 10 times as much as the place to its right and one-tenth as much as the place to its left.

| Thousands | | | Units | | | Decimals | | | |
|---|---|---|---|---|---|---|---|---|---|
| hundreds | tens | ones | hundreds | tens | ones | tenths | hundredths | thousandths | ten thousandths |
| | | | | 2 | 0 . | 3 | 9 | 8 | 1 |

*Multiply by 10 for each cell you move left*

*Divide by 10 for each cell you move right*

**B** When you compare numbers, the number with the most digits in the whole-number part is greater. For numbers with the same number of places, compare digits with the same place value. If the number of decimal places is different, insert zeros to fill the empty places. Use these symbols when comparing numbers:
= means equals
> means is greater than
< means is less than

43,041  <  43,062        14.359  <  14.370
0.285  >  0.231        12.46  >  11.59

The number 3,062.7 is read and written in words as **three thousand, sixty-two and seven tenths.** When rounded to the hundreds place, 3,062.7 is **3,100.**

### ✓ TEST-TAKING TIPS

Circle the digit you want to round. If the digit to its right is greater than or equal to 5, add 1 to the circled digit. If it is less than 5, do not change it. Change the digits to the right of the rounded digit to zeros.

1  On her income tax form, Val rounds her income to the nearest dollar. If her income is $30,237.59, what will she write on her income tax form?
   A  $30,237.00
   B  $30,237.50
   C  $30,237.60
 ↗ D  $30,238.00
   E  $30,240.00

**2** For a science experiment, Ryan measures the distance a ball travels when rolled down a ramp. His measurement is 3.27 meters. He needs to record his measurement to the nearest tenth of a meter. What measurement will he record?

  **A** 3 meters
  **B** 3.2 meters
  **C** 3.27 meters
  **D** 3.3 meters
  **E** 4 meters

**Directions:** Questions 3 and 4 are based on the information below.

The table below shows a sporting goods store's monthly sales for the first six months of the year.

| MONTHLY SALES | |
|---|---|
| January | $155,987 |
| February | $150,403 |
| March | $139,605 |
| April | $144,299 |
| May | $149,355 |
| June | $148,260 |

**3** In which month did the store have its highest sales?

  **A** January
  **B** February
  **C** March
  **D** April
  **E** May

**4** What sales trend can you determine?

  **A** People purchased the most sporting goods equipment during early spring.
  **B** Sales were at their highest in winter months.
  **C** Monthly sales remained the same from January through June.
  **D** People purchased more sporting goods as summer approached.
  **E** Sales suffered in the months immediately following the winter holidays.

**Directions:** Questions 5 and 6 are based on the information below.

Sliced deli meat is sold by the pound. Shana bought five different meats at the deli.

| DELI MEAT | WEIGHT |
|---|---|
| Chicken | 1.59 pounds |
| Turkey | 2.07 pounds |
| Ham | 1.76 pounds |
| Salami | 2.48 pounds |
| Roast beef | 2.15 pounds |

**5** Which package of deli meat weighed the least?

  **A** chicken
  **B** turkey
  **C** ham
  **D** salami
  **E** roast beef

**6** How many packages of deli meat weighed less than 2.25 pounds?

  **A** 1
  **B** 2
  **C** 3
  **D** 4
  **E** 5

**7** Colin is packing the car with three pieces of luggage. He packs the heaviest piece into the car first, then the next heaviest, and finally the lightest piece on top. The weights of the three pieces of luggage are 25.57 pounds, 24.30 pounds, and 25.98 pounds. According to the weights, in which order will he pack the luggage into the car?

  **A** 24.30 pounds, 25.57 pounds, 25.98 pounds
  **B** 25.98 pounds, 24.30 pounds, 25.57 pounds
  **C** 25.98 pounds, 25.57 pounds, 24.30 pounds
  **D** 25.57 pounds, 24.30 pounds, 25.98 pounds
  **E** 24.30 pounds, 25.57 pounds, 24.30 pounds

**Directions:** Questions 8 and 9 are based on the information below.

The seats in a large auditorium are identified by both numbers and letters. The range of numbers for each letter row is shown in the chart.

| SEAT NUMBERS | |
| --- | --- |
| LETTER OF ROW | NUMBER RANGE |
| A | 100–250 |
| B | 251–500 |
| C | 501–750 |
| D | 751–1000 |
| E | 1001–1250 |
| F | 1251–1500 |

8   In which row will you sit if your seat number is 1107?
  A   Row A
  B   Row B
  C   Row C
  D   Row E
  E   Row F

9   How many rows have seat numbers in the thousands?
  A   1
  B   2
  C   3
  D   4
  E   5

10   In a dive meet, Morgan finished in second place. The first-place diver scored 218.65 points. The third-place diver scored 218.15 points. Which of the following could be Morgan's score?
  A   218.00 points
  B   218.05 points
  C   218.45 points
  D   218.75 points
  E   218.85 points

**Directions:** Questions 11 and 12 are based on the information below.

In a dictionary, the following letters can be found on the following pages:
  **P**—pages 968–1096
  **Q**—pages 1097–1105
  **R**—pages 1105–1178
  **S**—pages 1178–1360
  **T**—pages 1360–1447

11   With which letter does a word found on page 1100 begin?
  A   P
  B   Q
  C   R
  D   S
  E   T

12   Words that begin with the letter S can be found on what pages?
  A   998–1045
  B   1046–1105
  C   1117–1165
  D   1234–1287
  E   1293–1376

13   Callie's job involves data entry. Which digits should she type in for the number twelve thousand, eight hundred two and twenty hundredths?
  A   1, 2, 8, 0, 2, 2, 0
  B   1, 2, 8, 2, 0, 2, 0
  C   1, 2, 0, 8, 0, 2, 2
  D   1, 2, 8, 0, 0, 2, 0, 2
  E   1, 2, 0, 8, 2, 0, 2, 0

**14** Romy's scores for her five social studies quizzes this semester are shown below.

$$98, 75, 84, 92, 95$$

What is the order of her social studies quiz scores from greatest to least?

A 75, 84, 92, 95, 98
B 75, 92, 95, 84, 98
C 95, 92, 98, 84, 75
D 98, 95, 84, 92, 75
E 98, 95, 92, 84, 75

**15** Calvin is writing an essay. He wants to write the sentence "150,218 people lived in the city in 2014," but a good rule for writing is to avoid beginning a sentence with a numeral. What is a better way for Calvin to write his sentence?

A One hundred fifty thousand, two eighteen people lived in the city in 2014.
B One hundred fifty, two hundred eighteen people lived in the city in 2014.
C One thousand fifty, two hundred eighteen people lived in the city in 2014.
D One hundred fifty thousand, twenty-one eight people lived in the city in 2014.
E One hundred fifty thousand, two hundred eighteen people lived in the city in 2014.

**16** A museum tracks its visitors each month. For its records, the museum rounds the number of monthly visitors to the nearest hundred. If 8,648 people visited the museum in July, what number will the museum record?

A 8,500
B 8,600
C 8,650
D 8,700
E 9,000

**17** Kate recorded the number of miles she commuted each day during one week. She listed the numbers of miles in order from least to greatest. Which shows her list?

A 38.3, 38.1, 37.8, 37.7, 37.5
B 37.5, 37.7, 37.8, 38.1, 38.3
C 37.5, 37.7, 38.1, 37.8, 38.3
D 38.3, 38.1, 37.8, 37.5, 37.7
E 37.7, 37.8, 38.1, 38.3, 37.5

**18** Natalia completed her balance beam routine in a gymnastics meet with a score of 15.975. The table shows the scores of four of Natalia's competitors. In what place did Natalia finish the competition?

| BALANCE BEAM SCORES | |
|---|---|
| GYMNAST | SCORE |
| Johnson | 15.995 |
| Hen | 15.98 |
| Kalesh | 15.97 |
| Ryder | 15.965 |

A first
B second
C third
D fourth
E fifth

**19** During jury selection, a clerk calls juror numbers. When a juror hears his or her number, he or she steps forward to speak with the judge. Bryan is juror number 807. When should he step forward to speak with the judge?

A when the clerk says "eighty-seven"
B when the clerk says "eight and seven"
C when the clerk says "eight hundred seven"
D when the clerk says "eight hundred seventy"
E when the clerk says "eighty hundred and seven"

**20** The table shows weekly sales of an automobile dealership over five weekends in the month of May.

| WEEKLY SALES IN MAY | |
|---|---|
| **WEEKEND** | **SALES** |
| Weekend 1 | $168,000 |
| Weekend 2 | $102,000 |
| Weekend 3 | $121,000 |
| Weekend 4 | $119,000 |
| Weekend 5 | $191,000 |

On which weekend did the automobile dealership most likely run a promotion to sell more cars?

A   Weekend 1
B   Weekend 2
C   Weekend 3
D   Weekend 4
E   Weekend 5

**21** Meredith wrote a check for $182 to pay a bill. How is 182 written in words?

A   one hundred eight-two
B   one hundred eighty-two
C   one hundred and eighteen-two
D   one-hundred eighty and two
E   one-hundred eight tens and two

**22** The scale at a doctor's office shows a person's weight to the thousandth of a pound. On a patient's chart, the nurse writes the weight to the nearest tenth. If a child weighs 42.468 pounds on the scale, what will the nurse record on the chart?

A   42.35
B   42.4
C   42.47
D   42.5
E   43.0

**23** Isaiah works in quality-control at an ice cream factory. Each half-gallon container of ice cream must weigh more than 1.097 kg and less than 1.103 kg. Which of the following containers of ice cream would Isaiah reject?

A   Sample A—1.099 kg
B   Sample B—1.101 kg
C   Sample C—1.121 kg
D   Sample D—1.098 kg
E   Sample E—1.102 kg

**Directions:** Questions 24 and 25 are based on the information below.

The batting averages of five baseball players are shown in the table.

| BATTING AVERAGES | |
|---|---|
| **PLAYER** | **BATTING AVERAGE** |
| A | 0.279 |
| B | 0.350 |
| C | 0.305 |
| D | 0.298 |
| E | 0.289 |

**24** Which player has the highest batting average?

A   Player A
B   Player B
C   Player C
D   Player D
E   Player E

**25** The players with the top three batting averages are first in the batting lineup. They will bat in order from the lowest batting average to the highest batting average. Which lists the correct batting order?

A   Player D, Player B, Player C
B   Player B, Player D, Player C
C   Player C, Player B, Player D
D   Player B, Player C, Player D
E   Player D, Player C, Player B

# Whole Numbers and Operations

## ① Learn the Skill

The four basic math operations are addition, subtraction, multiplication, and division. Add quantities to find a total, or **sum**. Use subtraction to find the **difference** between two quantities.

Multiply to find a **product** when you need to add a number many times. Divide when separating a quantity into equal groups. The initial quantity is the **dividend**. The **divisor** is the number by which you divide. The **quotient** is the answer.

## ② Practice the Skill

To successfully solve problems on high school equivalency mathematics tests, you must determine the correct operation(s) to perform and the proper order in which to perform them. Read the examples and strategies below. Use this information to answer question 1.

**A** Add the numbers in each column, working from right to left. If the sum of a column of digits is greater than 9, regroup to the next column on the left.

**B** To subtract, align the digits by place value. Subtract the numbers in each column, working from right to left. When a digit in the bottom number is greater than the digit in the top number, regroup.

**C** Multiply the ones digit of the bottom number by all the digits in the top number. Align each result, or partial product, under the digit you multiplied by. Use zeros as placeholders. When you have multiplied all digits in the bottom number by all digits in the top number, add the partial products.

**A Addition**

$$\begin{array}{r} \overset{1}{4}82 \\ + 208 \\ \hline \mathbf{690} \end{array}$$

**B Subtraction**

$$\begin{array}{r} 4\overset{7\,12}{8}\overset{}{2} \\ - 208 \\ \hline \mathbf{274} \end{array}$$

**C Multiplication**

$$\begin{array}{r} 482 \\ \times \quad 34 \\ \hline 1,928 \\ \times 14,460 \\ \hline \mathbf{16,388} \end{array}$$

**D Division**

$$\begin{array}{r} \mathbf{517\ R12} \\ 14\overline{)7250} \\ -70 \\ \hline 25 \\ -14 \\ \hline 110 \\ -98 \\ \hline 12 \end{array}$$

**D**

$$\begin{array}{r} \mathbf{517\ R12} \\ 14\overline{)7250} \\ 14 \times 5 = \ -70 \downarrow \\ \hline 25 \\ 14 \times 1 = \ -14 \downarrow \\ \hline 110 \\ 14 \times 7 = \ -98 \\ \hline 12 \end{array}$$

### ✔ TEST-TAKING TIPS

To determine which operation to use to solve a problem, look for key words or phrases. Words like *total* and *sum* indicate addition. Phrases like *how much is left*, *less than*, and *difference* indicate subtraction.

**1** Shirley has $1,256 in her bank account. She withdraws $340. How much money is left in her bank account?

A $816
B $916
C $926
D $996
E $1,006

2  Alex drove from Denver, Colorado, to Chicago, Illinois, in two days. The first day he drove 467 miles. The second day he drove 583 miles. What is the total distance that Alex drove?
   A  950 miles
   B  1,039 miles
   C  1,040 miles
   D  1,049 miles
   E  1,050 miles

3  During a word game, Alicia had 307 points. She was unable to use all of her letters, so she had to subtract 19 points at the end of the game. What was Alicia's final score?
   A  287
   B  288
   C  297
   D  298
   E  307

4  Juan works 40 hours per week. He earns $9 per hour. How much does Juan earn in one week?
   A  $32
   B  $36
   C  $320
   D  $360
   E  $400

5  Carl pays $45 per month for car insurance. How much does he spend on car insurance in 1 year?
   A  $550
   B  $540
   C  $530
   D  $450
   E  $440

6  Four friends went out for pizza. The total cost for appetizers, pizza, and drinks was $64. If the friends split the cost equally, how much did each friend pay?
   A  $13
   B  $14
   C  $15
   D  $16
   E  $17

7  Claire is purchasing bags of mulch to cover her vegetable garden. One bag of mulch will cover 12 square feet. How many bags of mulch will Claire need?

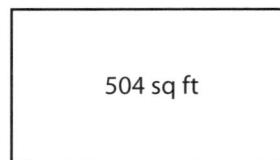

504 sq ft

   A  41
   B  42
   C  43
   D  44
   E  45

8  Each month, Anna pays $630 in rent. How much rent does she pay over the course of 18 months?
   A  $1,340
   B  $6,300
   C  $11,340
   D  $12,600
   E  $15,120

9  The quarterback on Scott's favorite football team is closing in on a 4,000-yard passing season. He has thrown for 3,518 yards with two games remaining. How many yards would the quarterback need to average over the final two games to reach his goal of 4,000 yards?
   A  221
   B  241
   C  271
   D  311
   E  482

**10** Alex saves $325 each month for college tuition. How much will he have saved after 6 months?

A  $331
B  $975
C  $1,300
D  $1,625
E  $1,950

**11** Angelo has paid $1,560 toward his car loan. If his loan is $2,750, how much does he still owe?

A  $190
B  $990
C  $1,190
D  $1,290
E  Not enough information is given.

**12** Tara pays the same amount for her electric bill each month. If she pays $72 per month, what is the total cost of her electricity for one year?

A  $84
B  $216
C  $720
D  $864
E  Not enough information is given.

**13** Four roommates equally share their monthly rent. Their monthly rent is $1,080. How much does each roommate pay per month?

A  $270
B  $250
C  $240
D  $216
E  $207

**14** A city budgets $567,800 for parks and recreation and $258,900 for facility maintenance. How much is spent on these two parts altogether?

A  $308,900
B  $358,900
C  $700,000
D  $801,900
E  $826,700

**15** Mara gave 22 shirts, 14 pairs of pants, and 12 scarves to charity. How many clothing items did she give away altogether?

A  26
B  30
C  32
D  48
E  54

**16** Eli is writing a 1,500-word essay for English class. He has written 892 words so far. How many words must he still write?

A  408
B  508
C  608
D  708
E  808

**17** The table shows the cost for five people to attend a professional football game.

| COST OF GAME | |
|---|---|
| PRODUCT/SERVICE | COST |
| Gasoline and parking | $50 |
| Tickets | $335 |
| Food | $80 |
| Souvenirs | $75 |

If five friends decided to evenly share the costs, how much would each person expect to pay?

A  $106
B  $107
C  $108
D  $109
E  $110

**Directions:** Questions 18 through 20 are based on the information below.

The table shows part of Antonio's monthly budget.

| MONTHLY BUDGET | |
|---|---|
| Rent | $825 |
| Utilities | $220 |
| Food | $285 |
| Recreation | $100 |
| Auto Loan | $179 |
| Auto Insurance | $62 |

18  What is the total amount included in Antonio's budget for rent, utilities, and food?
  A  $540
  B  $605
  C  $1,045
  D  $1,110
  E  $1,330

19  How much more money does Antonio allow in his budget for food than for his auto loan?
  A  $41
  B  $65
  C  $106
  D  $185
  E  $464

20  How much does Antonio pay toward his auto loan each year?
  A  $179
  B  $372
  C  $744
  D  $1,074
  E  $2,148

21  Annette works 5 days per week, 6 hours per day. She earns $13 per hour. How much does Annette earn in 4 weeks?
  A  $1,560
  B  $1,180
  C  $390
  D  $260
  E  $120

22  Andrew worked 54 hours one week and 39 hours the next week. He earns $11 per hour. How much did he earn in the 2 weeks?
  A  $429
  B  $594
  C  $945
  D  $1,023
  E  $2,106

23  In January, the Wilsons spent $458 on groceries. They spent $397 in February and $492 in March. What is the total amount the Wilsons spent on groceries for the three months?
  A  $800
  B  $855
  C  $950
  D  $1,250
  E  $1,347

24  A sports store ran a promotion on a specific tent. During the promotion, the store had tent sales of $23,870. If each tent cost $385, how many tents did the store sell during the promotion?
  A  59
  B  60
  C  61
  D  62
  E  63

**25** A pattern calls for 2 yards of material for a shirt. The same pattern for a dress calls for 5 yards of material. A seamstress makes five shirts and five dresses for a retail store. How many more yards of material does she use for the dresses than for the shirts?

A  5
B  10
C  15
D  20
E  25

**26** Joanne drives 37 miles round-trip each day commuting to and from work. She works Monday through Friday. How many miles does Joanne drive in 4 weeks?

A  148
B  185
C  370
D  740
E  1,480

**27** A charity wants to donate $12,500 to a food bank. It has already collected $4,020 in donations from local businesses and $2,902 in donations from individuals. About how much more does the charity need to collect to meet its goal?

A  $5,000
B  $6,000
C  $7,000
D  $9,000
E  $10,000

**28** Mr. and Mrs. Dale paid $1,445 to have a new floor installed in their kitchen. The company installed 289 square feet of flooring. How much did the Dales pay for each square foot of flooring?

A  $5
B  $57
C  $1,156
D  $1,734
E  Not enough information is given.

**29** Maggie purchased a used car. She financed the car through the auto dealership and owes a total of $13,392. She will make equal monthly payments on the car for 3 years. Which of the following expressions represents her monthly payment?

A  $13,392 ÷ (3 × 16)
B  $13,392 ÷ (3 × 12)
C  $13,392 ÷ (2 × 12)
D  $13,392 ÷ 12
E  $13,392 ÷ 3

**Directions:** Questions 30 and 31 are based on the information below.

The table shows the prices of various stocks for purchase.

| STOCK PRICES | |
|---|---|
| STOCK | PRICE PER SHARE |
| Computers4U | $30 |
| Sun Cell Phones | $23 |
| Online Airlines | $15 |
| Virtual Reality, Inc. | $42 |

**30** Mario wants to invest in an online travel service. How many shares of Online Airlines can he purchase for $270?

A  9
B  12
C  15
D  18
E  21

**31** Karissa purchased 25 shares of Virtual Reality, Inc., and then sold all of them three months later for a profit of $7 per share. How much profit did Karissa make on her sale of Virtual Reality, Inc.?

A  $175
B  $225
C  $275
D  $325
E  $375

# Word Problems

## 1 Learn the Skill

There are several steps to solving word problems. First, read the problem carefully. Make sure you understand all of the information. Next, determine the information that you need to solve the problem. There may be more information than you need. Then, solve the problem. Finally, check your answer. For some problems, you need only to estimate to find the answer.

## 2 Practice the Skill

Since many of the questions on mathematics tests are word problems, understanding how to quickly and accurately solve such problems will improve your ability to complete a test successfully. Read the example and strategies below. Use this information to answer question 1.

The table shows the prices of snacks at a snack shop.

| PRICES OF SNACK FOODS | |
|---|---|
| Hot Dog | C $3 |
| Pretzel | $2 |
| Granola | $1 |
| Yogurt | C $2 |

**B** **Step 2.** Determine the information you need to solve the problem. Remember, you may not need all of the information.

**D** **Step 4.** Solve the problem. Perform the necessary operations. Recall the **order of operations**.

**C** **Step 3.** Choose the operation or operations. This problem requires multiple steps. You must multiply the number of hot dogs by the price of each hot dog and the number of yogurts by the price of each yogurt. Then add the products.

**E** **Step 5.** Check your answer. Estimate to find an approximate answer. Compare your answer to your estimate. If they are close, your answer is reasonable.

**A** **Step 1.** Understand the problem. What are you asked to find?

### TEST-TAKING TIPS

When a multi-step problem contains more information than you need, highlight only the necessary information.

1 The snack shop sold 64 hot dogs and 38 yogurts. How much money did the snack shop make on hot dogs and yogurt?
A  $102
B  $192
C  $242
D  $268
E  $306

**2** Allison wants to buy a pair of shoes that originally cost $54. Today the shoes are on sale for $42. She uses a coupon for $5 off a pair of shoes. How much does Allison pay for the shoes?

A   $12
B   $37
C   $42
D   $47
E   $54

**3** Rasheed pays $645 each month for rent. He pays $78 each month for heat and electricity and $25 each month for phone service. Approximately how much does he pay each month for rent, heat and electricity, and phone service?

A   $670
B   $680
C   $690
D   $750
E   $800

**4** In the fall, 143 children signed up to play after-school sports at an elementary school. Sixty-seven children signed up for soccer, and thirty-three children signed up for softball. How many children signed up for a sport other than soccer?

A   76
B   110
C   176
D   210
E   Not enough information is given.

**Directions:** Questions 5 and 6 are based on the information below.

The following table shows the cost that a store pays for clothing and the price at which the store sells the clothing.

| PRICES OF CLOTHING | | |
|---|---|---|
| ITEM | AMOUNT STORE PAYS | PRICE STORE CHARGES |
| Women's T-shirt | $6 | $12 |
| Men's T-shirt | $6 | $10 |
| Women's sport shorts | $7 | $14 |
| Men's sport shorts | $9 | $16 |

**5** The store paid a bill to its supplier for a shipment of 144 women's T-shirts. What was the amount of the bill?

A   $720
B   $844
C   $864
D   $1,008
E   $1,728

**6** The profit the store makes on each item of clothing is the difference between what the store pays for the item and the price at which the store sells the item. If the store sells 508 men's sport shorts, what is the store's profit?

A   $3,556
B   $4,064
C   $4,572
D   $7,112
E   $8,128

**7** The Stein family budgeted $5,500 for their vacation to cover airfare, lodging, meals, and extras, such as souvenirs. Their total for airfare, lodging, and meals came to $5,000. Of the remainder, what was the average amount that each of the four family members could spend on souvenirs and remain within their budget?

A   $50
B   $125
C   $150
D   $250
E   $500

8  Juan is training for a triathlon. He needs to eat 3,500 calories per day to maintain his level of fitness. He ate 1,250 calories at breakfast. He ate another 780 calories at lunch. How many calories must he still eat today to maintain his level of fitness?

A  780
B  1,250
C  1,470
D  2,250
E  2,720

9  A large factory has 2,391 employees. Of these employees, 2,012 work on the production lines, 157 work in shipping, and the rest of the employees work in administration. How many employees work in administration?

A  157
B  222
C  379
D  2,548
E  4,403

10  Stella pays $126 per month on her car loan. She pays $57 per month for auto insurance. What is the cost of her car loan per year?

A  $183
B  $684
C  $828
D  $1,512
E  $2,196

11  Jackie saves $17.85 from her paycheck each week. She puts the money into a bank account. Estimate how much money she will have saved at the end of a year.

A  $200
B  $240
C  $928.20
D  $1000
E  Not enough information is given.

**Directions:** Questions 12 and 13 are based on the information below.

| GRACE'S WEEKLY TIME SHEET | | | | | |
|---|---|---|---|---|---|
| | HOURS WORKED PER DAY | | | | |
| Employee | Mon | Tues | Wed | Thu | Fri |
| Grace D. | 6 | 5 | 7 | 8 | 8 |

12  If Grace makes $8 per hour, how much money did she earn for the week?

A  $48
B  $144
C  $240
D  $264
E  $272

13  The following week, Grace worked 6 more hours than the week before. She also received a raise to $9 per hour. How much did Grace earn the following week?

A  $272
B  $304
C  $320
D  $360
E  $369

14  Theo and two friends went out for lunch. The bill was $30. They also left a tip. If they split the cost evenly, how much did each of them pay?

A  $12
B  $14
C  $16
D  $18
E  Not enough information is given.

15  Naomi has $913 in her checking account. She deposits $130, writes two checks for $75 apiece, and withdraws $50 with her bank card. How much money is left in Naomi's account?

A  $843
B  $893
C  $923
D  $973
E  $1,043

**16** The mileage on Marla's car was 41,868 when she last had her tires changed. She bought new tires that are advertised to last for 40,000 miles. Marla now has 97,634 miles on her car. How many miles over 40,000 has Marla driven on these tires?

A   15,766
B   15,776
C   15,867
D   16,766
E   16,776

**17** Charlotte can drive 400 miles on one tank of gas. Her gas tank holds 16 gallons of gas. If gas is $3 per gallon, how much will it cost her to drive 800 miles?

A   $48
B   $96
C   $133
D   $144
E   $1,200

**18** Karen owns three apartments. She rents each apartment for $895 per month. If she raises the rent to $950 per month, how much rent will she collect each month?

A   $165
B   $2,520
C   $2,550
D   $2,685
E   $2,850

**19** Trisha and her three siblings inherited $4,598 from an uncle. The lawyer's fees to settle the estate were $1,354. The remainder was split equally. How much money did Trisha and each of her siblings inherit?

A   $811
B   $1,081
C   $1,150
D   $1,622
E   $3,244

**Directions:** Questions 20 and 21 are based on the information below.

The table shows how many skeins of each color yarn that a craft store has in stock.

| SKEINS OF YARN | |
|---|---|
| YARN COLOR | NUMBER OF SKEINS |
| Blue | 3,156 |
| Red | 2,634 |
| Green | 1,920 |
| White | 4,208 |
| Off-white | 983 |
| Yellow | 732 |
| Orange | 531 |
| Brown | 1,828 |
| Purple | 935 |

**20** About how many more skeins of white yarn than off-white yarn does the store have in stock? Round to the nearest thousand.

A   1,000
B   2,000
C   3,000
D   4,000
E   5,000

**21** The white, blue, and red skeins are the craft store's most popular colors of yarn. Added together, the number of white, blue, and red skeins exceeds the combined amount of the other six skeins. How many more white, blue, and red skeins are there than the other skeins combined? Round the difference to the nearest hundred.

A   2,500
B   3,100
C   3,600
D   3,800
E   4,500

22 A furniture store is offering a payment plan for a dining table and chairs. The cost of the set is $1,620. The payment plan is $150 per month for 12 months. How much more will you pay if you choose the payment plan instead of purchasing the set up front?

A $1,800
B $1,620
C $180
D $135
E $0

23 Cody works at an electronics store. On Sunday, he sold four televisions for $757 each and three speaker systems for $533 each. He earns $50 commission for each television and $30 commission for each speaker system that he sells. How much money in commissions did Cody make on Sunday?

A $560
B $290
C $270
D $80
E Not enough information is given.

24 Angela bought $250 worth of fencing supplies, $207 worth of paving blocks, and 18 bags of mulch. What was the total cost of the mulch and paving blocks?

A $43
B $268
C $457
D $475
E Not enough information is given.

25 Hernando plans to make 12 monthly payments on his computer. The total cost of the computer plus interest is $1,476. How much will he pay per month?

A $112
B $123
C $131
D $139
E $147

**Directions:** Questions 26 and 27 are based on the information below.

Gretchen ordered T-shirts for the participants in a triathlon. The table shows how many boxes of each size T-shirt that she ordered. There are 35 T-shirts in each box of women's T-shirts and 25 T-shirts in each box of men's T-shirts.

| BOXES OF T-SHIRTS ORDERED | |
|---|---|
| SIZE | NUMBER OF BOXES |
| Women's Small | 5 |
| Women's Medium | 6 |
| Women's Large | 5 |
| Men's Medium | 3 |
| Men's Large | 8 |
| Men's Extra Large | 14 |

26 How many more men's T-shirts than women's T-shirts did Gretchen order?

A 9
B 10
C 65
D 475
E 1,185

27 How many shirts would Gretchen order if she only ordered medium and large sizes?

A 550
B 600
C 625
D 660
E 910

28 The Millersville Youth Athletic Center has children's sports leagues. Last fall, 460 children (242 boys, 218 girls) signed up to play basketball. In the spring, 540 children (295 boys, 245 girls) signed up to play soccer. How many more girls signed up to play soccer than signed up to play basketball?

A 17
B 27
C 37
D 47
E 57

# Integers and Operations

## ① Learn the Skill

**Integers** include positive whole numbers (1, 2, 3, . . .), their opposites or negative numbers (–1, –2, –3, . . .), and zero. Positive numbers show an increase and may be written with or without a plus sign. Negative numbers show a decrease and are written with a negative sign. Integers can be added, subtracted, multiplied, and divided. There are specific rules for adding, subtracting, multiplying, and dividing integers.

In some cases, you may need to determine an integer's **absolute value**, or its distance from 0. Absolute values are always greater than or equal to zero, never negative. So the absolute value of both 9 and –9 is 9.

## ② Practice the Skill

Many mathematics problems relating to real-world situations use integers. You must follow the rules for adding, subtracting, multiplying, and dividing integers to solve problems on mathematics tests. Read the examples and strategies below. Use this information to answer question 1.

**Ⓐ** If integers have like signs, add and keep the common sign. If integers have different signs, find the difference. Then use the sign of the number with the greater absolute value.

**Ⓑ** To subtract an integer, add its opposite. For example, the opposite of –5 is +5.

**Ⓒ** For multiplying or dividing integers: If the signs are the same, the answer will be positive. If the signs are different, the answer will be negative.

### Add Integers

$(+4) + (+7) = +11$      $(-5) + (-9) = -14$
$(-8) + (+4) = -4$      $(-5) + (+12) = +7$

### Subtract Integers

$(+8) - (-5) = (+8) + (+5) = 13$
$8 - 5 = 8 + (-5) = 3$

### Multiply and Divide Integers

$(4)(5) = +20$      $(-4)(5) = -20$
$(-4)(-5) = 20$      $(4)(-5) = -20$

$18 ÷ 9 = 2$      $(-18) ÷ (9) = -2$
$(-18) ÷ (-9) = 2$      $18 ÷ (-9) = -2$

### ✓ TEST-TAKING TIPS

It may be helpful to use a number line when solving problems that involve integers. To solve 12 – (–3), begin at –3 and count spaces to 12. You will see that the distance is +15.

1  In the morning, the temperature was –3°F. By mid-afternoon, the temperature was 12°F. What was the change in temperature between the morning and afternoon?

A   –15°F
B   –9°F
C   9°F
D   12°F
E   15°F

**Directions:** Questions 6 and 7 are based on the information below.

There were 3,342 students enrolled at a university. Of those students, 587 graduated in May. Over the summer, 32 students left the university, and 645 new students enrolled in the fall.

2   In a board game, Dora moves forward 3 spaces, back 4 spaces, and forward again 8 spaces in one turn. What is her net gain or loss of spaces?

A   1 space forward
B   7 spaces forward
C   1 space backward
D   7 spaces backward
E   15 spaces forward

3   Uyen has a balance of $154 in her savings account. She withdraws $40 from a cash machine. What is her new balance?

A   $94
B   $100
C   $104
D   $114
E   $194

4   Sasha's home is 212 feet above sea level. She participated in a scuba dive in which she descended to 80 feet below sea level. Which integer describes Sasha's change in position from her house to the lowest point of her dive?

A   −292
B   −132
C   132
D   292
E   302

5   Scott went golfing at a local course. He shot even-par over the first 9 holes, then shot +3 on each of the next three, −4 on each of the next three, and −1 on each of the final three holes. What was his score for the round?

A   6-under par
B   2-under par
C   even par
D   2-over par
E   7-over par

6   How many students were enrolled in the fall?

A   2,697
B   2,755
C   3,310
D   3,368
E   3,987

7   Which number describes the change in the number of students enrolled between May and the following fall?

A   −32
B   −26
C   26
D   32
E   58

8   Melanie played a game and kept track of her score. The table shows her points earned for each round.

| MELANIE'S POINTS SCORED | |
|---|---|
| ROUND | POINTS SCORED |
| 1 | 8 |
| 2 | −6 |
| 3 | −4 |
| 4 | 3 |
| 5 | 4 |

Melanie played a sixth round and scored −8 in that round. What was her overall score?

A   −13
B   −3
C   5
D   13
E   18

9  A football team has possession of the ball. On their first play, they gain 8 yards. On their second play, they lose 10 yards. On their third play, they gain 43 yards. How far have they gone?
   A  25 yards
   B  33 yards
   C  41 yards
   D  51 yards
   E  53 yards

10  The Dow Jones Industrial Average opened trading one morning at 11,498. It closed that day at 11,416. Which integer describes the change?
   A  +82
   B  +18
   C  −18
   D  −72
   E  −82

11  A mountain biker begins at the top of a mountain with an elevation of 8,453 feet. She rides 2,508 feet down the mountain before taking a break. She then rides another 584 feet up the mountain. At what elevation is she now?
   A  11,545 feet
   B  10,961 feet
   C  9,037 feet
   D  6,529 feet
   E  5,945 feet

12  A diver begins at a height of 3 meters above the water on a diving board. In her dive, she reaches a height of 2 meters above the board. From this point, she drops 8 meters. Which integer describes her position at this point, with regard to the surface of the water?
   A  3 meters
   B  2 meters
   C  −2 meters
   D  −3 meters
   E  −10 meters

13  In a card game, Deshon had −145 points. He then scored 80 points, and then 22 points. What is Deshon's score at this point?
   A  225
   B  167
   C  123
   D  −43
   E  −65

14  If −7 is added to a number, the result is 12. What is the number?
   A  −19
   B  −5
   C  5
   D  12
   E  19

**Directions:** Questions 15 and 16 are based on the information below.

Four friends played a game. Each player kept track of the number of points she scored in each round.

| POINTS SCORED EACH ROUND | | | |
|---|---|---|---|
| PLAYER | ROUND | | |
| | 1 | 2 | 3 |
| Nikki | 0 | 10 | −15 |
| Clara | −15 | 15 | 0 |
| Donna | 5 | −10 | −10 |
| Dorothy | 15 | 5 | 0 |

15  What was Donna's score after the end of Round 3?
   A  −25
   B  −15
   C  −10
   D  0
   E  25

16  How many more points did Dorothy score than Nikki?
   A  5
   B  10
   C  15
   D  20
   E  25

**17** A submarine is at 3,290 feet below sea level. It rises 589 feet before dropping another 4,508 feet. Which integer describes its current position with regard to sea level?

A 7,209
B 1,807
C −1,807
D −7,209
E −8,387

**18** A cogwheel train transports skiers to the top of a mountain. There are two stations where skiers can get on and off the train. Station A is 5,993 feet above sea level. Station B is 10,549 feet above sea level. The peak of the mountain is 872 feet above Station B. How tall is the mountain at its highest point?

A −16,542 feet
B −11,421 feet
C 6,865 feet
D 11,421 feet
E 16,542 feet

**19** Jordan had $890 in her bank account. In one week, she withdrew $45 three separate times. What was the balance of her account at the end of the week?

A $755
B $800
C $845
D $935
E $975

**20** A number is multiplied by −2. Then the product is increased by 2. The final result is 0. What is the original number?

A −2
B −1
C 0
D 1
E 2

**21** Four team members each have −120 points. How many points do they have together as a team?

A −480
B −360
C 240
D 360
E 480

**22** A group of rock climbers descended a rock face in three equal phases. They descended the same number of feet each time. If the rock face was 363 feet high, what number describes their change in height in each phase?

A −242 feet
B −121 feet
C 121 feet
D 242 feet
E 1,089 feet

**23** Erik biked 12 miles directly south from his home. He then turned around and biked 8 miles back toward his home before stopping to fix a flat tire. How far from home was he when he was fixing his flat tire?

A 4 miles
B 8 miles
C 12 miles
D 16 miles
E 20 miles

**24** Don played a game. He scored 3 points in the first round. After the second round, his total score was −10. How many points did Don score in the second round?

A −13
B −7
C 5
D 7
E 13

**25** The number –7 is multiplied by –1. The product is then multiplied by –1. Finally, this product is multiplied by –1. What is the final product?

A   –10
B   –7
C   –4
D   0
E   7

**26** Brenda has her health insurance premium of $156 automatically withdrawn from her checking account each month. Which integer describes the change in her bank account due to her health insurance premium in one year?

A   $1,872
B   $936
C   –$156
D   –$936
E   –$1,872

**27** Karin owes her sister $1,554. She has budgeted an equal amount of money over the next 6 months to pay her sister back. How much money will Karin pay her sister each month?

A   $257
B   $258
C   $259
D   $260
E   $261

**28** The absolute value of the difference between two numbers is the distance between the two numbers on the number line. What is the absolute value of the difference between *A* and *B*?

A   –11
B   –3
C   0
D   3
E   11

**29** Janet visited a skyscraper in Chicago. She entered the elevator on the ground floor and went up 54 floors. After looking at the city from the viewing area, she went back down 22 floors. She realized that she made a mistake and instead should have gotten out of the elevator 5 floors above, so she rides the elevator up. On what floor is Janet now?

A   27
B   32
C   37
D   59
E   76

**30** Cheryl receives $527 per month from her retirement fund. Which integer describes the amount she receives in 6 months?

A   –$3,689
B   –$3,162
C   $2,635
D   $3,162
E   $3,689

**31** In a year, Connor paid $3,228 for his car loan. He paid the same amount each month. Which integer describes the monthly change in his bank account after paying his monthly car payment?

A   –$269
B   –$239
C   –$229
D   $239
E   $269

**32** If –10 is subtracted from a number, the result is 6. What is the number?

A   –16
B   –4
C   0
D   4
E   16

# Fractions and Operations

## ① Learn the Skill

A **fraction** shows part of a whole or part of a group by separating two numbers with a fraction bar. The bottom number, the **denominator,** tells the number of equal parts in a whole. The top number, the **numerator,** tells the number of equal parts being considered.

Fractions and mixed numbers can be added, subtracted, multiplied, and divided. To add or subtract like fractions, simply add or subtract the numerators and keep the denominator. To add or subtract unlike fractions, first find a common denominator. Then add or subtract the same way as like fractions. Common denominators are not needed for multiplication or division.

## ② Practice the Skill

You must understand how to perform operations with fractions and mixed numbers to correctly solve problems on high school equivalency mathematics tests. Read the examples and strategies below. Use this information to answer question 1.

**Ⓐ** A proper fraction shows a quantity less than 1, such as $\frac{3}{4}$. An improper fraction has a numerator greater than the denominator.

**Ⓑ** To multiply fractions, simply multiply the numerators and multiply the denominators. Reduce the product to lowest terms. To divide, multiply by the reciprocal of the divisor. Write whole numbers as fractions by writing them over 1. For example, $9 = \frac{9}{1}$

**Ⓒ** To add mixed numbers, first find a common denominator. Then add the fractions. If the sum is an improper fraction, change it to a mixed number. Then add the sum of the fractions to the sum of the whole numbers.

**Add** $\quad \frac{3}{4} + \frac{5}{8} \rightarrow \frac{3 \times 2}{4 \times 2} = \frac{6}{8} \quad \frac{6}{8} + \frac{5}{8} = \frac{11}{8} = 1\frac{3}{8}$

**Subtract** $\frac{3}{4} - \frac{5}{8} \rightarrow \frac{3 \times 2}{4 \times 2} = \frac{6}{8} \quad \frac{6}{8} - \frac{5}{8} = \frac{1}{8}$

**Multiply** $\frac{3}{4} \times \frac{5}{8} \rightarrow \frac{3}{4} \times \frac{5}{8} = \frac{15}{32}$

**Divide** $\quad \frac{5}{9} \div \frac{2}{3} \rightarrow \frac{5}{9} \div \frac{2}{3} = \frac{5}{9} \times \frac{3}{2} = \frac{15}{18} = \frac{5}{6}$

**Add** $\quad 4\frac{5}{6} + 2\frac{1}{4}$

$4\frac{5}{6} + 2\frac{1}{4} = 4\frac{5 \times 2}{6 \times 2} + 2\frac{1 \times 3}{4 \times 3} = 4\frac{10}{12} + 2\frac{3}{12} = 6\frac{13}{12} = 7\frac{1}{12}$

**✓ TEST-TAKING TIPS**

To multiply and divide mixed numbers, first rename the mixed numbers as improper fractions.

**1** There are two containers of milk in Eric's refrigerator. One has $\frac{3}{5}$ gallon of milk. The other has $\frac{3}{4}$ gallon of milk. How many gallons of milk are in Eric's refrigerator?

A $\quad \frac{9}{20}$

B $\quad \frac{6}{11}$

C $\quad 1\frac{7}{20}$

D $\quad 1\frac{9}{20}$

E $\quad 1\frac{3}{5}$

2   Carly spent $3\frac{1}{3}$ hours organizing her room. She spent $1\frac{1}{2}$ hours cleaning her room. How many hours did she spend on her room altogether?

A   $1\frac{5}{6}$

B   $2\frac{1}{3}$

C   $4\frac{1}{6}$

D   $4\frac{2}{3}$

E   $4\frac{5}{6}$

3   Blake needs a wooden dowel that is $2\frac{1}{6}$ feet long. How much should he cut off the dowel shown below?

$$\longleftarrow 5\frac{1}{4} \text{ ft} \longrightarrow$$

A   $3\frac{1}{4}$ feet

B   $3\frac{1}{6}$ feet

C   $3\frac{1}{12}$ feet

D   $2\frac{1}{10}$ feet

E   $2\frac{1}{12}$ feet

4   An employee is packing shirts into boxes for shipping. On his inventory sheet, he marks that he is shipping $\frac{23}{4}$ boxes of shirts. What is $\frac{23}{4}$ written as a mixed number?

A   $4\frac{3}{4}$

B   $5\frac{1}{23}$

C   $5\frac{1}{4}$

D   $5\frac{1}{2}$

E   $5\frac{3}{4}$

5   Mr. White is sewing new curtains. He has a piece of material that is $11\frac{1}{4}$ yards long. He needs $2\frac{1}{4}$ yards for each curtain. How many curtains can he make from this material?

A   5

B   4

C   3

D   2

E   1

6   Colin has a stack of 8 books on his desk. If each book is $\frac{2}{3}$ of an inch thick, what is the height of the stack?

A   $5\frac{1}{3}$ inches

B   $5\frac{2}{3}$ inches

C   $6\frac{1}{3}$ inches

D   $8\frac{2}{3}$ inches

E   Not enough information is given.

7   Jane ran $3\frac{3}{8}$ miles on Tuesday, $1\frac{1}{4}$ miles on Wednesday, and $2\frac{1}{2}$ miles on Thursday. How many total miles did Jane run?

A   $6\frac{7}{8}$

B   $7\frac{1}{8}$

C   $7\frac{7}{8}$

D   $8\frac{1}{8}$

E   $8\frac{7}{8}$

8   A recipe calls for $1\frac{3}{4}$ cups of flour. If Liza cuts the recipe in half, how much flour will she use?

A   $\frac{7}{8}$ cup

B   1 cup

C   $1\frac{1}{4}$ cups

D   $2\frac{1}{4}$ cups

E   $3\frac{1}{2}$ cups

**9** Chandra has 8 sick days per year. If she has already used $3\frac{1}{2}$ sick days, how many sick days does she have left?

A $3\frac{1}{2}$

B 4

C $4\frac{1}{2}$

D 5

E $5\frac{1}{2}$

**10** Todd keeps track of the water he uses in his garden every week. One week he used $3\frac{3}{4}$ gallons, $5\frac{1}{2}$ gallons, and $4\frac{1}{4}$ gallons. How many gallons of water did he use in his garden that week?

A $4\frac{1}{2}$

B $8\frac{1}{4}$

C $9\frac{1}{2}$

D $12\frac{1}{2}$

E $13\frac{1}{2}$

**11** Carlos buys three different meats at the deli counter in the grocery store. The meats weigh $\frac{1}{10}$ pound, $\frac{7}{8}$ pound, and $2\frac{3}{16}$ pound. Rounding to the nearest whole number, which of the following expressions represents the best estimate for the total amount of meat Carlos buys?

A $0 + 1 + 3$

B $1 + 1 + 2$

C $1 + 1 + 3$

D $0 + 1 + 2$

E $0 + 0 + 2$

**12** In one year, Elias earns $29,400. His semi-monthly paycheck is $\frac{1}{24}$ of this amount. What is the amount of his semi-monthly paycheck?

A $12.25

B $122.50

C $1,225.00

D $12,250.00

E Not enough information is given.

**13** A chef uses $\frac{1}{4}$ pound of ground beef to make each dinner special. How many specials can the chef prepare from 24 pounds of ground beef?

A 48

B 64

C 80

D 96

E 112

**14** Mike owns a lawnmowing business. He can mow one lawn in about 15 minutes. If Mike works 30 hours per week, how many lawns can he mow?

A 30

B 60

C 90

D 120

E 150

**15** Amy works as an editor at a publishing house. It takes her 35 minutes to edit one textbook page. How many hours would it take Amy to edit a 300-page book?

A 150

B 175

C 200

D 225

E 250

**16** Forty people are part of a tour group in London. One day, $\frac{1}{2}$ of the group opted to go for a cruise on the River Thames. Of the $\frac{1}{2}$ that did not go on the cruise, $\frac{2}{3}$ visited the Tower of London. What fraction of the group chose to visit the Tower of London that day?

A $\frac{1}{6}$

B $\frac{1}{3}$

C $\frac{4}{6}$

D $\frac{2}{3}$

E $\frac{3}{4}$

**17** Ed purchased $25\frac{7}{8}$ yd of fencing for his yard. He only used $17\frac{5}{6}$ yd. How many yards of fencing does he have left?

A $8\frac{1}{48}$

B $8\frac{1}{24}$

C $9\frac{1}{8}$

D $9\frac{1}{12}$

E $9\frac{1}{2}$

**18** A group of 30 people attended a baseball game. Of those, $\frac{1}{5}$ sat in blue seats, $\frac{1}{2}$ sat in yellow seats, and the rest sat in red seats. What fraction of the people sat in red seats?

A $\frac{1}{3}$

B $\frac{3}{10}$

C $\frac{1}{4}$

D $\frac{1}{5}$

E $\frac{1}{10}$

**Directions:** Questions 19 and 20 are based on the information below.

In a water relay race, each team must fill a cup of water, race over to a bowl, and pour the water from the cup into the bowl. The relay is over when one team has filled its bowl to the top. The table below shows the results of the race.

| WATER RELAY RESULTS | |
| --- | --- |
| **Team** | **Bowl Capacity** |
| Team 1 | $\frac{1}{2}$ |
| Team 2 | $\frac{1}{1}$ |
| Team 3 | $\frac{3}{5}$ |
| Team 4 | $\frac{1}{3}$ |
| Team 5 | $\frac{4}{5}$ |

**19** Which team had a bowl less than $\frac{1}{2}$ full?
   A   Team 1
   B   Team 2
   C   Team 3
   D   Team 4
   E   Team 5

**20** Someone poured the contents of Team 4's bowl into Team's 1 bowl. Which of the following equations represents the result?

A $\frac{1}{3} \times \frac{1}{2} = \frac{1}{6}$

B $\frac{1}{3} + \frac{1}{2} = \frac{1}{5}$

C $\frac{1}{2} - \frac{1}{3} = \frac{1}{6}$

D $\frac{1}{3} - \frac{1}{2} = \frac{1}{1}$

E $\frac{1}{2} + \frac{1}{3} = \frac{5}{6}$

**21** The school chess club celebrated its recent championship. The 15-member team went out for dessert. Five members ordered pie, 4 ordered ice cream, 3 ordered cake, and 3 ordered milkshakes. What fraction of the chess club members ordered cake or milkshakes?

A  $\frac{0}{15}$

B  $\frac{1}{5}$

C  $\frac{1}{3}$

D  $\frac{2}{5}$

E  $\frac{11}{15}$

**22** Scott cut a board for a cabinet. The original board was $3\frac{7}{8}$ feet long. The board is now $3\frac{1}{4}$ feet long. If he cut the board $\frac{3}{8}$ of an inch too short, how many feet long is the board supposed to be?

A  $2\frac{5}{8}$

B  $3\frac{1}{4}$

C  $3\frac{5}{8}$

D  $7\frac{1}{8}$

E  Not enough information is given.

**23** Caroline has $50\frac{1}{2}$ feet of rope. She wants to divide it into 2 equal sections. Before she divides, she writes $50\frac{1}{2}$ as an improper fraction. Which shows the correct improper fraction form for $50\frac{1}{2}$?

A  $\frac{51}{2}$

B  $\frac{53}{2}$

C  $\frac{100}{2}$

D  $\frac{101}{2}$

E  $\frac{501}{2}$

**24** Riley and Maggie hiked a trail that is $4\frac{3}{4}$ miles long. If they hiked $2\frac{1}{4}$ miles each hour, how many hours did it take them to hike the trail?

A  $2\frac{1}{9}$

B  $2\frac{1}{8}$

C  $2\frac{1}{4}$

D  $2\frac{1}{3}$

E  $2\frac{1}{2}$

**25** Scott writes a 200-word blog in $\frac{3}{4}$ of an hour. How many hours would it take Scott to write a 500-word blog?

A  $1\frac{3}{8}$

B  $1\frac{5}{8}$

C  $1\frac{7}{8}$

D  $2\frac{1}{8}$

E  $2\frac{3}{8}$

**26** Mario needs to work $32\frac{5}{6}$ hours this week. He has worked $19\frac{7}{8}$ hours so far. How many more hours must Mario work this week?

A  $13\frac{23}{24}$

B  $13\frac{3}{4}$

C  $12\frac{23}{24}$

D  $12\frac{7}{8}$

E  $11\frac{7}{8}$

**27** It takes the drive-through staff $\frac{2}{3}$ of a minute to process a fast-food order. How many orders can they process in 18 minutes?

A  6

B  12

C  18

D  27

E  36

UNIT 1

# Ratios and Proportions

## 1 Learn the Skill

A **ratio** is a comparison of two numbers. You can write a ratio as a fraction, using the word *to*, or with a colon (:). A **proportion** is an equation with a ratio on each side. The ratios are equal. You can use proportions to solve problems involving equal ratios. Always write ratio answers in simplest form.

## 2 Practice the Skill

To succeed on mathematics tests, you must understand the concepts of rate and ratio and how to solve for each. Read the examples and strategies below. Use this information to answer question 1.

**A** A ratio is different from a fraction. The bottom or second number of a ratio does not necessarily represent a whole. Therefore, you do not need to rename improper fractions as mixed numbers. However, ratios should still be simplified.

**B** A **unit rate** is a ratio with the denominator of 1. It can be expressed using the word *per*.

**C** In a proportion, the **cross products** are equal. Use cross products to solve proportions. If one of the four terms is missing, cross-multiply and divide the product by the third number to find the missing number.

### Ratio

Jonathan earns $10 in 1 hour.

The ratio of dollars earned to hours is $\frac{10}{1}$, 10: ①, or 10 to 1. **A**

This also can be written as $10 per hour.

### Proportion

$\frac{3}{4} = \frac{6}{8}$   **Cross products:** $\frac{4 \times 6 = 24}{3 \times 8 = 24}$

$\frac{9}{12} = \frac{3}{x}$  ⟶  $12 \times 3 = 36$
$36 \div 9 = 4$
$x = 4$

### ✓ TEST-TAKING TIPS

When you write a proportion, the terms in both ratios need to be in the same order. In problem 1, the top numbers can represent gallons and the bottom numbers can represent cost.

1  Carleen bought 3 gallons of milk for $12. How much would 5 gallons of milk cost?
   A  $9
   B  $12
   C  $16
   D  $20
   E  $24

2   A store sold 92 pairs of pants and 64 shirts. What is the ratio of the number of pants sold to the number of shirts sold?
   A   92:64
   B   16:23
   C   64:92
   D   23:16
   E   16:92

3   Amanda traveled 558 miles in 9 hours. What is the unit rate that describes her travel?
   A   52 miles per hour
   B   61 miles per hour
   C   62 miles per hour
   D   71 miles per hour
   E   72 miles per hour

4   Jill mixed 2 cups of sugar with 10 cups of water to make lemonade. What ratio of sugar to water did she use?

   A   $\frac{1}{5}$

   B   $\frac{2}{10}$

   C   $\frac{5}{1}$

   D   $\frac{10}{2}$

   E   Not enough information is given.

5   Sarah can ride 4 miles in 20 minutes on her bike. How many miles can she bike in 120 minutes?
   A   12
   B   24
   C   48
   D   120
   E   480

6   The ratio of adults to children on a field trip is 2:7. If there are 14 adults on the trip, how many children are there?
   A   4
   B   5
   C   7
   D   28
   E   49

7   Sam averages 65 miles per hour on a road trip. How many hours will it take him to drive 260 miles?
   A   3
   B   4
   C   5
   D   6
   E   7

8   The ratio of cars to trucks at an auto dealership is $\frac{3}{2}$. If there are 144 cars at the dealership, how many trucks are there?
   A   288
   B   240
   C   216
   D   144
   E   96

9   The Jammers basketball team won 25 games and lost 5 games during their season. What was their ratio of wins to losses?
   A   4:1
   B   5:1
   C   6:1
   D   8:1
   E   25:5

10  The high school equivalency test preparation class has a teacher-to-student ratio of 1:12. If there are 36 students in the class, how many teachers are present?
   A   2
   B   3
   C   4
   D   6
   E   12

**11** On a swimming skills test, Olive performed 12 skills correctly. She performed 4 skills incorrectly. What is the ratio of incorrect skills to total skills?

A $\frac{1}{4}$

B $\frac{1}{2}$

C $\frac{2}{1}$

D $\frac{4}{1}$

E Not enough information is given.

**12** Joe's baseball team won 38 games and lost 4 games this season. What is the ratio of games lost to games won?

A 1:8
B 38:4
C 4:38
D 19:2
E 2:19

**13** A box of soup contains 8 cans. If the box costs $16, what is the unit rate?

A $0.50 per can
B $1 per can
C $2 per can
D $3 per can
E $4 per can

**14** A scale drawing of a living room has a scale of 1 inch : 3 feet. If one wall is 4 inches long on the drawing, how long is the actual wall?

A 4 inches
B 4 feet
C 7 feet
D 12 inches
E 12 feet

**15** Annie drove 96 miles on Monday and 60 miles on Tuesday. What is the ratio of miles she drove on Monday to miles she drove on Tuesday?

A $\frac{1}{2}$

B $\frac{2}{1}$

C $\frac{5}{8}$

D $\frac{8}{5}$

E $1\frac{3}{5}$

**16** A pancake recipe that serves 30 people calls for 12 eggs. Marti wants to make enough to serve only 10 people. What is ratio of eggs to servings for the reduced recipe?

A 2 to 5
B 3 to 5
C 2 to 3
D 3 to 4
E 4 to 5

**17** The ratio of wins to losses for the Wildcats rugby team was 8:3. If the team won 24 games, how many did they lose?

A 7
B 8
C 9
D 10
E 12

**18** There are 30 full-time and 12 part-time employees at the tire plant. What is ratio of full-time to part-time workers?

A 2:1
B 5:2
C 3:1
D 7:2
E 4:1

**19** A map scale states that 2 inches equal 150 miles. If two cities are 6 inches apart on a map, how many miles separate them?

A  154
B  300
C  450
D  600
E  900

**20** For each $5 given to a charity by an individual, the Bay Company will give $15 to that same charity. If individual contributions total $275, how many dollars will the Bay Company contribute?

A  $75
B  $825
C  $1,375
D  $4,125
E  Not enough information is given.

**21** One out of every five dogs that is brought to an animal shelter is adopted within one week. If 35 dogs arrive one week, how many dogs would you expect to be adopted within that week?

A  175
B  40
C  35
D  7
E  5

**22** Stuck in traffic, Trevor drove 48 miles in 3 hours. What was the unit rate of his speed?

A  16 miles per hour
B  24 miles per hour
C  45 miles per hour
D  48 miles per hour
E  144 miles per hour

**23** A recipe for a dessert sauce calls for 2 teaspoons of chocolate sauce and 3 teaspoons of caramel sauce. If Mary made 20 total teaspoons of dessert sauce, how many teaspoons of caramel did she use?

A  10
B  11
C  12
D  13
E  14

**24** There are 460 students in an elementary school. Of the students, $\frac{4}{5}$ ride the bus to school. How many students use a different method of transportation?

A  8
B  9
C  92
D  368
E  575

**25** Thirty people were surveyed about their type of work. Two of every five people work in a field related to education. How many of the people surveyed work in education?

A  5
B  8
C  12
D  15
E  30

**26** In a school, the ratio of students to teachers is 14 to 1. If there are 406 students, how many teachers work at the school?

A  14
B  29
C  406
D  570
E  5,684

**27** A person can burn about 110 calories by walking one mile. How many calories will a person burn by walking $4\frac{1}{2}$ miles?

A  415
B  435
C  455
D  475
E  495

**28** In a 30-person office, 16 people drive to work and the rest walk or ride their bicycles. What is ratio of people who drive to people who do not drive?

A  8:7
B  7:6
C  6:5
D  4:3
E  3:2

UNIT 1

**29** The ratio of cars to parking spots at a local business is 2:3. If there are 26 cars, how many parking spots are there?
A  33
B  36
C  39
D  42
E  45

**Directions:** Questions 30 and 31 are based on the information below.

The table shows the number of miles Leila drove each week on a full tank of gas.

| LEILA'S WEEKLY MILEAGE ||
|---|---|
| Week 1 | 420 miles |
| Week 2 | 414 miles |
| Week 3 | 389 miles |
| Week 4 | 421 miles |
| Week 5 | 396 miles |

**30** If Leila's gas tank holds 18 gallons of gasoline, how many miles per gallon did her car get during Week 2?
A  21
B  22
C  23
D  24
E  25

**31** What is the ratio of the number of miles Leila drove during Week 1 to the total number of miles she drove over the 5 weeks?
A  2:9
B  1:4
C  8:29
D  16:57
E  7:34

**32** The ratio of lifeguards to swimmers at a pool is 1:22. If there are 176 swimmers in the pool, how many lifeguards are there?
A  38
B  30
C  22
D  8
E  2

**33** The ratio of cats to people in a town is 3 to 8. How many people live in the town if there are 387 registered cats?
A  1,032
B  1,161
C  1,935
D  3,096
E  4,257

**34** Ayla bought an 8-pound turkey for $24. How much would she spend for a 12-pound turkey?
A  $4
B  $6
C  $12
D  $18
E  $36

**Directions:** Questions 35 and 36 are based on the information below.

The table shows job openings and the number of applicants at Booksmart Publishing.

| BOOKSMART PUBLISHING |||
|---|---|---|
| POSITION | OPENINGS | APPLICANTS |
| Graphic designer | 5 | 25 |
| Project manager | 2 | 15 |
| Art researcher | 3 | 27 |
| Staff writer | 4 | 48 |
| Editor | 2 | 12 |

**35** Which position shows a ratio of 12 applicants to 1 opening?
A  Graphic designer
B  Project manager
C  Art researcher
D  Staff writer
E  Editor

**36** Booksmart announced plans to hire a total of five, rather than three, art researchers. How many additional applicants can it expect to receive?
A  9
B  12
C  18
D  21
E  24

# Decimals and Operations

## 1 Learn the Skill

You can add, subtract, multiply, and divide decimal numbers. When you perform operations with decimals, you must pay close attention to the placement of the decimal point. For example, when you add or subtract, write the numbers so that the place values and decimal points align.

## 2 Practice the Skill

Problems on mathematics high school equivalency tests will require you to perform operations with decimals. Read the examples and strategies below. Use this information to answer question 1.

**A** Align the decimal points. Then add or subtract like you do with whole numbers.

**B** Multiply as you do with whole numbers. Then count the decimal places in the factors. The number of decimal places in the product will be the sum of these two. Count from the right of the product to place the decimal point.

**C** When dividing by a decimal, first make the divisor a whole number by moving the decimal place to the right. Then move the decimal point the same number of places to the right in the dividend. Add zeros to the dividend if necessary. For example,

$$0.08\overline{)12} \longrightarrow 8\overline{)1200}$$

**D** To multiply by 10, move the decimal one place to the right. To divide by 100, move the decimal two places to the left. The number of zeros shows how many spaces to move.

### Addition

$$\begin{array}{r} {}^{1}\ \\ 3.284 \\ +\ 5.681 \\ \hline \mathbf{8.965} \end{array}$$

### Subtraction

$$\begin{array}{r} {}^{17}\ \\ {}^{5\ 7\ 11}\ \\ 25.681 \\ -\ 3.284 \\ \hline \mathbf{22.397} \end{array}$$

### Division

$$\begin{array}{r} 12.283 \\ 8\overline{)98.264} \\ -8\phantom{0000} \\ \hline 18\phantom{000} \\ -16\phantom{000} \\ \hline 22\phantom{00} \\ -16\phantom{00} \\ \hline 66\phantom{0} \\ -64\phantom{0} \\ \hline 24 \end{array}$$

### Multiplication

$$\begin{array}{r} 5.61 \longleftarrow \text{2 decimal places} \\ \times\ 3.8 \longleftarrow \text{1 decimal place} \\ \hline 4488 \\ +\ 16830 \\ \hline \mathbf{21.318} \longleftarrow \text{3 decimal places} \end{array}$$

---

**TEST-TAKING TIPS**

Use estimation to help you determine the placement of the decimal point in a solution. Round each number to the nearest whole.

**1** Molly bought coffee for $2.95 and a muffin for $1.29. She paid with a $5 bill. How much change did she receive?

A   $0.76
B   $0.86
C   $2.05
D   $4.14
E   $4.24

**Directions:** Questions 2 through 5 are based on the information below.

Ben purchased groceries at Food U Eat. His receipt is shown below.

| FOOD U EAT | | |
|---|---|---|
| QUANTITY | ITEM | UNIT PRICE |
| 5 | Cereal | $3.85 |
| 6 | Milk | $3.50 |
| 2 | Butter | $1.29 |
| 4 | Bread | $2.33 |

2  **What amount did Ben pay for milk?**
A  $14.00
B  $17.50
C  $18.00
D  $21.00
E  $23.10

3  **How much more did Ben pay for the bread than for the butter?**
A  $9.32
B  $6.84
C  $6.74
D  $3.62
E  $3.00

4  **If Ben had purchased only milk and cereal, what would have been the amount of his bill?**
A  $80.85
B  $40.25
C  $40.00
D  $19.25
E  $7.35

5  **Ben has $12.25. If he spends his money only on cereal, about how many boxes could he buy?**
A  1
B  2
C  3
D  4
E  5

6  **Paper Plus sells reams of paper for $5.25 each. Discount Paper sells the same reams of paper for $3.99 each. How much would you save by purchasing 15 reams of paper at Discount Paper instead of at Paper Plus?**
A  $1.26
B  $18.90
C  $59.85
D  $78.75
E  $138.60

7  **David has a piece of rope that is 14.4 meters in length. If he divides the rope into 4 equal pieces, how many meters long will each piece be?**
A  57.6
B  18.4
C  10.4
D  3.6
E  2.2

**Directions:** Questions 8 and 9 are based on the information below.

Coach Steve needed to purchase new soccer equipment for the upcoming season.

| EQUIPMENT | PRICE | QUANTITY |
|---|---|---|
| Soccer ball | $12.95 | 6 |
| Shin guards | $10.95 | 12 sets |
| Knee pads | $8.95 | 12 sets |
| Uniforms | $17.00 | 12 sets |

8  **How much will Coach Steve spend on uniforms and soccer balls?**
A  $29.95
B  $47.95
C  $97.80
D  $211.77
E  $281.70

9  **How much more will Coach Steve spend on shin guards than knee pads?**
A  $2.00
B  $16.00
C  $24.00
D  $36.00
E  $48.00

10  Strawberries at a farmers market are $2.99 per pound. About how much would 5.17 pounds of strawberries cost?
   A  $8
   B  $10
   C  $12
   D  $13
   E  $15

11  A box of eight cans of baked beans sells for $5.89 at a discount warehouse. What is the price per can, rounded to the nearest penny?
   A  $0.73
   B  $0.74
   C  $0.75
   D  $0.76
   E  $0.77

12  Evan purchased a computer for $589.45, a keyboard for $82.32, and a mouse for $14.99. How much did he spend in all?
   A  $604.44
   B  $671.66
   C  $671.77
   D  $686.76
   E  $696.86

13  Alvarez Outdoor Furniture is selling Adirondack chairs for $89.79 each. In the Woods is selling the same chairs for $75.45 each. William buys 4 chairs at In the Woods. How much money did William save on 4 chairs by buying them at In the Woods instead of Alvarez Outdoor Furniture?
   A  $75.45
   B  $57.36
   C  $48.78
   D  $43.02
   E  $14.34

14  Six packages of buns at a bakery cost $7.62. What is the cost of a single package of buns?
   A  $1.09
   B  $1.27
   C  $1.52
   D  $1.91
   E  $2.54

15  Ariana has $37 in cash. She buys a book for $17.95 and a cup of coffee for $3.27. How much money does Ariana have left after her purchases?
   A  $33.73
   B  $21.22
   C  $19.05
   D  $15.78
   E  $14.68

16  Carmen has $163.60 deducted from her paycheck and placed into a retirement account over an 8-week period. How much money is deducted from her paycheck each week?
   A  $16.36
   B  $18.18
   C  $20.45
   D  $23.37
   E  $27.27

17  For lunch, Russell ordered a sandwich for $5.69, a salad for $3.98, and a drink for $1.99. How much did he spend on lunch?
   A  $11.66
   B  $10.46
   C  $9.86
   D  $9.67
   E  $7.68

18  Timothy pays $143 per year to subscribe to his local newspaper. What is the cost per week of his subscription?
   A  $2.75
   B  $2.80
   C  $2.86
   D  $2.92
   E  $2.98

**19** Cedric has $597.16 in his checking account. He deposits a check for $217.98. He takes out $45 in cash at the same time. Which of the following represents his new balance?

A  $597.16 − ($45 + $217.98)
B  $597.16 − $217.98 + $45
C  $597.16 − $217.98 − $45
D  $597.16 + $217.98 − $45
E  $597.16 + $217.98 + $45

**20** Salami at a deli costs $3.95 per pound. What is the cost, without sales tax, of 2.3 pounds of salami?

A  $9.09
B  $9.10
C  $9.85
D  $9.90
E  $10.08

**21** Sylvia bought a computer on a finance plan. She will make 12 equal payments altogether to pay for the computer. If the cost of the computer was $675.00, what is the amount of each month's payment?

A  $56.25
B  $57.30
C  $58.00
D  $58.25
E  $58.30

**22** At a home improvement store, Terese bought a new lamp for $14.89, a pack of light bulbs for $2.38, and a new light switch for $0.79. She paid with a $20 bill. How much change should be returned to her?

A  $1.94
B  $2.73
C  $4.32
D  $17.27
E  $18.06

**Directions:** Questions 23 and 24 are based on the information below.

The table shows the costs of various services at a hair salon.

| HAIR SALON SERVICES | |
|---|---|
| **SERVICE** | **COST** |
| Haircut | $22.95 |
| Shampoo and Style | $14.85 |
| Permanent Wave | $56.99 |
| Color | $66.25 |

**23** Lisa got a haircut and color at the hair salon. What was her total cost?

A  $43.30
B  $79.94
C  $81.10
D  $89.20
E  $123.24

**24** Mark is going to the hair salon before his wedding. He is trying to decide if he should get a haircut or just have his hair shampooed and styled. How much money would he save by only having his hair shampooed and styled?

A  $2.10
B  $6.85
C  $8.10
D  $9.20
E  $37.80

**25** Tim scored 95.75 on his first mathematics exam, 92.5 on his second exam, and 98.25 on his third exam. What is the combined total of his scores on the three exams?

A  285.95
B  286.05
C  286.5
D  286.75
E  287.5

**Directions:** Questions 26 and 27 are based on the information below.

The table shows the amount of Jonah's electric bill from July through December.

| MONTHLY ELECTRIC BILL COSTS | |
|---|---|
| **MONTH** | **AMOUNT OF BILL** |
| July | $124.53 |
| August | $118.92 |
| September | $95.41 |
| October | $88.73 |
| November | $85.04 |
| December | $86.29 |

**26** What is the difference between Jonah's highest and lowest electric bills during these 6 months?
A $61.57
B $39.49
C $38.24
D $33.88
E $5.61

**27** How much did Jonah pay for electricity from July through December?
A $386.84
B $474.39
C $510.24
D $512.63
E $598.92

**28** Angel hair pasta at Hometown Foods normally costs $2.29. This week, it is on sale for $2.05. Lorenzo bought 5 boxes on sale. Which of the following represents the amount of money he saved?
A 5 × ($2.29 − $2.05)
B ($2.29 − $2.05) ÷ 5
C 5 × $2.29 − $2.05
D 5 × $2.05
E 5 × $2.29

**29** A diver's score is calculated by adding the scores of three judges and then multiplying this sum by the degree of difficulty of the dive. Craig performed a dive with a degree of difficulty of 3.2. He received scores of 8, 8.5, and 7.5. What was his total score for the dive?
A 67.8
B 68.7
C 76.8
D 78.6
E 86.7

**30** A soft drink at a park costs $1.79. Dylan has $7.90. About how many drinks could Dylan buy at the park?
A 1
B 2
C 3
D 4
E 5

**31** Alexis bought 6 fruit smoothies for $2.65 each. She paid with a $20.00 bill. How much change should be returned to her?
A $10.35
B $18.60
C $7.35
D $4.10
E $2.35

**32** Walt took out a loan to buy his new car. He makes equal monthly payments. In a year, he pays $1,556.28. How much does he pay per month on his car loan?
A $128.29
B $129.69
C $130.99
D $131.09
E $132.89

**33** A car traveled at a speed of 60.2 miles per hour for 3.5 hours. How many miles did the car travel?
A 18.5
B 21.7
C 180.5
D 200.7
E 210.7

# Fractions, Decimals, and Percents

## ① Learn the Skill

As with fractions and decimals, **percents** show part of a whole. Recall that with fractions, a whole can be divided into any number of equal parts. With a decimal, the number of equal parts must be a power of 10. Percent always compares amounts to 100. The percent sign, %, means "out of 100."

## ② Practice the Skill

Understanding how to convert among fractions, decimals, and percent will help you to efficiently solve problems on mathematics tests. Read the examples and strategies below. Then complete the table of conversions. Use this information to answer question 1.

**A** To write a decimal as a percent, multiply by 100. Move the decimal point two places to the right and write the percent sign. To write a percent as a decimal, do the opposite.

| FRACTION | DECIMAL | PERCENT |
|---|---|---|
| $\frac{1}{5}$ | 0.2 | **20%** |
| $\frac{1}{4}$ | **0.25** | **25%** |
| $\frac{1}{2}$ | 0.5 | 50% |
| $\frac{3}{4}$ | 0.75 | |
| | 1.8 | |

**B** To write a fraction or mixed number as a decimal, divide the numerator by the denominator. To write a fraction or a mixed number as a percent, write it as a decimal and multiply by 100.

$\frac{1}{4} = 0.25; 0.25 \times 100 = 25\%$

To write a percent as a fraction or a mixed number, drop the percent sign and write the number as the numerator and 100 as the denominator. Then reduce the fraction.

$25\% = \frac{25}{100} = \frac{1}{4}$

**C** To rename a decimal as a fraction, write the number (without its decimal point) as the numerator. What is the place value of the last decimal digit? For 0.5, the place value is tenths. Write 10 as the denominator to show "five tenths."

$0.5 = \frac{5}{10} = \frac{1}{2}$

---

**✓ TEST-TAKING TIPS**

If a number is shown without a decimal point, such as 80%, assume that it lies directly to the right of the ones digit 80% = 80.0%.

**1** In a neighborhood, 27 of the 45 children are in elementary school. What percent of the children in the neighborhood are in elementary school?

- **A** 20%
- **B** 25%
- **C** 40%
- **D** 60%
- **E** 166%

---

2  Shelly's Boutique is advertising 25% off all merchandise. What fraction of the original price will customers save during the sale?

A  $\frac{2}{3}$

B  $\frac{1}{2}$

C  $\frac{1}{3}$

D  $\frac{1}{4}$

E  $\frac{1}{5}$

3  The unit price of a can of peas is $7\frac{1}{2}$ cents per ounce. To find the price of an 8-ounce can, Jeff first writes the mixed number as a decimal. What decimal does he write?
A  7.2
B  7.5
C  7.55
D  7.6
E  7.75

4  City Electric provides electricity for $\frac{3}{8}$ of the homes in Center City. For what percentage of the homes does City Electric provide electricity?
A  37.5%
B  37%
C  36.5%
D  36%
E  35.5%

5  In the spring, $\frac{1}{8}$ of the students at a community college participate in a work-study program. What decimal describes $\frac{1}{8}$?
A  0.08
B  0.105
C  0.125
D  0.8
E  8.0

6  In a survey, 0.22 of the respondents answered "Yes" to the question "Would you consider voting for a candidate from a third party?" What fraction of respondents answered "No"?

A  $\frac{11}{50}$

B  $\frac{22}{100}$

C  $\frac{22}{50}$

D  $\frac{39}{50}$

E  $\frac{78}{10}$

7  The Strikers girls soccer team won 9 of its 13 games. What percentage of games did the Strikers win?
A  61.5%
B  66.7%
C  69.2%
D  76.9%
E  84.6%

8  On a science test, Jarrod answered 88% of the questions correctly. If there were 25 questions on the test, how many did Jarrod correctly answer?
A  18
B  19
C  20
D  21
E  22

9  At Bright Minds Learning, 75% of employees work as instructors. If there are 300 employees at Bright Minds Learning, how many of them work as instructors?
A  150
B  175
C  200
D  225
E  250

**10** One-eighth of the children in first grade are dropped off at school by their parents. What percentage of first-graders are dropped off at school?

A  0.8%
B  8.5%
C  12%
D  12.5%
E  13%

**11** At the local college, $\frac{2}{50}$ of all students have a full-time job while attending college. What percentage of students attend college and work full-time?

A  0.4%
B  2%
C  4%
D  8%
E  40%

**12** Of 140 sixteen-year-olds, 85% have taken driver's education and earned their driver's licenses. What fraction of the students received their licenses?

A  $\frac{15}{100}$

B  $\frac{3}{20}$

C  $\frac{17}{100}$

D  $\frac{17}{27}$

E  $\frac{17}{20}$

**13** Each semester, about $\frac{3}{25}$ of students at a college study abroad. What percentage of students study abroad?

A  3%
B  9%
C  12%
D  21%
E  25%

**Directions:** Questions 14 and 15 are based on the information below.

Refrigerator
SALE

10% OFF

This promo is good while supplies last.

**14** What fraction of the original price will customers pay for the refrigerator if they buy it during the store's sale?

A  $\frac{1}{10}$

B  $\frac{3}{5}$

C  $\frac{3}{4}$

D  $\frac{4}{5}$

E  $\frac{9}{10}$

**15** Employees at the store receive a 30% discount on the original price. If the refrigerator ordinarily is priced at $580, how much will Sam, the store's manager, pay for it during the sale?

A  $348
B  $406
C  $435
D  $464
E  $522

**16** Marie needs 500 fliers to be printed at a cost of $2 per flier. Because of the size of the order, the print shop is asking Marie to pay 30% up front. How much of a down payment must Marie make on the fliers?

A  $300
B  $320
C  $330
D  $360
E  $400

**17** On a test, 45% of the questions are related to science. What fraction of the questions on the test are science-related?

A $\frac{1}{45}$

B $\frac{9}{20}$

C $\frac{55}{100}$

D $\frac{45}{55}$

E $\frac{55}{45}$

**18** The unit price of a can of soda is $22\frac{1}{2}$ cents. Aidan renames the mixed number as a decimal. What is the unit price in dollars when written as a decimal?

A $0.215
B $0.22
C $0.225
D $0.44
E $0.445

**19** Nina biked $54\frac{1}{2}$ miles in two days. If she biked 22.8 miles the first day, how many miles did she bike the second day?

A 32.2
B 31.7
C 31.6
D 28.2
E 25.3

**20** On Kennedy's income taxes, 5.5% of her income goes to pay her self-employment taxes. What fraction of her income is this?

A $\frac{1}{20}$

B $\frac{11}{200}$

C $\frac{1}{2}$

D $\frac{11}{20}$

E Not enough information is given.

**21** Eighty-two percent of the employees at a food processing plant belong to an employee's union. How is this percent expressed as a decimal?

A 0.82%
B 0.82
C 8.2
D 82.0%
E 82.0

**22** The Panthers won 22 of their 34 games. What percentage of the time did the Panthers win?

A 67.6%
B 64.7%
C 61.7%
D 58.8%
E 55.8%

**23** On a mathematics test, Ted correctly answered 41 of 50 questions. What percentage of questions did Ted answer correctly?

A 80%
B 81%
C 82%
D 83%
E 84%

**24** Bryon recently purchased a new laptop computer. He put 20 percent down on the purchase. If the computer cost $1,230, how much does Bryon owe after the down payment?

A $246
B $492
C $615
D $984
E $1,107

**25** Jim makes and sells denim knapsacks for $10.50 apiece. The knapsacks cost $7 apiece to produce. What is the percentage of profit that Jim makes on each knapsack?

A 50%
B 75%
C 100%
D 125%
E 150%

**Directions:** Questions 26 and 27 are based on the information below.

The Clothing Depot is having a two-day sale on its most popular items. The table shows the discount on each type of clothing in the sale.

| CLOTHING DISCOUNTS | |
| --- | --- |
| **ITEM** | **DISCOUNT** |
| Shirts | 30% off |
| Shorts | 25% off |
| Socks | 10% off |
| Jackets | 40% off |
| Vests | 25% off |

**26** Maya buys a shirt. What fraction of the original price does she pay for the shirt?

A $\frac{3}{10}$

B $\frac{2}{3}$

C $\frac{14}{21}$

D $\frac{7}{10}$

E $\frac{5}{6}$

**27** Carlos buys a jacket. What fraction of the original price will he save?

A $\frac{1}{5}$

B $\frac{1}{4}$

C $\frac{1}{3}$

D $\frac{2}{5}$

E $\frac{2}{3}$

**28** Delia earns $28,500 per year. She budgets $\frac{1}{6}$ of this for food and 0.35 of this for housing expenses. What fraction of her earnings is left to budget for other items?

A $\frac{6}{35}$

B $\frac{8}{26}$

C $\frac{29}{60}$

D $\frac{31}{60}$

E   Not enough information is given.

**29** Elliott sells 8x10 color photographs for $12 apiece. His friends receive a 10% discount. What fraction of the original price do Elliott's friends pay?

A $\frac{1}{10}$

B $\frac{1}{9}$

C $\frac{9}{10}$

D $\frac{9}{1}$

E $\frac{10}{1}$

**30** The Kickers soccer team played 24 matches and won 75 percent of them. How many matches did the Kickers win?
A   15
B   16
C   17
D   18
E   19

**31** Jason paid $5\frac{1}{2}$% interest on his computer loan. What is this interest rate expressed as a decimal?
A   0.055
B   0.512
C   0.55
D   5.12
E   5.5

# Percent Problems

## 1 Learn the Skill

There are three main parts of a percent problem—the base, the rate, and the part. The **base** is the whole amount. The **rate** tells how the base and whole are related. The **part** is a piece of the whole or base. The rate is always followed by a percent sign. You can use proportions to solve percent problems. You can also use the percent formula: **base × rate = part**. In the formula, write the rate as a decimal.

## 2 Practice the Skill

You must understand how the base, part, and rate are related to successfully solve percent problems on mathematics tests. As you practice percent problems, many common percents will become familiar to you, and eventually you will be able to use mental math to solve the problems. Read the examples and strategies below. Use this information to answer question 1.

**A** Set the rate over 100 equal to the part over the base.

**B** The variable $p$ = the amount of money borrowed, rate ($r$) is the percent charged, and time ($t$) is the time in years that you are borrowing money. Before you calculate, change the rate to a decimal and make sure the time is in years. Write months as a fraction or a decimal.

**C** You may be asked to find the percent of change. Subtract to find the difference between the original and new amounts. Then divide the difference by the original amount. Convert the decimal to a percent.

### Use a Proportion
Zach answered 86% of the questions on a math exam correctly. If there were 50 questions, how many questions did Zach answer correctly?

$$\frac{\text{Part}}{\text{Base}} = \frac{\text{Rate}}{100} \qquad \frac{?}{50} = \frac{86}{100}$$

$$50 \times 86 = 4300 \longrightarrow 4300 \div 100 = \textbf{43 questions}$$

### Find Percent Increase or Decrease
Last year, Kareem paid $750 per month for rent. This year he pays $820 a month. What is the percent of increase?

$\begin{cases} \$820 - \$750 = \$70.00 \\ \$70.00 \div \$750 = 0.09 \\ 0.09 \times 100 = \textbf{9\%} \end{cases}$

### Interest Problems
Kelly took out a $20,000 loan for four years at 3% interest. How much interest ($I$) will she pay on the loan?

$I = prt$
$I = \$20,000 \times 0.03 \times 4$
$I = \textbf{\$2,400}$

### ✓ TEST-TAKING TIPS

A proportion can be used to solve percent-of-change problems. Write the amount of change over the original amount. Set this ratio equal to the rate over 100.

**1** Kirsten borrowed $1,000 from her sister for six months. If she pays 5% interest, what is the total that she will owe her sister in six months?

A $1,000
B $1,025
C $1,050
D $1,075
E $1,300

UNIT 1

2  Tia earns $552 per week. Of this amount, 12% is deducted for taxes. What amount is deducted each week?
   A  $6.62
   B  $55.20
   C  $66.00
   D  $66.24
   E  $485.76

3  At a food packaging factory, 309 of the 824 employees work the third shift. What percentage of employees work the third shift?
   A  25%
   B  32.5%
   C  35%
   D  37.5%
   E  40%

4  Andrew received a raise from $24,580.00 per year to $25,317.40 per year. What percent raise did he receive?
   A  1%
   B  2%
   C  3%
   D  4%
   E  5%

5  Isabella paid $425 for a new bicycle, plus 6% sales tax. Which of the following represents the total amount she paid?
   A  0.06 × $425
   B  0.6 × $425
   C  $425 + (0.6 × $425)
   D  $425 + (0.06 × $425)
   E  $425 × 1.6

6  A computer company received 420 customer service calls in one day. Forty-five percent of the calls were about software issues. How many of the calls were about software?
   A  19
   B  189
   C  210
   D  229
   E  231

7  A furniture store is having a sale. A sofa is regularly priced at $659 but is on sale for 20% off. What is the sale price of the sofa?
   A  $649.00
   B  $527.20
   C  $450.80
   D  $427.50
   E  $131.80

8  Daria invested $5,000 in an account that earns 5% interest over nine months. How much interest will she earn over that time?
   A  $5,187.50
   B  $3,750.00
   C  $2,250.00
   D  $250.00
   E  $187.50

9  Fredrica's take-home pay is $2,250 per month. She spends 20% of this on her rent. How much does Fredrica spend each month on rent?
   A  $400
   B  $450
   C  $500
   D  $550
   E  $600

10 At a publishing house, Herb supervises 25 employees. His staff will grow by 40% next month. How many employees will Herb have on staff next month?
   A  35
   B  45
   C  55
   D  65
   E  75

**11** As of January 1, Theo's monthly rent increased from $585 to $615. What was the approximate percent increase of Theo's rent?

A 4%
B 5%
C 6%
D 7%
E 8%

**12** Pete paid $2.86 in sales tax on a purchase. If he paid 8% sales tax, what was the cost of the item?

A $12.50
B $15.36
C $22.88
D $28.60
E $35.75

**13** Ezra invested $3,000 for 18 months at 3% interest. How much interest will he earn on his investment?

A $90.00
B $135.00
C $180.00
D $900.00
E $1,350.00

**14** Dan paid 20% down on a new car that cost $16,584.00. He will pay the balance in 24 equal monthly installments. How much will he pay each month?

A $135.00
B $138.20
C $552.80
D $677.18
E $691.00

**15** Noelle bought a new jacket. The original price was $152.60. If the jacket was on sale for 40% off, how much did Noelle pay for the jacket?

A $146.50
B $122.08
C $112.60
D $91.56
E $61.04

**16** Last year, Cullen contributed $4\frac{1}{2}$% of his salary to charity. If he contributed $2,025.00 to charity, what was Cullen's salary?

A $4,500
B $9,125
C $45,000
D $91,125
E $450,000

**17** Remy took out a home-improvement loan to pay for new kitchen cabinets. The loan was for $10,000 at an annual interest rate of 5.6%. If Remy paid off the loan in 36 months, how much did she pay in all?

A $10,680
B $11,160
C $11,680
D $12,200
E $12,760

**18** Jae just bought her first car for $22,000. She put 20% down and financed the rest of the purchase over three years at 0% percent interest. What is the amount of Jae's monthly car payment?

A $427.77
B $488.89
C $586.66
D $611.11
E $651.30

**19** Jessica purchased a kayak in Florida, where the state sales tax is 6%. She paid $72 in sales tax. What was the retail price of the kayak?

A $1,050
B $1,100
C $1,150
D $1,200
E $1,250

**Directions:** Questions 20 and 21 are based on the information below.

A parts factory records the total number of each part sold in a monthly report. The report for May is shown below.

| PARTS SOLD IN MAY | |
|---|---|
| PART NUMBER | AMOUNT SOLD |
| A056284 | 120,750 |
| B057305 | 254,860 |
| P183456 | 184,340 |
| F284203 | 290,520 |
| Q754362 | 308,205 |

20 What percent of the parts sold were part number Q754362? Round to the nearest whole percent.
   A  23%
   B  24%
   C  25%
   D  26%
   E  27%

21 If combined, the sales of part numbers B057305 and F284203 equal what percentage of parts sold in May?
   A  37%
   B  39%
   C  43%
   D  47%
   E  51%

22 Ethan works as a waiter. A customer had a check that showed $38.00 for food and drinks, $2.66 in tax, and a $40.66 subtotal. He received a $6.10 tip based on the subtotal of the check. About what percent of the subtotal was his tip?
   A  10%
   B  15%
   C  17%
   D  18%
   E  20%

23 Grace used to earn $11.00 per hour at her job. She recently received a 4% raise. What is Grace's new hourly wage?
   A  $0.44
   B  $10.56
   C  $11.04
   D  $11.44
   E  $11.88

24 In March, the average number of passengers on a commuter train line was 5,478 per day. In April, this number dropped to 4,380 due to construction on the train line. What is the approximate percent decrease in ridership from March to April?
   A  15%
   B  18%
   C  20%
   D  22%
   E  25%

25 Michelle changed jobs this summer. Her new yearly salary is 15% more than her previous salary. If her new salary is $86,250.00, what was her previous salary?
   A  $97,500.00
   B  $75,000.00
   C  $73,312.50
   D  $63,750.00
   E  $11,250.00

26 There were 35 students signed up for an aerobics class. Then the class size increased by 20%. How many students are now in the class?
   A  38
   B  40
   C  42
   D  45
   E  48

27 Quinn borrows $10,500 for 3 years to pay college tuition. If she pays $787.50 in interest on the loan, what is the interest rate of the loan?
   A  1.5%
   B  2%
   C  2.5%
   D  3%
   E  3.5%

**28** Seventy-two percent of the members of a teacher's union voted to accept the terms of a new contract. If 324 teachers voted in favor of the new contract, how many voted against it?

A   126
B   198
C   233
D   252
E   278

**29** A clothing store has the following sign posted in the window.

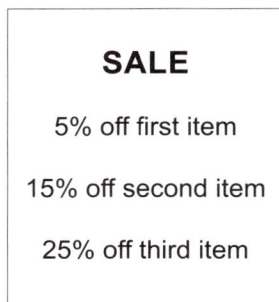

---

**SALE**

5% off first item

15% off second item

25% off third item

---

Adam buys 3 shirts. If each shirt has an original price of $12, what is the total that he pays for the three shirts on sale?

A   $9.00
B   $25.20
C   $27.00
D   $29.40
E   $30.60

**30** A cheese factory had 850 employees at the beginning of the first quarter. By the beginning of the second quarter, they had hired 8% more employees. Which of the following cannot be used to find the number of new employees the cheese factory hired by the beginning of the second quarter?

A   $\frac{8}{100} = \frac{x}{850}$

B   $0.08(850) = x$

C   $\frac{8}{x} = \frac{100}{850}$

D   $0.08(x) = 850$

E   $\frac{x}{850} = \frac{8}{100}$

**31** The number of students at a school changed from 756 last year to 711 to this year. What is the approximate percent of change in the student population?

A   increase of 6%
B   decrease of 6%
C   increase of 5%
D   decrease of 5%
E   increase of 4%

**32** Judith took out a loan for $4,300 over 6 months. What is the total amount she will owe in 6 months?

A   $4,300
B   $4,450
C   $4,600
D   $4,900
E   Not enough information is given.

**33** Dan is determining whether to do his holiday shopping in Maryland, where there is a 6% sales tax, or in Delaware, where there is no sales tax.

| PRICES OF GIFT ITEMS | | |
|---|---|---|
| **GIFT ITEM** | **DELAWARE PRICE** | **MARYLAND PRICE** |
| Women's jacket | $45 | $43 |
| Computer game | $32 | $30 |
| Hardcover book | $20.50 | $19.50 |
| Children's music CD | $10 | $9.50 |

For which item would Dan pay less if he purchased it in Maryland?

A   women's jacket
B   computer game
C   hardcover book
D   children's music CD
E   none of the items

# Squaring, Cubing, and Taking Roots

## ① Learn the Skill

When a number or a variable is multiplied by itself, the result is called the **square** of that number or variable. Squaring the number 5, for example, is finding the product $5 \times 5 = 25$; this product is written as $5^2$, where the $^2$ indicates that the product is composed of two factors of 5.

When a number or variable is multiplied by itself an additional time, the result is called the **cube** of the number or variable. For example the cube of 5 is $5 \times 5 \times 5 = 125$; this product is written as $5^3$.

To find the **square root** of a number, find the number that, when squared, equals the given number. The **cube root** of a number is that number which, when cubed, equals the given number. Square and cube roots are indicated by root signs, such as $\sqrt{25}$ and $\sqrt[3]{125}$, respectively.

## ② Practice the Skill

By practicing the skills of squaring, cubing, and taking the corresponding roots of quantities, you will improve your study and test-taking abilities, especially as they relate to mathematics tests. Read the examples and strategies below. Use this information to answer question 1.

**A** The square of a negative number is positive; if two numbers differ only in their sign, their squares are both positive and equal. The square root of a positive number can, as a result, have two values. Since there are no real numbers that, when multiplied by themselves, give a negative number, square roots of negative numbers are undefined when dealing with real numbers.

**B** The cube of a negative number is negative. As a result, the cube root of a negative number exists, is negative, and is equal in magnitude to the cube root of the absolute value of the number.

$$1^2 = 1 \times 1 = 1 \qquad 1^3 = 1 \times 1 \times 1 = 1$$

$$2^2 = 2 \times 2 = 4 \qquad 2^3 = 2 \times 2 \times 2 = 8$$

$$(-1)^2 = (-1) \times (-1) = 1 \qquad (-1)^3 = (-1) \times (-1) \times (-1) = -1$$

$$(-2)^2 = (-2) \times (-2) = 4 \qquad (-2)^3 = (-2) \times (-2) \times (-2) = -8$$

$$\sqrt{1} = 1, -1 \qquad \sqrt[3]{1} = 1$$

$$\sqrt{4} = 2, -2 \qquad \sqrt[3]{8} = 2$$

$$\sqrt{9} = 3, -3 \qquad \sqrt[3]{27} = 3$$

$$\sqrt{-9} = \text{undefined} \qquad \sqrt[3]{-27} = -3$$

### ✓ TEST-TAKING TIPS

Taking the square root of a number is different from dividing a number by 2. When finding the square root of $x$, consider: *what number times itself equals x?* When dividing $x$ by 2, think: *what number plus itself equals x?*

**1** The length of a square can be determined by finding the square root of its area. If a square has an area of 81 m², what is the length of the square?

   **A** 7.5 m

   **B** 8.0 m

   **C** 8.5 m

   **D** 9.0 m

   **E** 9.5 m

UNIT 1

## ③ Apply the Skill

**Directions:** Questions 2 and 3 are based on the information below.

The diagram shows the area of Meredith's square garden.

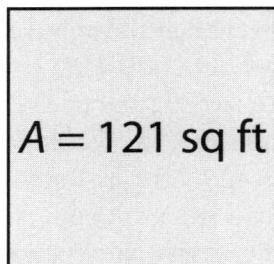

$$A = 121 \text{ sq ft}$$

**2** Use the formula for the area of a square. What is the length of one side of Meredith's garden?
- **A** 12 ft
- **B** 11 ft
- **C** 10 ft
- **D** 9 ft
- **E** 8 ft

**3** Meredith decides to double the length of each side of her garden. What will its new area be?
- **A** 88 sq ft
- **B** 242 sq ft
- **C** 363 sq ft
- **D** 484 sq ft
- **E** 605 sq ft

**4** Carlos completed $x^3$ squats as part of his football workout. If $x = 5$, how many squats did he complete?
- **A** 10
- **B** 15
- **C** 25
- **D** 125
- **E** 243

**5** The length of a side of a cube can be determined by finding the cube root of its volume. If a cube has a volume of 64 cm³, what is the length of a side?
- **A** 4.0 cm
- **B** 8.0 cm
- **C** 16.0 cm
- **D** 32.0 cm
- **E** 262,144 cm

**6** To determine the length of yarn needed for a project, Josie must solve $\frac{\sqrt{x}}{4}$ for $x = 64$. What is the solution?
- **A** 2
- **B** 4
- **C** 6
- **D** 8
- **E** 16

**7** Mark multiplied a number by itself. He found a product of 30. What is the number, rounded to the nearest tenth?
- **A** 4.5
- **B** 5.4
- **C** 5.5
- **D** 10.0
- **E** 15.0

**8** A square has an area of 50 square feet. What is the perimeter of the square, rounded to the nearest foot?
- **A** 7 ft
- **B** 14 ft
- **C** 28 ft
- **D** 49 ft
- **E** 100 ft

**9** A student is told that a moving car travels a distance $x$, where $x$, expressed in miles, is the solution of the equation $(8 - x)^2 = 64$. The student argues that the car can't be moving, since $x$ must equal zero. Which of the following describes the distance, $x$?
- **A** The student is correct; $x$ must equal zero.
- **B** $x$ is less than 0.
- **C** $x$ is greater than zero, but less than 10.
- **D** $x$ is between 10 and 20.
- **E** $x$ is greater than 20.

**Lesson 10 | Squaring, Cubing, and Taking Roots**

**10** A math website gives the number of questions in its daily quiz as a square root. Today, there are $\sqrt{144}$ questions. How many questions are there?

A  10
B  11
C  12
D  13
E  14

**11** Amanda used her calculator to find $\sqrt{7,788}$. What is this number rounded to the nearest hundredth?

A  88.24
B  88.25
C  89.24
D  89.25
E  90.25

**12** The length of a cube is 29 cm. What is the volume of the cube in cubic centimeters?

A  87
B  841
C  24,389
D  78,586
E  707,281

**13** The area of a square is 6.7 sq ft. What is the length of a side of the square to the nearest tenth of a foot?

A  2.5
B  2.6
C  2.9
D  3.3
E  3.4

**14** $\sqrt{33}$ is between which two numbers?

A  4 and 5
B  5 and 6
C  6 and 7
D  7 and 8
E  8 and 9

**15** Hannah buys carpet for her room, which has an area of 216 square feet and is 50% longer than it is wide. What are the dimensions of the room?

A  width = 12 feet; length = 18 feet
B  width = 6 feet; length = 9 feet
C  width = 12 feet; length = 12 feet
D  width = 18 feet; length = 12 feet
E  width = 9 feet; length = 24 feet

**16** Which set of integers solves the equation $x^2 = 16$?

A  4, −2
B  4, −4
C  2, 8
D  −2, −8
E  −256, 256

**17** Jon's monthly car payment is the square of 19 minus the square root of 169. What is Jon's monthly car payment?

A  $188
B  $348
C  $361
D  $374
E  $530

**18** A soccer field for a youth league has an area of 4,000 square yards, and is 60% longer than it is wide. What is the length of the field?

A  50 yards
B  60 yards
C  70 yards
D  80 yards
E  1,538 yards

**19** A glass table is a cube with a side length of 4 feet. The center of the table is a hollow cube with a side length of 1.5 feet. What volume of glass was used to make the table?

A  3.375 cubic feet
B  60.625 cubic feet
C  61.75 cubic feet
D  64 cubic feet
E  67.375 cubic feet

**20** The length of each side of a rectangular prism is increased by a factor of three. By what factor does the volume increase?

A  3
B  9
C  18
D  27
E  81

**21** The length of each side of a rectangle increases by a factor of 2. By what factor does the area increase?

A  2
B  4
C  6
D  8
E  16

**22** The equation $x^2 = 25$ has two different solutions. What is the product of these two solutions?

A  −25
B  −5
C  0
D  5
E  25

**23** The equation $(x − 1)^2 = 64$ has two different solutions. What is the product of these two solutions?

A  81
B  64
C  49
D  −49
E  −63

**24** How many times greater is the cube of the positive square root of 64 than the square of the cube root of 64?

A  4
B  8
C  16
D  32
E  The values are equal.

**25** The equation $(x − 6)^2 = 4$ has two different solutions. What is the product of these two solutions?

A  −64
B  −32
C  4
D  32
E  64

**26** The area of one side of a cube is 30.25 square inches. What is the volume of the cube, rounded to the nearest cubic inch?

A  121
B  166
C  242
D  915
E  27,681

**27** A square table has an area of 2,000 square inches. What is the length of each side, rounded to the nearest inch?

A  40 inches
B  44 inches
C  45 inches
D  50 inches
E  500 inches

**28** A small refrigerator is cubic in shape and has an outside width of 18 inches. What is the volume of the space the refrigerator takes up, expressed as cubic *feet*, to the nearest tenth of a cubic foot?

A  1.4
B  2.3
C  3.3
D  3.4
E  4.5

**29** A child has a set of play blocks that are cubic in shape and measure 2 inches on each side. How many blocks does the child need to make a cubic stack with a volume of 1 cubic foot?

A  18
B  36
C  144
D  216
E  1,728

**30** A relationship between the temperature measured in degrees Fahrenheit ($F$) and the temperature in degrees Celsius ($C$) is $25(F - 32)^2 = 81C^2$. If the temperature is 41 degrees Fahrenheit, what is the temperature in degrees Celsius?

A  5
B  9
C  23
D  25
E  45

**31** Which number solves the equation $x^2 = -16$?

A  8
B  4
C  -4
D  -8
E  It is undefined.

**32** Which number solves the equation $x^3 = -64$?

A  8
B  4
C  -4
D  -8
E  It is undefined.

**33** Chris is laying floor tiles in his kitchen, which has a width of $\frac{3}{4}$ its length, and an area of 192 square feet. What are the dimensions of the room?

A  8 feet by 24 feet
B  12 feet by 15 feet
C  12 feet by 16 feet
D  16 feet by 18 feet
E  20 feet by 24 feet

**34** Kelly makes a box that has a width equal to its depth, a length equal to three times the depth, and a volume of 192 cubic inches. What are the dimensions of the box?

A  depth = 4 in.; width = 4 in.; length = 12 in.
B  depth = 2 in.; width = 4 in.; length = 8 in.
C  depth = 4 in.; width = 4 in.; length = 8 in.
D  depth = 12 in.; width = 4 in.; length = 4 in.
E  depth = 3 in.; width = 3 in.; length = 9 in.

**35** For which list of numbers is the following expression undefined: $\sqrt{(x^2 - 1.5)}$ ?

A  1.5, -1.5, -2
B  1.4, 1.6, -1.8
C  1.3, 2, -3
D  -1, 0, 1
E  -2, 1, 3

**36** Rob refinanced his monthly student loan so that it equals the square of 14. Rob pays in monthly rent the square of 21 minus the square root of 49. How much does Rob pay each month in student loans and rent?

A  $396
B  $445
C  $630
D  $644
E  $686

**37** For which equation is $x$ undefined?

A  $x^3 = 8$
B  $x^3 = -27$
C  $x^2 = -49$
D  $x^2 = 121$
E  $x^3 = 0$

# Exponents and Scientific Notation

## ① Learn the Skill

**Exponents** are used when a number, the **base**, is multiplied by itself many times. The exponent shows the number of times that the base appears in the product. When a quantity is given an exponent of $n$, it is said to be raised to the $n$th **power**; for example, $2^5$ is the same as 2 raised to the 5th power. There are rules for adding, subtracting, multiplying, and dividing quantities with exponents.

**Scientific notation** uses exponents and powers of 10 to write very small and very large numbers in a compact form that simplifies calculations. Scientific notation requires that the decimal point be located just to the right of the first non-zero digit.

## ② Practice the Skill

By practicing the skill of working with exponents and scientific notation, you will improve your study and test-taking abilities, especially as they relate to mathematics tests. Read the examples and strategies below. Use this information to answer question 1.

**A** A number or quantity raised to the first power equals itself. A number or quantity (except zero) raised to the zero power equals one. When a number is raised to a negative power, write the reciprocal and change the negative exponent to a positive.

**B** Terms can be added and subtracted if they are alike, meaning they must have the same variable raised to the same exponent.

**C** To multiply terms with the same base, keep the base and add the exponents. Do the opposite for division. If the bases are not the same, simplify using the order of operations.

$$5^1 = 5 \qquad 5^0 = 1 \qquad 5^{-2} = \frac{1}{5^2} = \frac{1}{25}$$

$$2x^2 + 4x^2 + 1 = 6x^2 + 1 \qquad 4x^2 - x^2 = 3x^2$$

$$(3^2)(3^3) = (3)^{2+3} = 3^5 \qquad \frac{6^5}{6} = 6^{5-1} = 6^4$$

$$4.2 \times 10^7 = 42{,}000{,}000 \qquad 5{,}800{,}000 = 5.8 \times 10^6$$

$$3.7 \times 10^{-5} = 0.000037 \qquad 0.000052 = 5.2 \times 10^{-5}$$

**D** To write a number shown in scientific notation as a number in expanded form, look at the power of 10. The exponent tells how many places to move the decimal point—right for positive, left for negative. To write a number in scientific notation, place the decimal point directly after the ones digit. Next, count the number of places you need to move. Then drop the zeros at the ends.

### ✔ TEST-TAKING TIPS

Simplifying algebraic expressions typically involves practices such as expanding terms, combining like terms, looking for common factors among terms, and canceling factors.

**1** The distance between the sun and Mercury is about 58,000,000 km. What is this distance written in scientific notation?

A $5.8 \times 10^6$

B $5.8 \times 10^7$

C $58 \times 10^6$

D $58 \times 10^7$

E $5.8 \times 10^8$

**2** There are 25,400,000 nanometers in an inch. What is this number written in scientific notation?

A  $2.54 \times 10^6$

B  $2.54 \times 10^7$

C  $2.54 \times 10^8$

D  $2.54 \times 10^9$

E  $2.54 \times 10^{10}$

**3** The width of a rectangle is $2^6$, and the length is $2^5$. What is the area of the rectangle?

A  $2^1$

B  $2^{11}$

C  $2^{30}$

D  $4^1$

E  $4^{11}$

**4** The width of a certain strand of human hair is about $1.5 \times 10^{-3}$ cm. What is the width of $2.0 \times 10^5$ of these hairs placed next to each other?

A  $3.5 \times 10^8$ cm

B  $3.0 \times 10^{-2}$ cm

C  $3.0 \times 10^8$ cm

D  $3.0 \times 10^2$ cm

E  $5.0 \times 10^{-8}$ cm

**5** Which has the same value as $5^1 + 4^0$?

A  9

B  8

C  6

D  5

E  4

**6** Which of the following expressions is equivalent to $5(7^2 7^2) + 5(7^4 7^{-4}) - (7^8 7^{-4})$?

A  $4(7^4) + 5$

B  $6(7^8) + 1$

C  $10(7^4) - (7^{-2})$

D  $10(7^4) - (7^2)$

E  $5(7^4) + 5(7^{-16}) - (7^{-32})$

**7** Which has the same value as $6(2^{-3}) + (5)(2^{-4}) + (4)(2^{-5})$?

A  $\dfrac{15}{2}$

B  $\dfrac{19}{8}$

C  $\dfrac{19}{16}$

D  $-256$

E  $256$

**8** Which of the following expressions is equivalent to $(3x^2 + 3x + 2) + 2(x^2 - 5x - 2)$?

A  $5x^2 - 2x$

B  $5x^2 - 13x - 6$

C  $5x^2 + 13x + 6$

D  $5x^2 - 7x + 6$

E  $5x^2 - 7x - 2$

**9** Which of the following expressions is equivalent to $(3x^2 + 3x + 2) - 2(x^2 - 5x - 2)$?

A  $x^2 + 13x + 6$

B  $x^2 + 13x - 2$

C  $x^2 - 7x - 2$

D  $x^2 + 8x + 4$

E  $x^2 - 2x$

**10** Which of the following expressions is equivalent to $\dfrac{[(6x^2 + 4) - 2(2 - 3x^2)]}{4x}$?

A  $\dfrac{x}{2} + \dfrac{5}{2x}$

B  $3x$

C  $\dfrac{2}{x}$

D  $3x^2$

E  $0$

**11** Max says that $x^2$ is always greater than $x^{-2}$. Which value of $x$ shows that Max is incorrect?

A  $-2$

B  $\dfrac{1}{3}$

C  $\sqrt{3}$

D  $3$

E  $30$

**12** The North American feather-winged beetle is one of the world's smallest beetles. It is less than 0.0005 meters long. What is this length written in scientific notation?

A $5.0 \times 10^{-4}$

B $5.0 \times 10^{-3}$

C $5.0 \times 10^{-2}$

D $5.0 \times 10^{3}$

E $5.0 \times 10^{4}$

**13** The Smithsonian Institute has about $3^4$ items in its Division of Old World Archeology Collection. About how many items are in this collection?

A 30

B 60

C 70

D 80

E 90

**14** There are $3^4$ students in one math class and $2^6$ students in another math class. Which expression represents the total number of students in the two classes?

A $5^{10}$

B $6^{10}$

C $3^4 + 2^6$

D $3^4 \times 2^6$

E $34 + 26$

**15** Olivia wrote $4^0$ for the number of siblings she has. What is another way to write the number of siblings using exponents?

A $5^0$

B $5^1$

C $5^{-1}$

D $4^1$

E $3^1$

**16** The expression $4x(x^2 + 2y)$ is equal to which of the following expressions?

A $4x^3 + 2y$

B $4x^2 + 8xy$

C $4x^3 + 2xy$

D $4x^3 + 8y$

E $4x^3 + 8xy$

**Directions:** Questions 17 through 20 are based on the table below.

| PLANET MASSES | |
|---|---|
| **Planet** | **Mass (kg)** |
| Mercury | $3.3 \times 10^{23}$ |
| Venus | $4.87 \times 10^{24}$ |
| Earth | $5.97 \times 10^{24}$ |
| Mars | $6.42 \times 10^{23}$ |
| Jupiter | $1.899 \times 10^{27}$ |
| Saturn | $5.68 \times 10^{26}$ |
| Uranus | $8.68 \times 10^{25}$ |
| Neptune | $1.02 \times 10^{26}$ |

**17** Which planet has the greatest mass?

A Venus

B Earth

C Jupiter

D Uranus

E Neptune

**18** Which planet has the least mass?

A Mercury

B Venus

C Earth

D Mars

E Jupiter

**19** About how many times greater is the mass of Jupiter than the mass of Mars?

A $3.0 \times 10^2$

B $3.0 \times 10^3$

C $3.0 \times 10^4$

D $3.0 \times 10^5$

E $3.0 \times 10^6$

**20** About what is the sum of the masses of the planets?

A $2.7 \times 10^{27}$

B $2.7 \times 10^{28}$

C $3.8 \times 10^{24}$

D $3.8 \times 10^{27}$

E $3.8 \times 10^{28}$

21 Which of the following expressions is equivalent to $5x^5 - 15x^4 + 10x^3$?

A $5x^5(1 - 3x + 2x^2)$

B $5x^3(x^2 - 3x + 2)$

C $5x^3(x^2 + 3x - 2)$

D $5x^{-3}(x^2 - 3x + 2)$

E $5x^{-5}(1 - 3x + 2x^2)$

22 Which of the following expressions is equivalent to $\dfrac{(x^4 + 5x^3) + x^2(x^2 - 2x)}{x^3}$?

A $\dfrac{2x^3 + 5x^3 - 2}{x^2}$

B $x^3(2x + 7)$

C $x^6(2x + 7)$

D $x^6(2x + 3)$

E $2x + 3$

23 Which of the following expressions is equivalent to $3(2x^2 - 1) - (5x^2 + x + 3)$?

A $x^2 - x - 6$

B $x^2 - 4x + 3$

C $-x(2x + 1)$

D $x(x - 1)$

E $x(x + 1)$

24 Which best describes the value of the expression $\dfrac{x^3}{x^3 - 1}$ when $x = -1$?

A $< 0$

B $= 0$

C $> 0$

D undefined

E a negative number

25 Which best describes the value of the expression $\dfrac{x^3}{x^3 + 1}$ when $x = -1$?

A $< 0$

B $= 0$

C $> 0$

D undefined

E a decimal number

26 Which best describes the value of the expression $\dfrac{(x - 1)^4}{x^2 - 1}$ when $x = -1$?

A $< 0$

B $= 0$

C $> 0$

D undefined

E a square of a number

**Directions:** Questions 27 and 28 are based on the information below.

The table shows the distances of planets from the sun.

| Planet | Distance from the Sun (km) |
|---|---|
| Mercury | $5.79 \times 10^7$ |
| Venus | $1.082 \times 10^8$ |
| Earth | $1.496 \times 10^8$ |
| Mars | $2.279 \times 10^8$ |
| Jupiter | $7.786 \times 10^8$ |
| Saturn | $1.4335 \times 10^9$ |
| Uranus | $2.8725 \times 10^9$ |
| Neptune | $4.4951 \times 10^9$ |

27 Find the distance between Saturn and the sun. What is this number written in standard notation?

A 14,335,000 km

B 14,350,000 km

C 143,350,000 km

D 1,433,500,000 km

E 14,335,000,000 km

28 How many kilometers farther from the sun is Jupiter than Venus?

A $6.704 \times 10^7$

B $6.704 \times 10^8$

C $6.704 \times 10^9$

D $6.704 \times 10^{10}$

E $6.704 \times 10^{16}$

**29** The approximate number of people who visited an amusement park in July can be written as $10^5$. About how many people attended the park in July?
  A   1,000
  B   10,000
  C   100,000
  D   1,000,000
  E   10,000,000

**30** Which expression is equivalent to $b^{-4}$?
  A   $b^4$

  B   $\dfrac{1}{b^4}$

  C   $-b^4$

  D   $\dfrac{1}{b^{-4}}$

  E   $(-b)(-b)(-b)(-b)$

**31** One of the largest genomes (sets of genetic material) is that of the marbled lungfish. Even so, this genome weighs only $1.3283 \times 10^{-10}$ grams. What is this number expressed in standard notation?
  A   0.000013283
  B   0.0000013283
  C   0.00000013283
  D   0.000000013283
  E   0.00000000013283

**32** The perihelion of Pluto is 4,435,000,000 km. What is this distance written in scientific notation?
  A   $4.435 \times 10^7$
  B   $4.435 \times 10^8$
  C   $4.435 \times 10^9$
  D   $4.435 \times 10^{10}$
  E   $4.435 \times 10^{11}$

**33** Which has the same value as $4.404 \times 10^9$?
  A   $0.4404 \times 10^8$
  B   $0.4404 \times 10^9$
  C   $0.4404 \times 10^{10}$
  D   $44.04 \times 10^9$
  E   $44.04 \times 10^{10}$

**34** If $3^x = 81$, what is the value of $x$?
  A   1
  B   2
  C   3
  D   4
  E   5

**35** The number of students who attend Shadyside High School can be written as $2^9$. Sunnyside High School has 3 times the number of students as Shadyside. How many students attend Sunnyside High School?
  A   171
  B   512
  C   1,536
  D   4,096
  E   Not enough information is given.

**36** Which statement is true about the expression $(-5)^x$?
  A   If $x$ equals an even number, the answer will be negative.
  B   If $x$ equals an even number, the answer will be positive.
  C   If $x$ equals an odd number, the answer will be positive.
  D   If $x$ equals zero, the answer will be zero.
  E   If $x$ equals an even number, the answer could be positive or negative.

**37** If $(3^x)(3^x) = 3$, what must the value of $x$ be?
  A   $-1$
  B   0
  C   $\dfrac{1}{2}$
  D   1
  E   3

**38** A supercomputer can perform $5 \times 10^{13}$ operations per second. How many operations can the computer perform in one hour?
  A   $1.8 \times 10^{15}$
  B   $1.8 \times 10^{17}$
  C   $3 \times 10^{14}$
  D   $3 \times 10^{15}$
  E   $6 \times 10^{14}$

# Unit 1 Review

The Unit Review is structured to resemble mathematics high school equivalency tests. Be sure to read each question and all possible answers very carefully before choosing your answer.

To record your answers, fill in the lettered circle that corresponds to the answer you select for each question in the Unit Review.

Do not rest your pencil on the answer area while considering your answer. Make no stray or unnecessary marks. If you change an answer, erase your first mark completely.

Mark only one answer space for each question; multiple answers will be scored as incorrect.

**Sample Question**

Rodrigo pays $165.40 per month on his car loan. How much does he pay on his loan in 1 year?

A $661.60
B $719.13
C $992.40
D $1,984.80
E $3,969.60

Ⓐ Ⓑ Ⓒ ● Ⓔ

1 Thirty-five percent of residents surveyed were in favor of building a new road. The remaining residents objected. If 1,200 people were surveyed, how many objected to the new road?

A 35
B 360
C 420
D 600
E 780

Ⓐ Ⓑ Ⓒ Ⓓ Ⓔ

2 Dina purchased a new dining room table for $764.50 and four new chairs for $65.30 each. What was the cost of the whole set?

A $699.20
B $829.80
C $926.30
D $1,025.70
E $1,091.00

Ⓐ Ⓑ Ⓒ Ⓓ Ⓔ

3 The Martins drove 210.5 miles on the first day of their trip and 135.8 miles the second day. How many more miles did they drive the first day than the second day?

A 60.0
B 74.7
C 149.4
D 271.6
E 346.3

Ⓐ Ⓑ Ⓒ Ⓓ Ⓔ

4 Erin must add $4\frac{1}{2}$ cups of flour to her cookie batter using a $1\frac{1}{2}$-cup measuring cup. How many times will she need to fill the measuring cup with flour?

A one
B two
C three
D four
E five

Ⓐ Ⓑ Ⓒ Ⓓ Ⓔ

**Directions:** Questions 5 and 6 are based on the information below.

The table shows the breakdown of after-school options for students at Oak Ridge Elementary School.

| WHAT STUDENTS DO AFTER SCHOOL | |
|---|---|
| OPTION | NUMBER OF STUDENTS |
| Parent pickup | 118 |
| Walk | 54 |
| Bus | 468 |
| After-school programs | 224 |

**5** What fraction of the students walk home?

A $\frac{1}{216}$

B $\frac{9}{216}$

C $\frac{1}{24}$

D $\frac{1}{16}$

E $\frac{1}{3}$

Ⓐ Ⓑ Ⓒ Ⓓ Ⓔ

**6** What fraction of the students take the bus or stay after school?

A $\frac{197}{432}$

B $\frac{468}{864}$

C $\frac{117}{216}$

D $\frac{173}{216}$

E $\frac{13}{24}$

Ⓐ Ⓑ Ⓒ Ⓓ Ⓔ

**7** Kara invested $1,250 in the production of a friend's music CD. Her friend paid her back at 6% annual interest after 36 months. How much money did Kara get back?

A  $225
B  $500
C  $1,250
D  $1,025
E  $1,475

Ⓐ Ⓑ Ⓒ Ⓓ Ⓔ

**8** The population of a city grew from 43,209 to 45,687 in just five years. What was the percent increase in the population to the nearest whole percent?

A  18%
B  17%
C  6%
D  5%
E  1%

Ⓐ Ⓑ Ⓒ Ⓓ Ⓔ

**9** The number of students at a large university can be written as $8^4$. How many students are at the university?

A  512
B  4,096
C  10,024
D  32,028
E  32,768

Ⓐ Ⓑ Ⓒ Ⓓ Ⓔ

**10** Tracy bought two pretzels for $1.95 each and two soft drinks for $0.99 each. If she paid with a $10 bill, how much change did she receive?

A  $4.12
B  $5.12
C  $6.10
D  $7.06
E  $8.02

Ⓐ Ⓑ Ⓒ Ⓓ Ⓔ

**Directions:** Questions 11 and 12 are based on the information below.

A number of women participate in five different intramural college sports. The fraction of women who participate in each sport is shown in the table.

| WOMEN'S INTRAMURAL SPORTS | |
|---|---|
| **SPORT** | **FRACTION OF WOMEN** |
| Basketball | $\frac{1}{6}$ |
| Volleyball | $\frac{1}{20}$ |
| Soccer | $\frac{1}{3}$ |
| Ultimate frisbee | $\frac{1}{5}$ |
| Lacrosse | $\frac{1}{4}$ |

**11** In which sport do the greatest number of women participate?
   **A** basketball
   **B** volleyball
   **C** soccer
   **D** ultimate frisbee
   **E** lacrosse

Ⓐ Ⓑ Ⓒ Ⓓ Ⓔ

**12** What fraction of women participates in lacrosse and basketball?

   **A** $\frac{2}{12}$

   **B** $\frac{2}{10}$

   **C** $\frac{5}{12}$

   **D** $\frac{5}{8}$

   **E** $\frac{3}{4}$

Ⓐ Ⓑ Ⓒ Ⓓ Ⓔ

**13** If $x^2 = 36$, then $2(x + 5)$ could equal which of the following numbers?
   **A** 6
   **B** 11
   **C** 12
   **D** 22
   **E** 28

Ⓐ Ⓑ Ⓒ Ⓓ Ⓔ

**14** A skier takes a chairlift 786 feet up the side of a mountain. He then skis down 137 feet and catches a chairlift 542 feet up the mountain. What is his position when he gets off the chairlift relative to where he began?
   **A** −1,191 feet
   **B** −649 feet
   **C** +679 feet
   **D** +1,191 feet
   **E** +1,465 feet

Ⓐ Ⓑ Ⓒ Ⓓ Ⓔ

**15** Benjamin drove a distance of 301.5 miles in 4.5 hours. If Benjamin drove at a constant rate, how many miles per hour did he drive?
   **A** 63
   **B** 64
   **C** 65
   **D** 66
   **E** 67

Ⓐ Ⓑ Ⓒ Ⓓ Ⓔ

**16** Scarlett purchased 20 shares of AD stock at $43 per share. She sold the 20 shares at $52 per share. How much money did Scarlett make on her investment?
   **A** $80
   **B** $180
   **C** $200
   **D** $280
   **E** $860

Ⓐ Ⓑ Ⓒ Ⓓ Ⓔ

**17** A group of 426 people is going to a rally. Each bus can take 65 people. What is the minimum number of buses needed?

- A 4
- B 5
- C 6
- D 7
- E 8

ⒶⒷⒸⒹⒺ

**18** Alice typed her income into tax-preparation software. If her income was fifty-six thousand, two hundred, twenty-eight dollars, what digits did she type?

- A 5, 6, 2, 2, 0, 8
- B 5, 0, 6, 2, 2, 8
- C 5, 6, 2, 2, 8
- D 5, 6, 2, 0, 8
- E 5, 6, 2, 8

ⒶⒷⒸⒹⒺ

**19** Earth averages 149,600,000 kilometers from the sun. What is this distance written in scientific notation?

- A $1.496 \times 10^{-7}$ km
- B $1.496 \times 10^{-8}$ km
- C $1.496 \times 10^{8}$ km
- D $1.496 \times 10^{9}$ km
- E $1.496 \times 10^{10}$ km

ⒶⒷⒸⒹⒺ

**20** Each day for three days, Emmit withdrew $64 from his account. Which number shows the change in his account after the three days?

- A −$192
- B −$128
- C −$64
- D $128
- E $192

ⒶⒷⒸⒹⒺ

**Directions:** Questions 21 through 23 are based on the information below.

Kurt and his family went to the state fair. They ate lunch at a wild game restaurant. The menu is shown below.

| ITEM | PRICE |
|---|---|
| Walleye fillet | $5.89 |
| Elk sandwich | $9.65 |
| Wild boar barbecue | $9.19 |
| Salmon on a stick | $5.45 |
| Kid's buffalo platter | $3.50 |

**21** What is the most expensive item on the menu?

- A Walleye fillet
- B Elk sandwich
- C Wild boar barbecue
- D Salmon on a stick
- E Kid's buffalo platter

ⒶⒷⒸⒹⒺ

**22** Kurt ordered 1 wild boar barbecue, 1 walleye fillet, and 3 kid's buffalo platters. If he brought $50 with him to the fair, how much does he have left?

- A $18.58
- B $24.42
- C $25.58
- D $26.42
- E $31.42

ⒶⒷⒸⒹⒺ

**23** How much more do 2 elk sandwiches cost than 3 kid's platters?

- A $2.39
- B $6.15
- C $7.88
- D $8.80
- E $15.80

ⒶⒷⒸⒹⒺ

**24** The ratio of men to women in a chorus is 2:3. If there are 180 women in the chorus, how many men are in the chorus?

A   80
B   100
C   120
D   160
E   180

Ⓐ Ⓑ Ⓒ Ⓓ Ⓔ

**25** Anna can knit a scarf in $1\frac{2}{3}$ hours. How many scarves can she knit in 4 hours?

A   $2\frac{2}{5}$

B   $2\frac{2}{3}$

C   $3\frac{1}{5}$

D   $3\frac{2}{3}$

E   $3\frac{3}{5}$

Ⓐ Ⓑ Ⓒ Ⓓ Ⓔ

**26** Eighty-four percent of student athletes attended a preseason meeting. If there are 175 student athletes, how many attended the meeting?

A   84
B   128
C   137
D   147
E   149

Ⓐ Ⓑ Ⓒ Ⓓ Ⓔ

**27** Two-thirds of Mrs. Jensen's class passed the math exam. If there are 24 students in her class, how many passed the exam?

A   12
B   13
C   14
D   15
E   16

Ⓐ Ⓑ Ⓒ Ⓓ Ⓔ

**Directions:** Questions 28 and 29 are based on the information below.

During an election year, 200 people were surveyed about their political affiliation. The results are shown in the table.

| VOTERS' POLL | |
| --- | --- |
| PARTY AFFILIATION | NUMBER OF PEOPLE |
| Democratic | 78 |
| Republican | 64 |
| Independent | 46 |
| Green | 10 |
| Libertarian | 2 |

**28** What is the ratio of Green party supporters to Libertarian party supporters?

A   5 to 1
B   10 to 1
C   1 to 10
D   1 to 5
E   2 to 10

Ⓐ Ⓑ Ⓒ Ⓓ Ⓔ

**29** If 400 people were surveyed, how many would you expect to affiliate themselves with the Democratic Party?

A   278
B   156
C   78
D   39
E   Not enough information is given.

Ⓐ Ⓑ Ⓒ Ⓓ Ⓔ

# Unit 2

## Unit Overview

Each time you step on a scale, plan a trip, or cook a meal, you are using skills related to measurement. Similarly, you are surrounded by geometric figures. The skills developed in geometry are used to solve many everyday problems using points, lines, and angles, as well as figures such as squares, rectangles, triangles, circles, and solids.

Measurement and geometry skills are important both to your everyday life as well as to your success on mathematics high school equivalency tests. As with other subject areas, measurement and geometry comprise about 25 percent of questions on the mathematics test. In Unit 2, you will study different measurement systems and forms of measurement, along with lines and angles, a variety of regular and irregular figures, and circles and solids. Developing a solid understanding of these concepts will help you prepare for high school equivalency tests.

## Table of Contents

## Key Terms

**acute triangle:** a triangle in which all three angles are less than 90°

**adjacent angles:** two angles that share a common side and have the same vertex

**alternate angles:** a pair of congruent angles formed by two parallel lines cut by a transversal; may be exterior, located outside the parallel lines, or interior, located inside the parallel lines

**angle:** two rays that share the same endpoint ($\angle$)

**area:** the amount of space within a two-dimensional figure; measured in square units

**base (geometry):** the bottom line segment of a figure; forms a 90° angle with a height segment

**circumference:** the distance around a circle

**complementary angles:** two angles for which their sum equals 90° (a right angle)

**congruent:** exactly the same, such as line segments that have the same length or angles that have the same measurement

**congruent figures:** two or more figures that have the same shape and size

**corresponding angles:** congruent angles that are in the same relative location (i.e., both are to the left of the intersecting line and on top of a parallel line)

**diameter:** the distance across the center of a circle, from one side to the other

**equilateral triangle:** a triangle with three congruent sides

**exterior angles:** angles that fall outside two parallel lines cut by a transversal

**height:** a perpendicular line segment from a vertex to its opposite side; forms a 90° angle when paired with a base segment

**hemisphere:** half of a sphere

**hypotenuse:** the longest side of a right triangle

**indirect measurement:** using proportions and corresponding parts of similar figures to find a measurement that you cannot find directly

**interior angles:** angles that fall between two parallel lines cut by a transversal

**irregular figures:** figures made up of several plane or solid shapes

**isosceles triangle:** a triangle with two congruent sides

**legs:** the two shorter sides of a right triangle

**line:** in geometry, a line is straight and continues forever in both directions; represented as —

**obtuse triangle:** a triangle in which the largest angle is greater than 90°

**parallel:** two lines on a plane that never meet; similar line segments that are the same distance apart

**parallelogram:** a four-sided figure with opposite pairs of congruent and parallel sides

**perimeter:** the distance around a polygon (a multi-sided figure with straight sides)

**prism:** a solid figure with two parallel bases

**pyramid:** a three-dimensional figure that has a polygon as its single base and triangular faces

**Pythagorean theorem:** a description of the relationship between the sides of a right triangle; the sum of the squares of the two smaller legs equals the square of the hypotenuse; $a^2 + b^2 = c^2$

**quadrilateral:** a closed four-sided figure with four angles

**radius:** the distance from the center of a circle to any point on the perimeter; half the diameter

**ray:** a part of a line that has one endpoint and extends forever in the opposite direction

**rhombus:** a parallelogram with four equal sides

**right triangle:** a triangle in which the largest angle is 90°

**scalene triangle:** a triangle with no congruent sides

**similar figures:** figures that have the same shape and congruent angles, but have proportional sides

**solid figure:** a three-dimensional figure, such as a cube, pyramid, or cone

**square pyramid:** a pyramid that has a square base and four congruent triangular faces

**surface area:** the sum of the area of a figure's two bases and its lateral surfaces

**supplementary angles:** two angles that produce a sum equal to 180° (a straight line)

**transversal:** a line crossing parallel lines

**trapezoid:** a four-sided figure with only one pair of parallel sides

**vertex:** the endpoint shared by two rays; the points at which the faces of a three-dimensional figure connect

**vertical angles:** congruent, nonadjacent angles that share a vertex formed by intersecting lines

**volume:** the amount of space inside a three-dimensional figure, measured in cubic units

# Measurement Systems and Units of Measure

## 1 Learn the Skill

When solving measurement problems, you will use either the **U. S. customary system** or the **metric system**. **Units of measure** in the U. S. customary system include inch and foot (length), ounce and pound (weight), and pint and quart (capacity). Units of measure in the metric system include centimeter and meter (length), gram and kilogram (mass), and milliliter and liter (capacity).

## 2 Practice the Skill

When you convert and rename a unit in the metric system, multiply or divide by 10, 100, or 1,000. These prefixes can help in making metric conversions.

*milli-* means 1/1000        *centi-* means 1/100        *kilo-* means 1,000

---

### U.S. CUSTOMARY UNITS OF MEASURE

**Length**
1 foot (ft) = 12 inches (in.)
1 yard (yd) = 3 feet
1 mile (mi) = 5,280 feet
1 mile = 1,760 yards

**Liquid Capacity**
1 cup (c) = 8 fluid ounces (fl oz)
1 pint (pt) = 2 cups
1 quart (qt) = 2 pints
1 gallon (gal) = 4 quarts

**Weight**
1 pound (lb) = 16 ounces (oz)
1 ton (tn) = 2,000 pounds

### METRIC UNITS OF MEASURE

**Length**
1 kilometer (km) = 1,000 meters (m)
1 meter (m) = 100 centimeters (cm)
1 centimeter (cm) = 10 millimeters (mm)

**Capacity**
1 kiloliter (kL) = 1,000 liters (L)
1 liter (L) = 100 centiliters (cL)
1 centiliter (cL) = 10 milliliters (mL)

**Mass**
1 kilogram (kg) = 1,000 grams (g)
1 gram (g) = 100 centigrams (cg)
1 centigram (cg) = 10 milligrams (mg)

**ELASPED TIME**
Time from 1:15 P.M. to 4:35 P.M.
1:15 to 2:00 = 45 min; 2:00 to 4:35 = 2 hr 35 min
Add: 45 min + 2 hr 35 min = 2 hr 80 min
Rename: 80 min = 1 hr 20 min
Add: 2 hr + 1 hr 20 min = 3 hr 20 min

---

☑ **TEST-TAKING TIPS**

First identify the measurement system and the units of measure that are being converted. If you are converting from a lesser unit to a greater unit, divide. If you are converting from a greater unit to a lesser unit, multiply.

**1**  Dante mixes 30 milliliters of one liquid with 2 centiliters of a second liquid. How many centiliters of liquid does he have altogether?

A   5 cL
B   32 cL
C   50 cL
D   302 cL
E   500 cL

**Directions:** Questions 2 and 3 are based on the schedule below.

| TRAINS FROM CENTRAL STATION | | |
|---|---|---|
| TIME | DESTINATION | GATE |
| 11:52 A.M. | Ridley | 31 |
| 1:05 P.M. | Harrison | 28 |
| 2:35 P.M. | Ridley | 32 |
| 3:20 P.M. | Harrison | 29 |

**2** If a passenger misses the 11:52 train to Ridley, how long must he or she wait for the next train to Ridley?

A  1 hr 5 min
B  1 hr 13 min
C  2 hr 35 min
D  2 hr 43 min
E  3 hr 15 min

**3** The journey to Harrison takes 1 hour 45 minutes. If Lance takes the 3:20 train, what time can he expect to arrive at Harrison?

A  1:45 P.M.
B  4:45 P.M.
C  5:05 P.M.
D  5:45 P.M.
E  Not enough information is given.

**4** Mr. Trask wants to fill his four hummingbird feeders with liquid food. Two feeders hold 6 fluid ounces each. One larger feeder holds 1 cup of liquid. The largest feeder holds 1 pint. How many fluid ounces of liquid food does Mr. Trask need to fill the four bird feeders?

A  8 fl oz
B  14 fl oz
C  28 fl oz
D  30 fl oz
E  36 fl oz

**Directions:** Questions 5 through 7 are based on the information below.

Five students in Ms. Craig's chemistry class conducted an experiment using different amounts of the same powdered chemicals.

| CHEMICAL AMOUNTS FOR EXPERIMENTS | | | |
|---|---|---|---|
| STUDENT | CHEMICAL A | CHEMICAL B | CHEMICAL C |
| Tory | 2 cg | 5 cg | 2 g |
| Shantell | 5 mg | 8 cg | 60 cg |
| Diego | 15 mg | 50 cg | 1 g |
| Janice | 15 mg | 2 cg | 50 cg |
| Dana | 5 mg | 3 cg | 0.5 g |

**5** How many centigrams of Chemical A were used by the five students?

A  6 cg
B  15 cg
C  42 cg
D  60 cg
E  402 cg

**6** How much greater was the mass of Chemical C than the mass of Chemical A in Shantell's experiment?

A  10 mg
B  55 mg
C  100 mg
D  550 mg
E  595 mg

**7** Dana used a total of 535 mg of chemicals for her experiment. How much greater was the mass of chemicals used by Diego?

A  10 mg
B  170 mg
C  980 mg
D  1,765 mg
E  8,965 mg

**Directions:** Questions 8 through 10 are based on the information below.

Cedric is making punch to take to a family reunion. He wants to triple the following recipe for Pineapple Punch.

*Pineapple Punch*
2 cups tea
$\frac{1}{2}$ cup lemon juice
2 cups orange juice
1 cup sugar
5 pints ginger ale
4 pints carbonated water
8 slices canned pineapple

| Liquid Capacity |
| --- |
| 1 cup (c) = 8 fluid ounces (fl oz) |
| 1 pint (pt) = 2 cups |
| 1 quart (qt) = 2 pints |
| 1 gallon (gal) = 4 quarts |

8   How many pints of orange juice will Cedric need?
   **A**   1 pt
   **B**   2 pt
   **C**   3 pt
   **D**   4 pt
   **E**   6 pt

9   How many quarts of carbonated water will Cedric need?
   **A**   2 qt
   **B**   3 qt
   **C**   4 qt
   **D**   6 qt
   **E**   12 qt

10   How many cups of ginger ale will Cedric need?
   **A**   7.5 c
   **B**   10 c
   **C**   13 c
   **D**   15 c
   **E**   30 c

**Directions:** Questions 11 through 13 are based on the information below.

Sabrina is marking the route for an upcoming cross-country foot race. She keeps a record of the number of meters she marks each day.

| METERS MARKED FOR A CROSS-COUNTRY RACE | | | | |
| --- | --- | --- | --- | --- |
| DAY | 1 | 2 | 3 | 4 |
| METERS MARKED | 700 m | 600 m | 800 m | 1,000 m |

| Metric Units of Length |
| --- |
| 1 kilometer (km) = 1,000 meters (m) |
| 1 meter (m) = 100 centimeters (cm) |
| 1 centimeter (cm) = 10 millimeters (mm) |

11   How many kilometers of the cross-country route did Sabrina mark on Day 1?
   **A**   0.0007 km
   **B**   0.007 km
   **C**   0.07 km
   **D**   0.7 km
   **E**   7.0 km

12   How many kilometers of the route had Sabrina marked by Day 4?
   **A**   0.31 km
   **B**   3.1 km
   **C**   31 km
   **D**   310 km
   **E**   3,100 km

13   On Day 5, Sabrina marked another 900 meters. How many kilometers must the runners race?
   **A**   0.04 km
   **B**   0.4 km
   **C**   4 km
   **D**   40 km
   **E**   400 km

**Directions:** Questions 14 through 17 are based on the table below.

> **Standard Measures of Length**
> 1 foot (ft) = 12 inches (in.)
> 1 yard (yd) = 3 feet
> 1 mile (mi) = 5,280 feet
> 1 mile = 1,760 yards

**14** During a school track meet, Jason threw the javelin 63 feet. Hector threw the javelin 54 feet. What is the difference between the two throws?
A   3 yd
B   9 yd
C   21 yd
D   33 yd
E   39 yd

**15** Hannah walks 875 yards on Monday, on Wednesday, and on Friday. She walks 2,625 yards on Tuesday and again on Thursday. Approximately how many miles has she walked by the end of the day Friday?
A   1.5 mi
B   4 mi
C   4.5 mi
D   7.9 mi
E   78.8 mi

**16** Each place setting requires 18 inches of ribbon. If Mara makes 24 place settings, how many feet of ribbon will she need?
A   12 ft
B   36 ft
C   144 ft
D   432 ft
E   5,184 ft

**17** George ran for a school-record of 2,640 yards. About how many miles did George run during the football season?
A   0.5 mi
B   1.5 mi
C   3 mi
D   5 mi
E   26.4 mi

**Directions:** Questions 18 through 20 are based on the table below.

> **Metric Measures of Capacity**
> 1 kiloliter (kL) = 1,000 liters (L)
> 1 liter (L) = 100 centiliters (cL)
> 1 centiliter (cL) = 10 milliliters (mL)

**18** Kyle is decorating his restaurant with tropical fish tanks. Three of the tanks each have a capacity of 448 L of water. Two of the tanks each have a capacity of 236 L of water. How many kiloliters of water will Kyle need to fill all of the tanks?
A   0.0186 kL
B   0.186 kL
C   1.816 kL
D   18.16 kL
E   1,816 kL

**19** Microbiologists gathered 15 vials of pond water. They filled 5 vials with 10 milliliters of pond water each, 5 vials with 1 milliliter of pond water each, and 5 vials with 0.1 milliliter of pond water each. How many centiliters of pond water did they gather in all?
A   0.6 cL
B   1.11 cL
C   11.1 cL
D   5.55 cL
E   55.5 cL

**20** The soccer team drank a combined 17 liters of water during the game. How many kiloliters did the team consume?
A   0.017 kL
B   0.17 kL
C   1.7 kL
D   170 kL
E   1,700 kL

**Directions:** Questions 21 through 23 are based on the tables below.

Students in an ecology class are tracking the growth of saplings over the summer.

| SAPLINGS' GROWTH OVER 4 MONTHS | | | | |
|---|---|---|---|---|
| | MAY | JUNE | JULY | AUGUST |
| MAPLE | 55 cm | 62 cm | 83 cm | 101 cm |
| CHERRY | 15 cm | 18 cm | 21 cm | 23 cm |
| OAK | 91 cm | 98 cm | 105 cm | 121 cm |
| ASH | 33 cm | 38 cm | 45 cm | 57 cm |

> **Metric Measures of Length**
> 1 kilometer (km) = 1,000 meters (m)
> 1 meter (m) = 100 centimeters (cm)
> 1 centimeter (cm) = 10 millimeters (mm)

**21** What was the height, in meters, of the tallest sapling in August?
- **A** 1.01 m
- **B** 1.21 m
- **C** 2.3 m
- **D** 5.7 m
- **E** 12.1 m

**22** What is the difference between the ash's height in May and its height in August?
- **A** 0.24 m
- **B** 2.4 m
- **C** 24 m
- **D** 240 m
- **E** 2,400 m

**23** The students measured the saplings' heights the following summer. How many more meters had the maple sapling grown than the cherry sapling?
- **A** 0.38 m
- **B** 0.46 m
- **C** 3.8 m
- **D** 4.6 m
- **E** Not enough information is given.

**Directions:** Questions 24 and 25 are based on the table below.

> **Metric Measures of Mass**
> 1 kilogram (kg) = 1,000 grams (g)
> 1 gram (g) = 100 centigrams (cg)
> 1 centigram (cg) = 10 milligrams (mg)

**24** One serving of breakfast cereal contains 45 grams of carbohydrates. If there are 8 servings per box of cereal, how many kilograms of carbohydrates does one box contain?
- **A** 0.036 kg
- **B** 0.36 kg
- **C** 3.6 kg
- **D** 36 kg
- **E** 360 kg

**25** A can of tuna contains 250 mg of sodium. If Jim eats half a can, how many centigrams sodium will he consume?
- **A** 12.5 cg
- **B** 25 cg
- **C** 100 cg
- **D** 125 cg
- **E** 1,250 cg

**26** Kyle took 1 hr 34 min to finish a 25-mile bicycle race for charity. If the race started at 12:30 P.M., at what time did Kyle cross the finish line?
- **A** 1:34 P.M.
- **B** 1:39 P.M.
- **C** 1:47 P.M.
- **D** 1:53 P.M.
- **E** 2:04 P.M.

**27** One Sunday, Nick starts his daily run at 11:35 A.M. and runs 5 miles. If he maintains a 7 mph rate, what time will he finish?
- **A** 12:15 P.M.
- **B** 12:18 P.M.
- **C** 12:20 P.M.
- **D** 12:25 P.M.
- **E** 12:42 P.M.

# Length, Perimeter, and Area

## ① Learn the Skill

The distance around a polygon, such as a triangle or rectangle, is called the **perimeter**. Determine the perimeter of a polygon by measuring and adding the **lengths** of all its sides.

**Area** is the amount of space that covers a two-dimensional figure. Square units are used to measure area. In this lesson, you will find the area of rectangles, triangles, parallelograms, and trapezoids. A **parallelogram** is a four-sided figure with opposite pairs of congruent and parallel sides. A **trapezoid** is a four-sided figure with only one pair of parallel sides.

The formulas for finding the area may require you to identify the **base** and **height** of a figure. The base and height form a right angle. A trapezoid has two bases. To find the bases of a trapezoid, look for the two parallel sides.

## ② Practice the Skill

When finding area and perimeter, you will practice using formulas and simplifying expressions. Read the example and strategies below. Use this information to answer question 1.

**A** Since the door is square, you know that each side is 12 ft. The length and width of a square are equal. The formula for the area of a square is $A = s \times s$, where $s$ represents the length of one side.

**B** The units used to express the area of a figure are based on the unit of length. In this problem, the length is measured in feet, so the area is measured in square feet (sq ft or ft²).

Jorge is in charge of painting the front wall of a storage facility. The facility has a large square metal door that does not need to be painted.

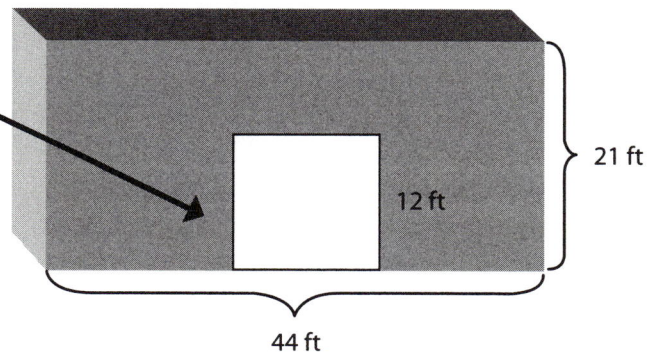

21 ft

12 ft

44 ft

### ✔ TEST-TAKING TIPS

Some word problems require several steps to find the answer. Finding the area of the square door is not sufficient, nor is finding the area of the wall including the door. You must find the area of the wall and the area of the door. Then you must perform a third step.

**1** What is the area of the front wall of the storage facility that Jorge must paint?
A  144 sq ft
B  288 sq ft
C  672 sq ft  **B**
D  780 sq ft
E  924 sq ft

**Directions:** Questions 2 through 4 are based on the information below.

A home-building company wants to divide a large plot of land into three separate parcels. Parcel B and Parcel C will have the same perimeter. The plot of land is shown below.

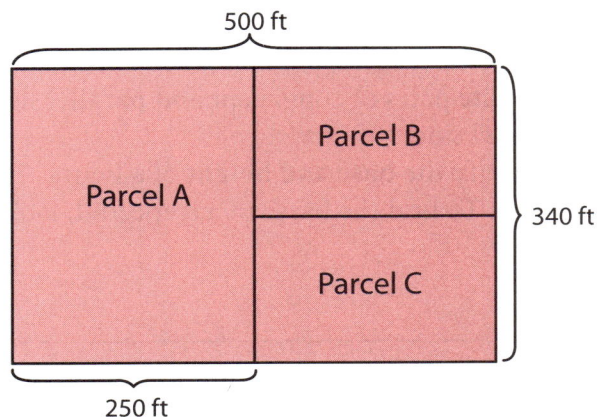

**Directions:** Questions 5 through 7 are based on the information below.

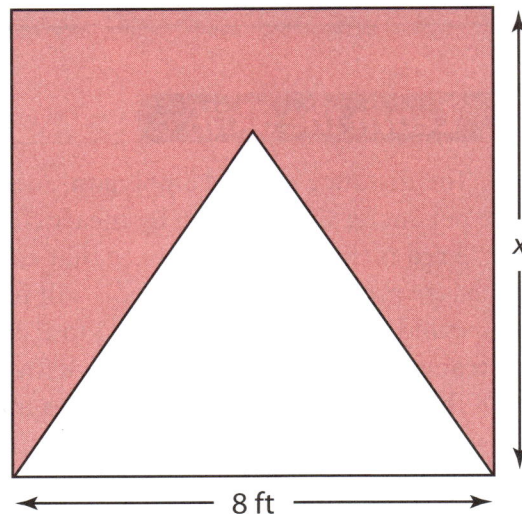

The triangle inside the rectangle is equilateral. The length of one side of the triangle is 8 feet. The perimeter of the rectangle is 36 feet.

2   **What is the perimeter of Parcel A?**
   A   500 ft
   B   840 ft
   C   1,000 ft
   D   1,090 ft
   E   1,180 ft

5   **What is the perimeter of the triangle?**
   A   18 ft
   B   24 ft
   C   28 ft
   D   32 ft
   E   36 ft

3   **What is the perimeter of Parcel C?**
   A   420 ft
   B   590 ft
   C   840 ft
   D   1,340 ft
   E   1,680 ft

6   **What is the length of x?**
   A   8 ft
   B   10 ft
   C   16 ft
   D   20 ft
   E   28 ft

4   **What effect would the deletion of Parcel C have on the overall perimeter of the land?**
   A   It would increase.
   B   It would decrease.
   C   It would remain the same.
   D   It first would increase and then decrease.
   E   It first would decrease and then increase.

7   **What is the perimeter of the shaded area?**
   A   12 ft
   B   24 ft
   C   36 ft
   D   44 ft
   E   Not enough information is given.

**Directions:** Questions 8 through 10 are based on the information below.

Melanie is having new hardwood floors installed in her bedroom and closet. Her bedroom and closet are both rectangular.

**MELANIE'S BEDROOM**

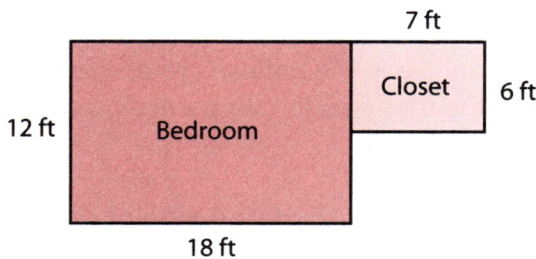

**8** Each square foot of flooring costs $6. How much will it cost to cover the bedroom and closet?
- **A** $516
- **B** $648
- **C** $1,152
- **D** $1,374
- **E** $1,548

**9** How much would it cost if Melanie chose to install flooring in only her bedroom?
- **A** $252
- **B** $504
- **C** $756
- **D** $1,296
- **E** $1,548

**10** Melanie decides to install baseboard around her bedroom and closet. How many feet of baseboard will she need?
- **A** 68 ft
- **B** 74 ft
- **C** 80 ft
- **D** 86 ft
- **E** 258 ft

**11** Vanessa designed the logo for a neighborhood walkathon. Each strip has the same area.

**WALK FOR OUR PARK**

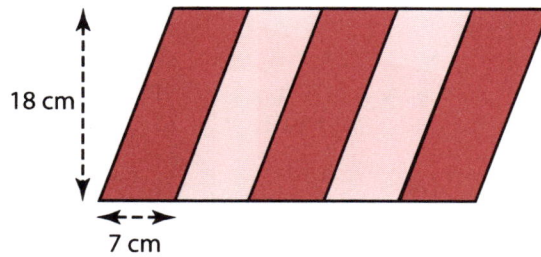

What is the area of the logo that Vanessa covered with red ink?
- **A** 25 sq cm
- **B** 75 sq cm
- **C** 126 sq cm
- **D** 378 sq cm
- **E** 630 sq cm

**12** Quentin built a triangular pen for his dog Toro. The pen is shown below.

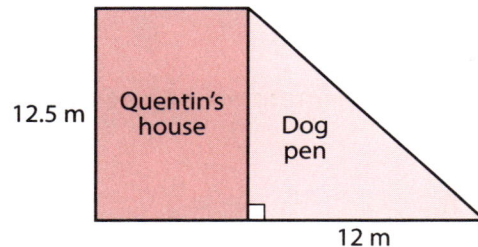

What is the area in which Toro can run?
- **A** 75 sq m
- **B** 150 sq m
- **C** 162.5 sq m
- **D** 600.25 sq m
- **E** Not enough information is given.

**VIEW OF JULIO'S YARD WITH DIAGONAL FENCE**

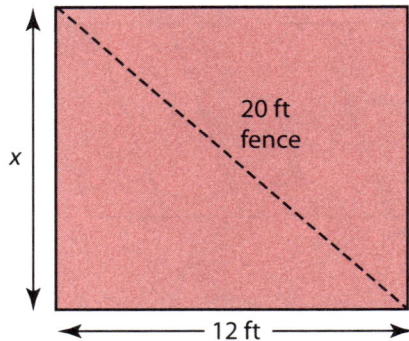

The width of the rectangle is 12 feet. The diagonal of the rectangle is 20 feet. The perimeter of the rectangle is 56 feet.

**13  What is the value of $x$?**
- A  12 ft
- B  16 ft
- C  20 ft
- D  32 ft
- E  44 ft

**14  What is the perimeter of one of the triangles?**
- A  32 ft
- B  44 ft
- C  48 ft
- D  68 ft
- E  96 ft

**15  What is the area of Julio's yard?**
- A  120 ft²
- B  144 ft²
- C  192 ft²
- D  240 ft²
- E  384 ft²

**16  A rectangle has a perimeter of 54 cm. The length is 16 cm. What is the width of the rectangle?**
- A  11 cm
- B  19 cm
- C  38 cm
- D  140 cm
- E  864 cm

**KAYLA'S FABRIC PATTERN**

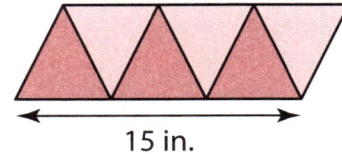

The triangles in the pattern are all the same size. Together, the triangles form a parallelogram with a length of 15 inches.

**17  What is the length of one side of one of the equilateral triangles?**
- A  2 in.
- B  3 in.
- C  5 in.
- D  10 in.
- E  15 in.

**18  What is the perimeter of one of the triangles?**
- A  5 in.
- B  10 in.
- C  15 in.
- D  20 in.
- E  30 in.

**19  What is the perimeter of the parallelogram formed by the triangles?**
- A  20 in.
- B  30 in.
- C  35 in.
- D  40 in.
- E  45 in.

**20  Kayla decided to add two more triangles to the pattern. What is the new perimeter of the parallelogram formed by the triangles?**
- A  40 in.
- B  44 in.
- C  50 in.
- D  54 in.
- E  60 in.

**Directions:** Questions 21 through 23 are based on the information below.

Leon has a piece of wood in the shape of a parallelogram. He wants to cut two triangular pieces off each end.

**LEON'S PLAN FOR CUTTING**

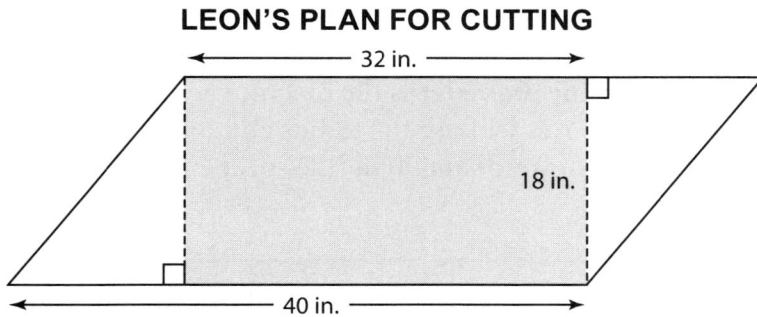

**21 What is area of the resulting rectangular piece of wood?**
- **A** 100 in.$^2$
- **B** 324 in.$^2$
- **C** 432 in.$^2$
- **D** 576 in.$^2$
- **E** 756 in.$^2$

**22 What is the combined area of the resulting two triangular pieces of wood?**
- **A** 36 in.$^2$
- **B** 72 in.$^2$
- **C** 108 in.$^2$
- **D** 144 in.$^2$
- **E** 288 in.$^2$

**23 What is the area of the original piece of wood?**
- **A** 116 in.$^2$
- **B** 576 in.$^2$
- **C** 720 in.$^2$
- **D** 864 in.$^2$
- **E** 1,280 in.$^2$

**24 A triangle has an area of 33.84 cm$^2$. The base of the triangle is 14.1 cm. What is the height of the triangle?**
- **A** 2.4 cm
- **B** 4.8 cm
- **C** 238.6 cm
- **D** 477.1 cm
- **E** Not enough information is given.

**Directions:** Questions 25 through 28 are based on the information below.

The Miller family wants to pave a section of their backyard. The portion to be paved is shaded.

**THE MILLER FAMILY'S BACKYARD**

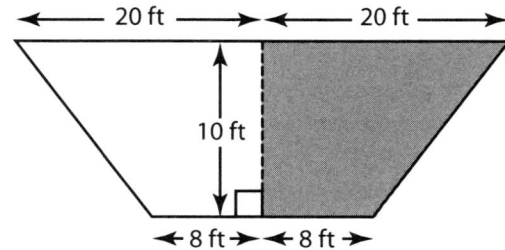

**25 What is the total area of the Millers' backyard?**
- **A** 140 ft$^2$
- **B** 180 ft$^2$
- **C** 240 ft$^2$
- **D** 280 ft$^2$
- **E** 560 ft$^2$

**26 What is the area of the lot that the Millers want to pave?**
- **A** 140 ft$^2$
- **B** 180 ft$^2$
- **C** 240 ft$^2$
- **D** 280 ft$^2$
- **E** 560 ft$^2$

**27 The cost for pavers is $3 per square foot. How much will the Miller family spend to pave the shaded area of their backyard?**
- **A** $400
- **B** $420
- **C** $600
- **D** $800
- **E** $840

**28 The cost for pavers is $3 per square foot. How much would the Miller family spend to pave their entire backyard?**
- **A** $400
- **B** $420
- **C** $600
- **D** $800
- **E** $840

# Circles

## ① Learn the Skill

In any **circle**, all of the points on the circle are the same distance from the center of the circle. The distance from the center to any point is called the **radius**. The **diameter** is the distance across a circle through its center. The diameter is always twice the radius. You will use the radius and diameter to solve problems relating to the **circumference** (the distance around) and **area** (the space in) of circles.

## ② Practice the Skill

By mastering the skills involving circles, their parts, and how to use formulas to find circumference and area, you will be able to successfully solve problems on mathematics tests. Read the examples and strategies below. Use this information to answer question 1.

**(A)** To find the area of a circle, you must know the radius of the circle. In some problems, the diameter will be given instead of the radius. In those cases, divide the diameter by 2 to get the radius. Likewise, multiply the radius by 2 to get the diameter.

**(B)** To find the circumference of a circle, use the formula $C = \pi d$, where $\pi \approx 3.14$. The formula $C = 2\pi r$ can also be used because the diameter is twice the radius. Circumference is measured in units.

**(C)** To find the area of a circle, use the formula $A = \pi r^2$. Area is measured in square units.

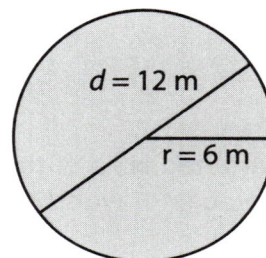

$d = 12$ m

$r = 6$ m

$C = \pi d$
$C = 3.14(12)$
$C = 37.68$ m

$A = \pi r^2$
$A = 3.14(6)^2$
$A = 3.14 \times 6 \times 6$
$A = 113.04$ sq m

### ☑ TEST-TAKING TIPS

Use estimation to check your answers to problems involving circumference and area. When multiplying by $\pi$, round 3.14 to 3 to estimate before finding the product.

1  The diameter of the smaller circle is equal to the radius of the larger circle, which is 9 inches. What is the area of the larger circle?

   A  28.26 in.²
   B  56.52 in.²
   C  63.59 in.²
   D  190.76 in.²
   E  254.34 in.²

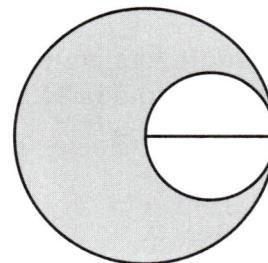

2   Alisha is painting a perfectly round sun as part of a mural on the side of a building. If the diameter of her sun is 15 cm, what is its area in square centimeters?
  A   4.78 cm²
  B   7.1 cm²
  C   176.63 cm²
  D   225 cm²
  E   706.5 cm²

3   A circle has a diameter of 25 inches. What is its circumference?
  A   39.25 inches
  B   78.5 inches
  C   156.25 inches
  D   490.63 inches
  E   1,092.5 inches

4   Jon and Gretchen are laying a circular brick patio in their backyard. The patio is shown in the diagram below.

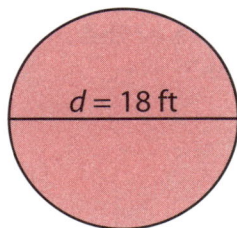

$d = 18$ ft

If the pavers charge $1.59 per square foot, how much will the pavers charge for the whole patio?
  A   $56.52
  B   $89.87
  C   $404.40
  D   $1,017.36
  E   $1,617.60

**Directions:** Questions 5 through 7 are based on the information below.

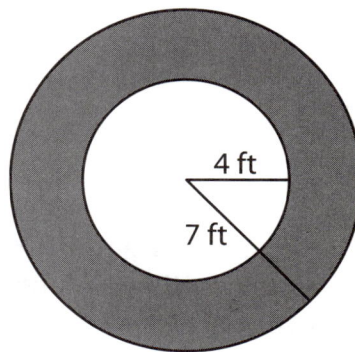

4 ft

7 ft

5   Henry wants to add a fringed edge around the outside of the rug. About how many feet of edging should he buy to go around the outside edge of the rug?
  A   69
  B   44
  C   28
  D   25
  E   13

6   The rug is divided into a white interior and a gray border. What is the area of the interior of the rug in square feet?
  A   5.1
  B   16.4
  C   28.3
  D   50.2
  E   153.9

7   What is the area of the entire area rug to the nearest square foot?
  A   50
  B   103
  C   104
  D   105
  E   154

8   One circle has a radius of 5.5 cm. Another circle has a diameter of 12.5 cm. What is the difference in area between the two circles?
  A   4.7 sq cm
  B   7 sq cm
  C   27.7 sq cm
  D   110.7 sq cm
  E   395.6 sq cm

UNIT 2

**9** A birdbath has a diameter of 30 inches. What is the circumference of the birdbath in inches?

A 47.1 in.
B 94.2 in.
C 188.4 in.
D 706.5 in.
E 2,826.0 in.

**10** Lydia is using an old tractor tire to build a sandbox. She is cutting a piece of plywood to use as a cover for the sandbox.

132 cm

Which of the following expressions represent the number of square centimeters of plywood needed to cover the sandbox if the cover goes to the edge of the tire?

A $\pi \times \frac{132}{2}$
B $\pi \times 132$
C $2 \times \pi \times 132$
D $\pi \times \left(\frac{132}{2}\right)^2$
E $\pi \times (132)^2$

**11** Mr. Dunn is painting a large circle in the middle of a gymnasium. The diameter of the circle is 12 feet. What is the area of the circle?

A 18.84 ft²
B 37.68 ft²
C 75.36 ft²
D 113.04 ft²
E 452.16 ft²

**Directions:** Questions 12 through 15 are based on the information below.

The circular mirror below has a frame that is 2 inches wide. The diameter of the mirror and frame together is 11 inches.

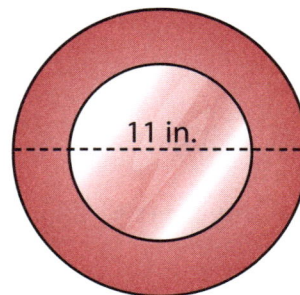

11 in.

**12** What is the radius of the mirror without the frame?

A 3.5 in
B 4.5 in.
C 7 in.
D 9 in.
E 11 in.

**13** What is the area of the mirror only in square inches?

A 34.5
B 38.5
C 40.8
D 95.0
E 379.9

**14** What is the area of the frame only in square inches?

A 284.9
B 60.5
C 56.5
D 54.2
E 6.3

**15** What is the circumference of the mirror with the frame in inches?

A 34.54
B 35.64
C 47.10
D 94.99
E 176.62

**Directions:** Questions 16 and 17 are based on the information below.

A circular tablecloth is shown in the diagram below.

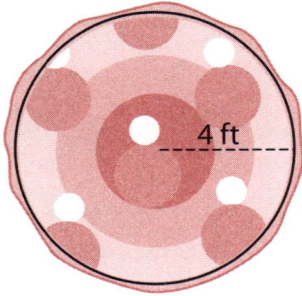

**16 How many square feet is the tablecloth?**
- A  12.56
- B  25.12
- C  50.24
- D  100.48
- E  200.96

**17 Tomás places the tablecloth on a circular table that has a diameter of 5 feet. How many square feet of the tablecloth will hang over the edge of the table?**
- A  19.63
- B  30.62
- C  50.24
- D  78.50
- E  Not enough information is given.

**18 Jonna is sewing a front cover for a circular pillow. The pillow has a diameter of 15 inches. To sew the front cover, she must cut the fabric two inches wider all the way around. What is the minimum area, in square inches, of the piece of fabric she will use?**
- A  1,133.5
- B  706.5
- C  314.0
- D  283.4
- E  226.9

**19 What is the difference in square inches between the two circles?**

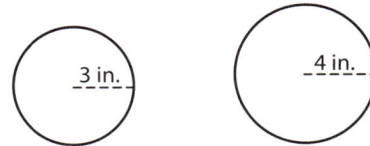

- A  1.0
- B  3.14
- C  6.28
- D  15.0
- E  21.98

**Directions:** Questions 20 and 21 are based on the information below.

Ava hung a circular sunshield in a window in her living room to block the afternoon sun.

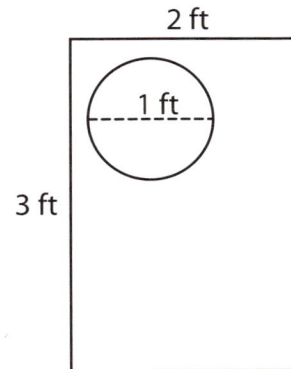

**20 What is the area of the sunshield in square feet?**
- A  6.0
- B  3.14
- C  1.57
- D  1.0
- E  0.79

**21 About what percent of the area of her window is covered by the sunshield?**
- A  13%
- B  17%
- C  25%
- D  50%
- E  78%

**22** A large circle has a radius of 6 meters. A smaller circle has a radius of 2 meters. How many small circles are needed to cover the large circle?

A 4.0
B 9.0
C 12.6
D 100.5
E 125.6

**23** The area of the circular garden shown in the diagram is 144$\pi$ square feet.

Russell is building a fence around the garden to keep out animals. What will be the circumference of the fence in feet?

A 37.68
B 47.12
C 56.52
D 75.36
E 16,277.76

**24** The area of a circle is 19.625 square centimeters. What is the diameter, in centimeters, of the circle?

A 2.5
B 3.1
C 5.0
D 6.3
E Not enough information is given.

**25** Dillon is building a model airplane. The front tires have a diameter of 3.5 inches. What is the total circumference of both tires?

A 5.50 in.
B 9.62 in.
C 10.99 in.
D 19.23 in.
E 21.98 in.

**Directions:** Questions 26 and 27 are based on the information below.

A hotel swimming pool is in the shape of the number eight.

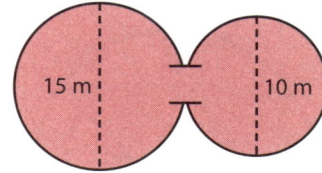

**26** What is the approximate area of the swimming pool in square meters?

A 40 sq m
B 78 sq m
C 255 sq m
D 705 sq m
E 1019 sq m

**27** If the diameter of the smaller section is increased by 45 cm, what will be the area of the entire enlarged pool?

A 93.27 square meters
B 198.46 square meters
C 219.85 square meters
D 262.35 square meters
E 342.90 square meters

**28** Linda is cutting circle shapes for a bulletin board. If she wants to cut a circle that has a circumference of 25.12 inches, what should the radius be in inches?

A 4
B 8
C 12
D 16
E 20

**29** Mary used about 56.5 cm of yarn to form a circle. What could be the diameter of the circle?

A 9.0 cm
B 18.0 cm
C 88.7 cm
D 177.4 cm
E Not enough information is given.

# Lines and Angles

## ① Learn the Skill

A **line** continues forever in both directions. A **ray** is a part of a line. It has one **endpoint** and extends in the other direction without end. The first figure contains $\overrightarrow{AB}$ and $\overrightarrow{AC}$. The arrow above the letters means "ray." Two rays that share the same endpoint form an **angle**. The shared endpoint is called the **vertex**. In this figure, the vertex is point $A$. The angle in the first figure can be named $\angle BAC$, $\angle CAB$, $\angle A$, or $\angle 1$.

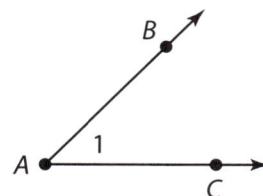

Two angles that have the same measure are **congruent**. If the sum of the measures of the angles is 90°, the angles are **complementary**. The two angles shown directly to the right are complementary. If the sum of the measures of two angles is a straight angle, or 180°, the angles are **supplementary**. The notation for the "measure of angle $A$" is $m\angle A$.

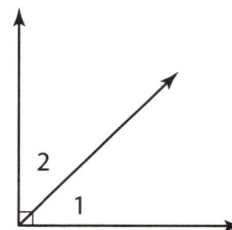

## ② Practice the Skill

Read the example and strategies below. Use this information to answer question 1.

**A** Angles that share a common side and have the same vertex are **adjacent angles**. In this figure, $\angle 1$ and $\angle 2$ are adjacent and supplementary. Angles formed by intersecting lines and that are not adjacent are called **vertical angles**. Vertical angles are congruent. In this figure, $\angle 5$ and $\angle 7$ are vertical angles. Which other pairs of angles are vertical angles?

**B** In this figure, $\angle 3$, $\angle 4$, $\angle 5$, and $\angle 6$ are **interior angles**. Angles 1, 2, 7, and 8 are **exterior angles**. Angles 3 and 5 are **alternate interior angles**. They are congruent. The same is true of $\angle 4$ and $\angle 6$. There are two pairs of **alternate exterior angles**: $\angle 1$, $\angle 7$ and $\angle 2$, $\angle 8$. These pairs of angles also have the same measure. **Corresponding angles** are also congruent. Angles 2 and 6 are corresponding because they are in the same relative location.

The figure below shows two parallel lines intersected by $\overleftrightarrow{JK}$. This line is called a **transversal** (a line that intersects parallel lines).

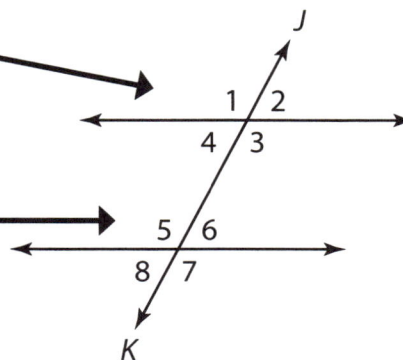

### ✓ TEST-TAKING TIPS

In the example, $\angle 1$ and $\angle 5$ are congruent because they are corresponding, and $\angle 6$ and $\angle 5$ are supplementary. Find $180 - 115$ to find $m\angle 6$.

**1** If $m\angle 1$ is 115°, what is the measure of $\angle 6$?

  **A**  65°
  **B**  85°
  **C**  115°
  **D**  180°
  **E**  Not enough information is given.

## ③ Apply the Skill

**Directions:** Questions 2 and 3 are based on the information below.

Two lines intersect as shown in the diagram below.

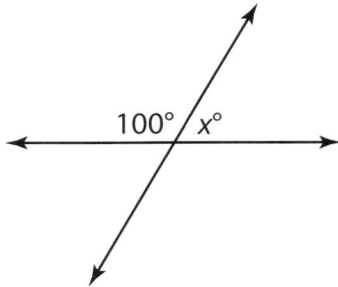

**2** What is the value of *x*?
A 80
B 90
C 100
D 180
E Not enough information is given.

**3** What is the measure of the angle that is vertical to the angle measuring 100°?
A 80°
B 90°
C 100°
D 180°
E Not enough information is given.

**4** Two angles are complementary. If one angle has a measure of 25°, then what is the measure of the other angle?
A 25°
B 65°
C 75°
D 155°
E 165°

**Directions:** Questions 5 and 6 are based on the information below.

This figure shows two parallel lines that are transversed, or crossed, by a third line. Remember that parallel lines are equidistant and never intersect.

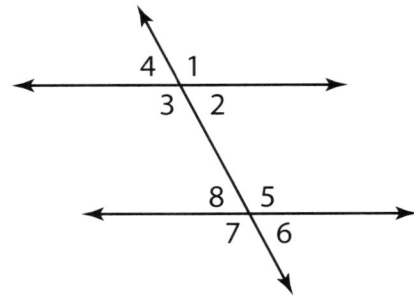

**5** Which two angles have the same measure?
A ∠1 and ∠4
B ∠2 and ∠5
C ∠7 and ∠8
D ∠1 and ∠3
E ∠1 and ∠6

**6** Which three angles have the same measure?
A ∠1, ∠2, and ∠3
B ∠2, ∠5, and ∠6
C ∠2, ∠4, and ∠8
D ∠5, ∠7, and ∠8
E ∠6, ∠8, and ∠3

**7** If ∠*PMR* is a right angle and *m*∠*LMN* is 35°, what is the sum of the measures of ∠*LMP* and ∠*RMP*?

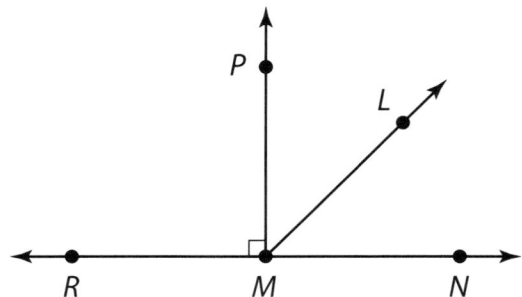

A 55°
B 90°
C 125°
D 145°
E 180°

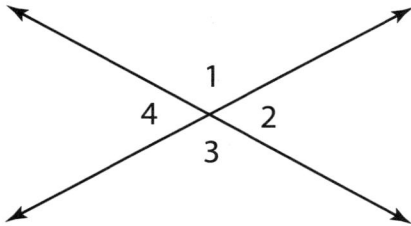

**8** If ∠2 measures 55°, what is the measure of ∠4?

- **A** 35°
- **B** 55°
- **C** 110°
- **D** 125°
- **E** 180°

**9** What kind of angles are ∠1 and ∠3?

- **A** right
- **B** supplementary
- **C** complementary
- **D** adjacent
- **E** vertical

**10** If angle 4 has a measure of 52°, then what is the measure of angle 3?

- **A** 38°
- **B** 48°
- **C** 52°
- **D** 128°
- **E** 138°

**11** Two angles are congruent and complementary. What is the measure of the supplement of one of the angles?

- **A** 45°
- **B** 90°
- **C** 125°
- **D** 135°
- **E** Not enough information is given.

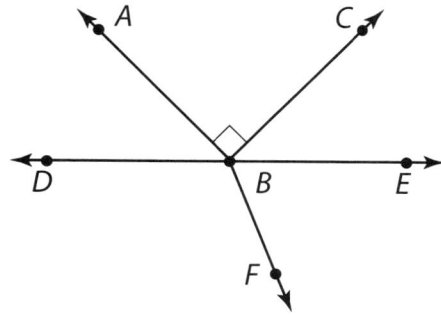

**12** If ∠DBF measures 115°, what is the measure of ∠EBF?

- **A** 45°
- **B** 65°
- **C** 70°
- **D** 75°
- **E** Not enough information is given.

**13** If ∠CBE and ∠ABD are congruent, what is the measure of each angle?

- **A** 15°
- **B** 25°
- **C** 45°
- **D** 55°
- **E** 90°

**14** If ∠CBE and ∠ABD are congruent, what is the measure of ∠CBD?

- **A** 45°
- **B** 90°
- **C** 100°
- **D** 135°
- **E** 155°

**15** Two angles are supplementary. Angle *A* is 20 degrees larger than three times the size of angle *B*. What is the measure of angle *A*?

- **A** 40°
- **B** 60°
- **C** 120°
- **D** 140°
- **E** 180°

UNIT 2

**Directions:** Questions 16 through 19 are based on the information below.

The figure shows two parallel lines intersected by a third line.

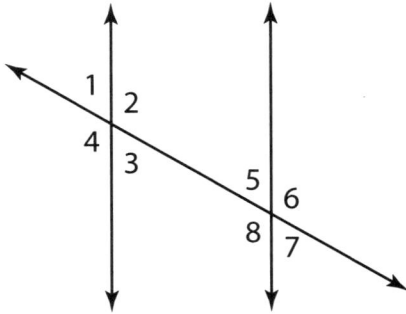

**16 Which two angles are both adjacent and supplementary?**

- **A** ∠1 and ∠3
- **B** ∠3 and ∠5
- **C** ∠4 and ∠5
- **D** ∠2 and ∠3
- **E** ∠5 and ∠7

**17 If ∠7 measures 60°, what is the measure of ∠8?**

- **A** 60°
- **B** 90°
- **C** 120°
- **D** 180°
- **E** Not enough information is known.

**18 Which angle is alternate exterior to ∠4?**

- **A** ∠1
- **B** ∠2
- **C** ∠6
- **D** ∠7
- **E** ∠8

**19 Which angles are corresponding angles?**

- **A** ∠6 and ∠8
- **B** ∠4 and ∠7
- **C** ∠2 and ∠5
- **D** ∠1 and ∠7
- **E** ∠3 and ∠7

**Directions:** Questions 20 and 21 are based on the information below.

The figure shows several rays with vertex *E* in common.

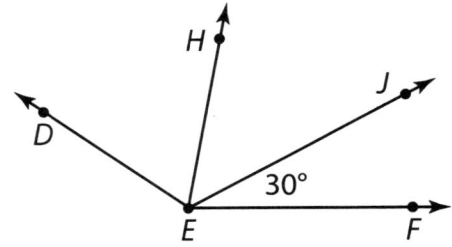

**20 Which angle has a 30° measure?**

- **A** ∠JEF
- **B** ∠FEH
- **C** ∠DEJ
- **D** ∠HED
- **E** ∠JEH

**21 What is the measure of ∠HED?**

- **A** 30°
- **B** 45°
- **C** 60°
- **D** 90°
- **E** Not enough information is given.

**22 The figure shows three rays that have a common endpoint.**

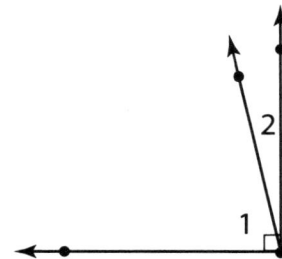

**What kind of relationship do ∠1 and ∠2 have?**

- **A** vertical
- **B** complementary
- **C** congruent
- **D** supplementary
- **E** straight

**Directions:** Questions 23 through 25 are based on the figure below.

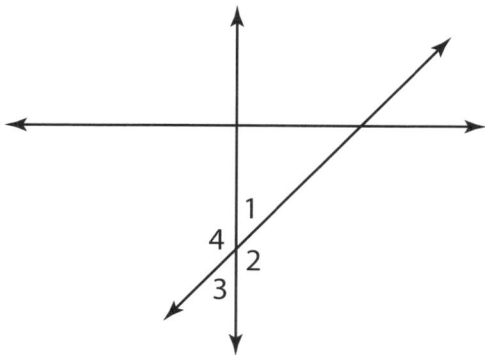

**23** If ∠1 has a measure of 35°, what is the sum of the measures of ∠2 and ∠4?
A 70°
B 145°
C 180°
D 290°
E Not enough information is given.

**24** What is the sum of the measures of ∠1, ∠2, ∠3, and ∠4?
A 140°
B 180°
C 270°
D 360°
E Not enough information is given.

**25** If ∠3 has a measure of 30°, then what is the measure of ∠4?
A 30°
B 60°
C 90°
D 150°
E 180°

**26** Perpendicular lines are lines that meet at right angles. What is the sum of the measures of any two adjacent angles formed when two perpendicular lines intersect?
A 45°
B 90°
C 180°
D 360°
E Not enough information is given.

**Directions:** Questions 27 through 30 are based on Figures A–E below.

A

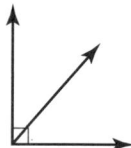

B

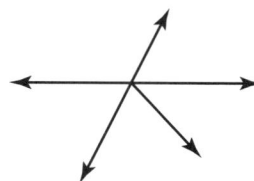

C

D

E

**27** Which figure shows two congruent and supplementary angles?
A A
B B
C C
D D
E E

**28** Which figure contains complementary angles?
A A
B B
C C
D D
E E

**29** Which figure includes a pair of vertical angles?
A A
B B
C C
D D
E E

**30** Which figure shows two supplementary angles that are not congruent?
A A
B B
C C
D D
E E

# Triangles and Quadrilaterals

## ① Learn the Skill

A **triangle** is a closed three-sided figure with three angles. The sum of the three interior angles of any triangle is always 180°. To name a triangle, use the symbol △ followed by the names of the vertices in any clockwise or counterclockwise order. For example, two names for the triangle to the right include △CBA and △CAB.

A **quadrilateral** is a closed four-sided figure with four angles. The sum of the four interior angles of any quadrilateral is always 360°. The sides of a quadrilateral may or may not be parallel. This quadrilateral is a trapezoid, which has one pair of parallel sides. Other quadrilaterals are the parallelogram, rectangle, rhombus, and square.

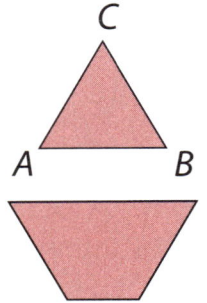

## ② Practice the Skill

By mastering your understanding of triangles and quadrilaterals, you will improve your study and test-taking skills, especially as they relate to mathematics high school equivalency tests. Read the example and strategies below. Use this information to answer question 1.

Ⓐ Triangles can be classified by their largest angle: **right** (90°), **acute** (less than 90°), or **obtuse** (greater than 90°). They also can be classified by their sides:
 **equilateral**: three congruent sides
 **isosceles**: at least two congruent sides
 **scalene**: no congruent sides
This is a scalene triangle. No two sides are the same length, which means that no two angles have the same measure.

Ⓑ In a rectangle and square, all angles are right angles and are congruent. In a parallelogram and rhombus, opposite angles and opposite sides are congruent. You can use this information to find the measure of an unknown angle. In this parallelogram, ∠D and ∠B are congruent, and ∠A and ∠C are congruent.

The first figure is a triangle and the second figure is a parallelogram.

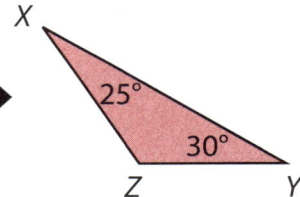

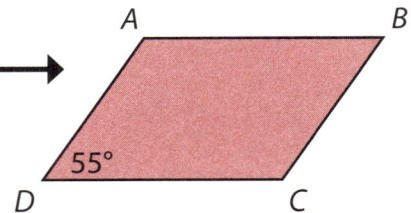

1 **What is the measure of ∠A in the parallelogram above?**
 **A** 55°
 **B** 110°
 **C** 125°
 **D** 180°
 **E** 250°

**Directions:** Questions 2 and 3 are based on the information below.

Supplementary angles are two angles whose measures add to 180°. Use this information and the figure below to answer the questions.

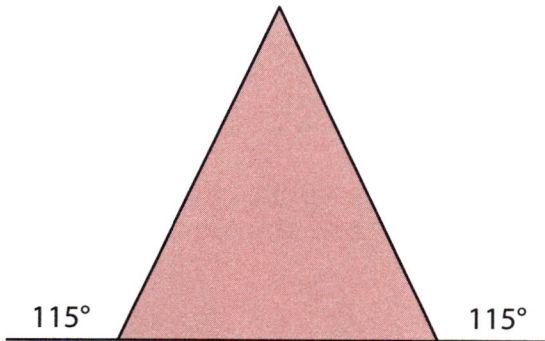

2   **What is the measure of the smallest acute angle within the triangle?**
   A   50°
   B   65°
   C   90°
   D   115°
   E   130°

3   **Which statement best describes this isosceles triangle?**
   A   All angles are congruent.
   B   All sides are congruent.
   C   One angle is obtuse.
   D   Two sides are congruent.
   E   All sides and angles are congruent.

4   **In a certain right triangle, the measure of one acute angle is twice the measure of the other acute angle. What is the measure of the smaller angle?**
   A   10°
   B   25°
   C   30°
   D   55°
   E   60°

**Directions:** Questions 5 and 6 are based on the information below.

Quadrilateral *RSUV* is a parallelogram.
Quadrilateral *RSTV* is a trapezoid.

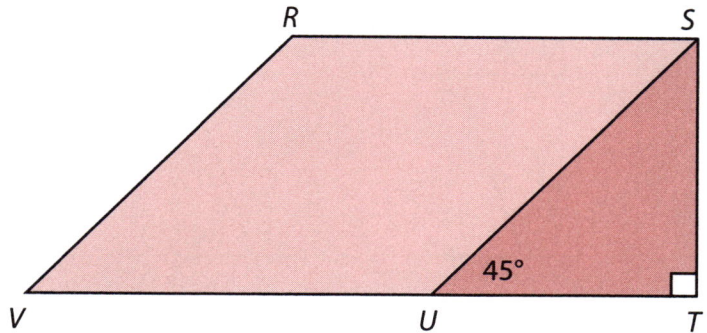

5   **What is the measure of ∠R?**
   A   45°
   B   90°
   C   135°
   D   145°
   E   Not enough information is given.

6   **What is the measure of ∠RST?**
   A   22.5°
   B   45°
   C   65°
   D   90°
   E   Not enough information is given

7   **Angle *H* in the rhombus shown is 35°.**

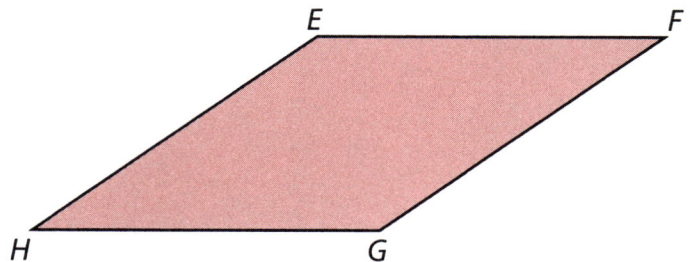

**What is the measure of each obtuse angle in the rhombus?**
   A   35°
   B   45°
   C   90°
   D   135°
   E   145°

**Directions:** Questions 8 through 10 are based on the information below.

The figure shows two transversals intersecting two parallel lines to form a parallelogram.

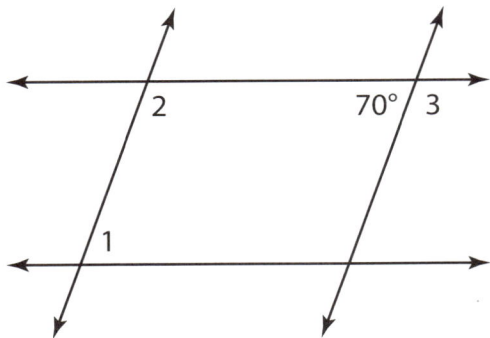

8  What is the measure of ∠1?
   A  20°
   B  70°
   C  90°
   D  110°
   E  Not enough information is given.

9  What is the measure of ∠2?
   A  70°
   B  90°
   C  110°
   D  140°
   E  Not enough information is given.

10  What is the measure of the unlabeled angle in the parallelogram?
   A  110°
   B  90°
   C  70°
   D  20°
   E  10°

**Directions:** Questions 11 through 13 are based on the information below.

An architect is designing a triangular eating area beside an outdoor pool.

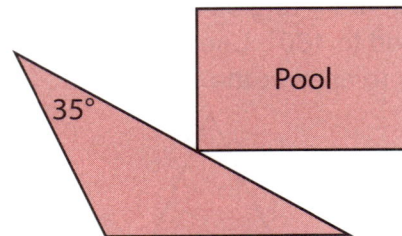

11  If the other acute angle of the eating area is 45°, what is the measure of the obtuse angle?
   A  45°
   B  80°
   C  90°
   D  100°
   E  Not enough information is given.

12  If the shape of the pool is a rectangle, what is the measure of each angle?
   A  35°
   B  90°
   C  180°
   D  360°
   E  Not enough information is given.

13  If the two acute angles of the eating area are congruent, what is the measure of the obtuse angle?
   A  55°
   B  73°
   C  110°
   D  145°
   E  180°

14  An acute angle in a right triangle has a measure of 40°. What is the measure of the other acute angle in the triangle?
   A  40°
   B  50°
   C  90°
   D  140°
   E  Not enough information is given.

UNIT 2

**15** In a certain right triangle, one acute angle is one-half the measure of the right angle. What is the measure of the other acute angle?

   **A** 35°
   **B** 45°
   **C** 60°
   **D** 90°
   **E** Not enough information is given.

**Directions:** Questions 16 and 17 are based on the information below.

A section of a square park has been roped off so that the park attendants can plant grass seed. The roped-off area forms △XYZ.

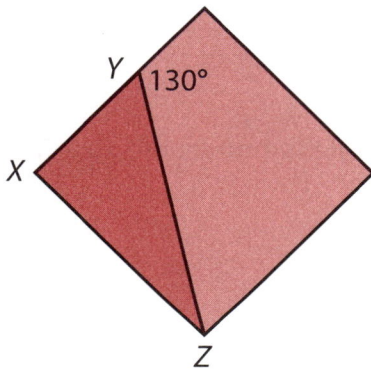

**16** What is the measure of the larger of the two acute angles of the triangle?

   **A** 30°
   **B** 40°
   **C** 50°
   **D** 90°
   **E** Not enough information is given.

**17** What is the measure of the angle that is complementary to ∠YZX?

   **A** 35°
   **B** 40°
   **C** 45°
   **D** 50°
   **E** Not enough information is given.

**Directions:** Questions 18 and 19 are based on the information below.

Quadrilateral *HIJK* is a trapezoid. Quadrilateral *HIJL* is a parallelogram.

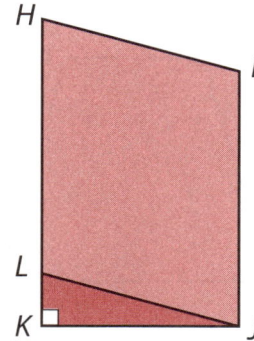

**18** If ∠L of △JKL measures 60°, what is the measure of the other acute angle in the triangle?

   **A** 30°
   **B** 35°
   **C** 45°
   **D** 60°
   **E** Not enough information is given.

**19** If ∠L of △JKL measures 60°, what is the measure of ∠H in quadrilateral *HIJK*?

   **A** 30°
   **B** 35°
   **C** 45°
   **D** 55°
   **E** 60°

**20** Which two sides of this quadrilateral are parallel?

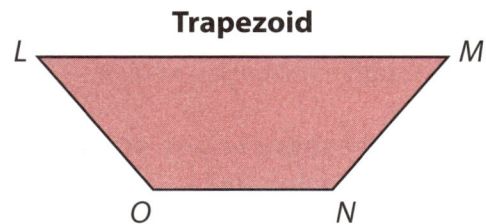

   **A** $\overline{LM}$ and $\overline{MN}$
   **B** $\overline{NO}$ and $\overline{OL}$
   **C** $\overline{OL}$ and $\overline{MN}$
   **D** $\overline{LM}$ and $\overline{NO}$
   **E** $\overline{OL}$ and $\overline{NO}$

**Directions:** Questions 21 and 22 are based on the information below.

Quadrilateral *QRST* is a trapezoid. The measures of ∠*T* and ∠*S* are equal. The measures of ∠*Q* and ∠*R* are equal.

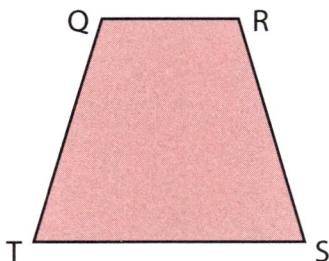

**21  What is the sum of angles *Q, R, S,* and *T*?**
  A  90°
  B  180°
  C  270°
  D  360°
  E  Not enough information is given.

**22  If the measure of ∠*R* is 105°, what is the measure of ∠*S*?**
  A  37.5°
  B  52.5°
  C  75°
  D  150°
  E  Not enough information is given.

**23  What kind of angle would represent the sum of the interior angles of a triangle?**
  A  right angle
  B  straight angle
  C  acute angle
  D  obtuse angle
  E  Not enough information is given.

**24  What is the measure of one acute angle in an isosceles right triangle?**
  A  30°
  B  45°
  C  60°
  D  90°
  E  Not enough information is given.

**Directions:** Questions 25 through 27 are based on the information below.

The figure shows △*BCD*, △*BFE* and quadrilateral *CDEF*. Segments *EF* and *DC* are parallel.

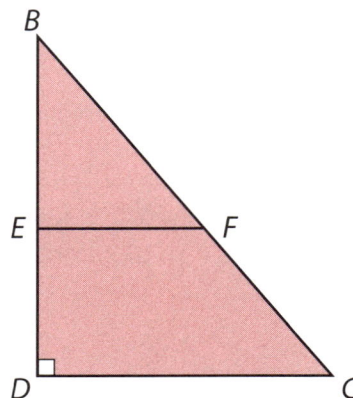

**25  Which angle is congruent to ∠*BDC*?**
  A  ∠*BEF*
  B  ∠*BCD*
  C  ∠*DCF*
  D  ∠*BFE*
  E  ∠*EFB*

**26  If the measure of ∠*FCD* is 60°, what is the measure of ∠*BFE*?**
  A  30°
  B  60°
  C  90°
  D  120°
  E  Not enough information is given.

**27  If the measure of ∠*B* is 30°, what is the measure of the obtuse angle in quadrilateral *CDEF*?**
  A  180°
  B  150°
  C  145°
  D  120°
  E  Not enough information is given.

# Pythagorean Theorem

## ① Learn the Skill

A **right triangle** is a triangle with a right (90°) angle. The **legs** (shorter sides) and **hypotenuse** (longest side) of a right triangle have a special relationship. The **Pythagorean theorem** describes this relationship: In any right triangle, the sum of the squares of the lengths of the legs is equal to the square of the length of the hypotenuse. In equation form, the Pythagorean theorem is $a^2 + b^2 = c^2$.

## ② Practice the Skill

Whenever the lengths of two sides of a right triangle are known, you can use the Pythagorean theorem to find the length of the third side. Be sure to use the Pythagorean theorem for right triangles only. Read the examples and strategies below. Use this information to answer question 1.

**UNIT 2**

**A** The Pythagorean theorem is written:
$$a^2 + b^2 = c^2$$
where $a$ and $b$ are the leg lengths and $c$ is the length of the hypotenuse. The hypotenuse is always the side opposite the right angle. After solving for the missing side length, check to make sure that the hypotenuse has the greatest measure. Here, solve for $c$ using the values for $a$ and $b$.

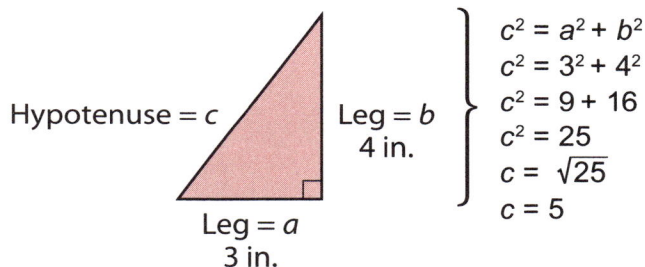

Hypotenuse = $c$     Leg = $b$
                    4 in.

Leg = $a$
3 in.

$c^2 = a^2 + b^2$
$c^2 = 3^2 + 4^2$
$c^2 = 9 + 16$
$c^2 = 25$
$c = \sqrt{25}$
$c = 5$

**B** Use the Pythagorean theorem to find the distance between points $A$ and $B$. Draw a triangle with segment $AB$ as the hypotenuse. Count units to find the lengths of the legs and use the Pythagorean theorem to solve for the length of the hypotenuse.

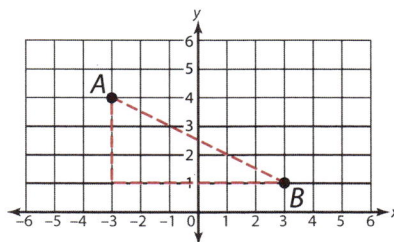

$c$ = distance between $A$ and $B$
$c^2 = 3^2 + 6^2$
$c^2 = 45$
$c \approx 6.7$

✔ **TEST-TAKING TIPS**

If the sides of a triangle make the equation $a^2 + b^2 = c^2$ true, then the triangle proves to be a right triangle.

**1** The bottom of a ladder is resting 5 feet from the wall of a garage. The wall and the ground form a right angle. If the ladder is 10 feet long, how far up the wall does it reach?

A   5.0 ft
B   6.4 ft
C   7.5 ft
D   8.7 ft
E   11.2 ft

**Directions:** Questions 2 through 4 are based on the information below.

A telephone pole is 30 feet tall. A cable attached to the top of the pole is anchored to the ground 15 feet away from the base of the pole.

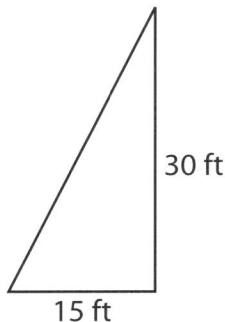

30 ft

15 ft

**2** **What is the length of the cable to the nearest tenth of a foot?**
- **A** 26.0
- **B** 30.7
- **C** 32.2
- **D** 33.5
- **E** 45.1

**3** **If a 35-foot cable were run from the top of the pole and anchored to the ground at a distance from the pole, about how far away from the pole would it be anchored?**
- **A** 16 feet
- **B** 18 feet
- **C** 30 feet
- **D** 38 feet
- **E** 46 feet

**4** **If the telephone pole were 2 feet taller, and the cable was still anchored in 15 feet away from the pole, how would the change in pole height affect the length of the cable?**
- **A** The cable would be exactly 2 feet longer.
- **B** The length of the cable would not change.
- **C** The cable would be about 1.8 feet longer.
- **D** The cable would be about 2.2 feet longer.
- **E** The cable would be about 1.9 feet longer.

**Directions:** Questions 5 and 6 are based on the information below.

The river is 120 meters wide. Sara starts out swimming across the river. The current pushes her, so she ends up 40 meters downriver from where she started.

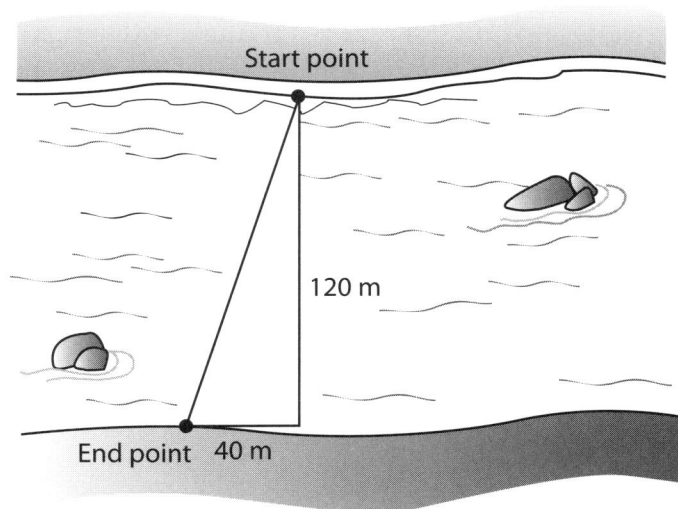

Start point

120 m

End point   40 m

**5** **To the nearest meter, how many meters did Sara actually swim?**
- **A** 80
- **B** 113
- **C** 105
- **D** 126
- **E** 160

**6** **If the current had not been so strong and had swept Sara only 20 meters downstream, about how many meters would she actually have swum? Round to the nearest whole number.**
- **A** 122
- **B** 118
- **C** 116
- **D** 106
- **E** 100

**7** **What is the distance between points $M(-4, 5)$ and $N(4, 3)$? Round to the nearest tenth.**
- **A** 6.3
- **B** 6.5
- **C** 7.1
- **D** 7.3
- **E** 8.2

**Directions:** Questions 8 through 10 are based on the graph below.

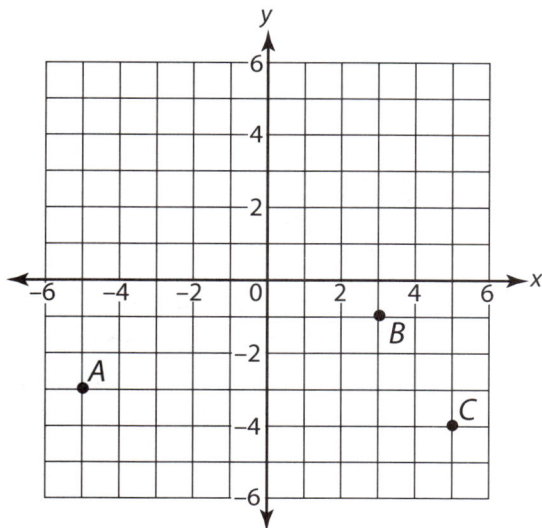

**8** What is the distance between points *A* and *B*? Round to the nearest hundredth.
A   6.32
B   7.07
C   7.28
D   7.75
E   8.25

**9** What is the distance between points *B* and *C*? Round to the nearest hundredth.
A   3.46
B   3.61
C   3.74
D   3.87
E   Not enough information is given.

**10** What is the perimeter of the triangle with vertices *A*, *C*, and (–5, –4)? Round to the nearest hundredth.
A   26.18
B   22.20
C   21.05
D   20.95
E   20.00

**Directions:** Questions 11 and 12 are based on the information below.

Dana designed the quilt square shown below.

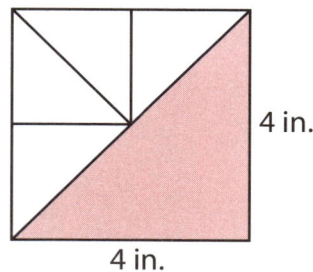

**11** What is the length of the diagonal of the outside square in inches? Round to the nearest hundredth.
A   5.66
B   4.00
C   2.83
D   2.01
E   1.41

**12** Notice the small square in the upper left corner of the large square. One side of this small square is half the length of one side of the large square. What is the length of the diagonal of the small square? Round to the nearest hundredth.
A   2.00 in.
B   2.83 in.
C   3.25 in.
D   3.82 in.
E   5.66 in.

**13** Ella incorrectly determined the length of one of the legs of a right triangle. Her work is shown below. Which answer choice best describes why Ella's answer is incorrect?

$$4^2 + a^2 = 10^2$$
$$16 + a^2 = 100$$
$$a^2 = 116$$
$$a \approx 10.8$$

A   She incorrectly found the square root of 116.
B   She incorrectly squared 10.
C   She incorrectly squared 4.
D   She incorrectly added 16 when she should have subtracted.
E   She incorrectly set up the equation to find the value of *a* when she should have found the value of *c*.

**14** A computer monitor is listed as measuring 21 inches. This is the distance across the diagonal of the screen. If the screen is 16 inches wide, what is the height of the screen to the nearest tenth of an inch?

   **A**  4.5
   **B**  9.1
   **C**  13.6
   **D**  27.2
   **E**  37.1

**15** A surveyor wants to find the width of the pond. She placed stakes at points *A, B,* and *C.* She knows that △*ABC* is a right triangle. If the distance between *A* and *C* is 75 feet and the distance between *B* and *C* is 63 feet, what is the approximate width of the pond between points *A* and *B*?

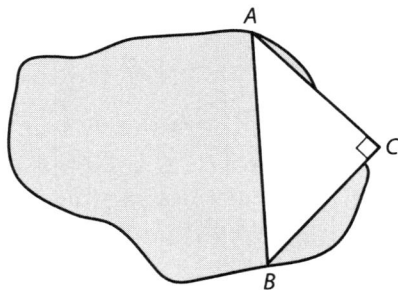

   **A**  12.2 feet
   **B**  40.7 feet
   **C**  54.5 feet
   **D**  97.9 feet
   **E**  144.2 feet

**16** A right triangle has a leg with length 48 and a hypotenuse with length 50. Which equation could be used to determine the length, *x,* of the missing third side?

   **A**  $48^2 + 50^2 = x^2$
   **B**  $48 = \sqrt{x^2 + 50^2}$
   **C**  $48 = \sqrt{50^2 - x^2}$
   **D**  $50^2 + x^2 = 48^2$
   **E**  $x = \sqrt{48^2 + 50^2}$

**Directions:** Questions 17 and 18 are based on the information below.

**17** A 7.9-foot ramp runs from the back of a moving truck to the ground. If the ramp meets the ground 6.5 feet away from the truck, about how many feet off of the ground is the ramp?

   **A**  2.0
   **B**  4.5
   **C**  7.1
   **D**  10.2
   **E**  14.4

**18** If the ramp on the back of the truck were 5 feet off the ground and touched the ground at a point 8 feet away from the back of the truck, about how many feet long would the ramp be?

   **A**  3.6
   **B**  5.3
   **C**  9.0
   **D**  9.1
   **E**  9.4

**19** Which group of segments are very close to forming a right triangle?

   **A**  6.5 m, 2.7 m, 7.04 m
   **B**  4.25 m, 3.8 m, 5.07 m
   **C**  11.5 m, 13.6 m, 16.81 m
   **D**  12.75 m, 8.6 m, 14.45 m
   **E**  7.25 m, 3.5 m, 10.75 m

**20** A 15-foot ladder is placed against the side of a building so that it reaches 12 feet up the side of the building. How far away from the building is its base?

   **A**  8 feet
   **B**  9 feet
   **C**  10 feet
   **D**  11 feet
   **E**  12 feet

Henry is building a walkway through a rectangular garden as shown in the diagram below.

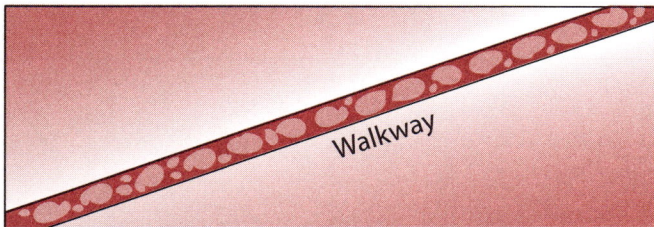

**21** If the length of the garden is 30 yards and the width of the garden is 17 yards, what is the approximate length of the walkway in yards?
   **A** 34.5
   **B** 24.7
   **C** 21.4
   **D** 15.9
   **E** 13.3

**22** If the length of the walkway is 40 yards and the width of the garden is 12 yards, what is the approximate length of the garden in yards?
   **A** 12.5
   **B** 28.4
   **C** 38.2
   **D** 41.8
   **E** 52.3

**23** Which group of segments with the given side lengths cannot form a right triangle?
   **A** 7, 24, 25
   **B** 9, 12, 15
   **C** 10, 24, 26
   **D** 15, 33, 39
   **E** 24, 45, 51

The front view of a dollhouse is shown below.

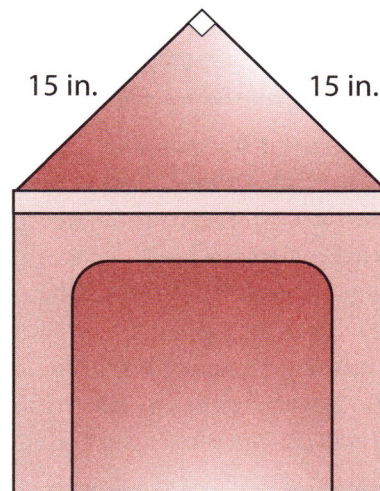

**24** What is the width of the dollhouse to the nearest tenth of an inch?
   **A** 15.0
   **B** 17.5
   **C** 18.6
   **D** 19.2
   **E** 21.2

**25** What is the area of the triangular section of the roof in square inches?
   **A** 51.2
   **B** 112.5
   **C** 159.0
   **D** 225.0
   **E** 318.0

**26** If the height of the dollhouse, not including the triangular roof, is 18 inches, what is the perimeter of the front view?
   **A** 69.2 inches
   **B** 87.2 inches
   **C** 90.4 inches
   **D** 108.4 inches
   **E** Not enough information is given.

# Congruent and Similar Figures

## 1 Learn the Skill

**Congruent figures** have the same shape and size. They have corresponding angles of equal measure and corresponding sides of equal length. The figures at the right are congruent figures.

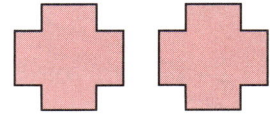

**Similar figures** have the same shape and congruent angles, but have proportional sides. The similar triangles to the right have corresponding angles of equal measure but do not have corresponding sides of equal measure. If the base of the larger triangle is three times greater than the base of the other, the same relationship exists for the heights.

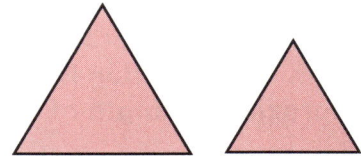

## 2 Practice the Skill

If you know that two figures are similar, then you know that the corresponding angles are congruent and the sides are proportional. Therefore, if you know that two figures have congruent corresponding angles and proportional sides, then you know that the figures are similar. Read the example and strategies below. Use this information to answer question 1.

**A** When angles or line segments of two figures correspond, they are in the same position. Angle *C* of △*ABC* corresponds to ∠*T* of △*RST*. Angle *B* corresponds to ∠*S*. Similarly, $\overline{AC}$ corresponds to $\overline{RT}$, and $\overline{BC}$ corresponds to $\overline{ST}$.

**B** The symbol ≅ means "is congruent to." The symbol ~ means "is similar to." When saying that two figures are congruent or similar, name corresponding parts in the same order. For example, △*BCA* ~ △*STR*.

Triangles *ABC* and *RST* are similar.

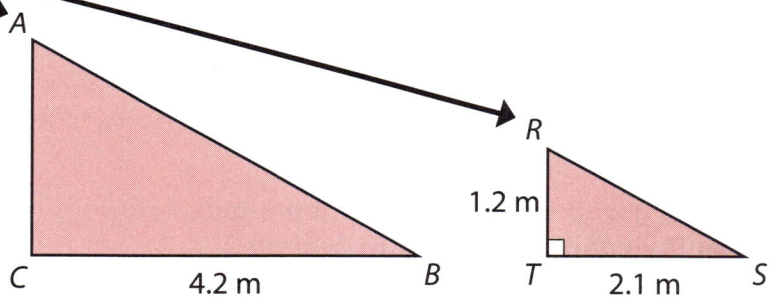

1.2 m
4.2 m
2.1 m

### ☑ TEST-TAKING TIPS

In question 1, we are told that the triangles are similar. A proportion can be written to solve for the missing length. A sample proportion for the figures is $\frac{x}{1.2} = \frac{4.2}{2.1}$.

**1** What is *AC*, which is another way to say "the length of $\overline{AC}$"?

- **A** 1.2 m
- **B** 2.1 m
- **C** 2.4 m
- **D** 3.2 m
- **E** 4.2 m

**Directions:** Questions 4 and 5 are based on the information below.

Triangle *ABC* and triangle *FGH* are similar figures.

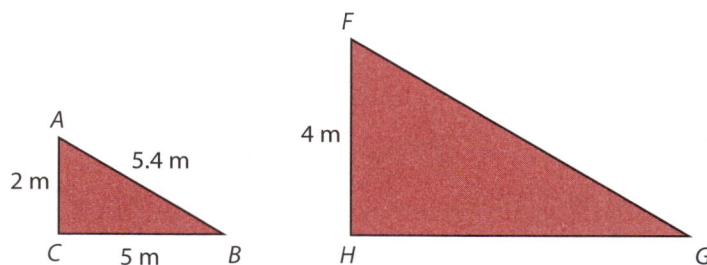

**2** Refer to the following figures.

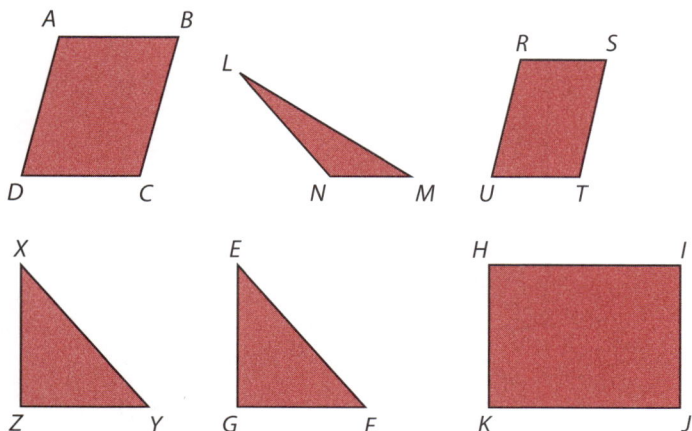

Which statement appears to be true?

A   parallelogram *ABCD* ~ rectangle *HIJK*
B   triangle *LMN* ~ △*EFG*
C   parallelogram *ABCD* ≅ rectangle *HIJK*
D   triangle *XYZ* ≅ △*EFG*
E   parallelogram *RSTU* ~ rectangle *HIJK*

**3** Triangles 1 and 2 shown below are congruent. The lengths of two sides of triangle 1 are given.

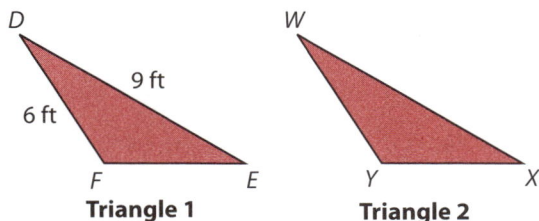

If the perimeter of triangle 1 is 19 ft, what is the length of $\overline{XY}$?

A   4 ft
B   6 ft
C   9 ft
D   19 ft
E   Not enough information is given.

**4** What is the length of $\overline{FG}$?

A   5.4 m
B   10.0 m
C   10.8 m
D   12.4 m
E   Not enough information is given.

**5** What is the perimeter of △*FGH*?

A   12.4 m
B   14.8 m
C   24.0 m
D   24.8 m
E   Not enough information is given.

**6** Figure *MNOP* and figure *XYZW* are congruent.

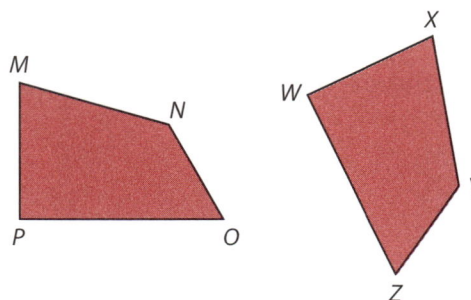

Which line segment of quadrilateral *XYZW* is congruent to $\overline{MN}$?

A   $\overline{WX}$
B   $\overline{XY}$
C   $\overline{YZ}$
D   $\overline{ZW}$
E   Not enough information is given.

**Directions:** Questions 7 through 9 are based on the information below.

The corresponding angles and sides of quadrilateral *CDEF* and quadrilateral *MJKL* are congruent. $\overline{CD}$ and $\overline{FE}$ are parallel. Angle *F* and angle *E* are congruent.

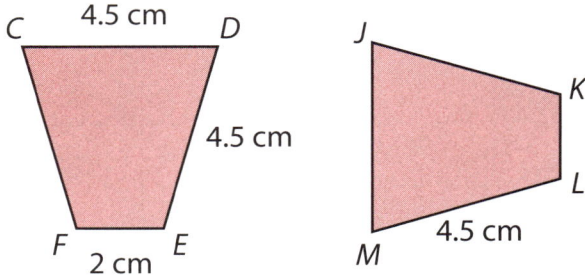

**Directions:** Questions 10 through 12 are based on the information below.

The following right triangles are similar. $\overline{CB}$ corresponds to $\overline{GF}$.

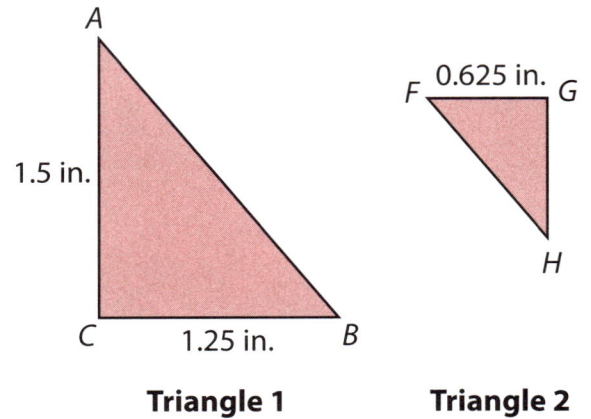

Triangle 1          Triangle 2

7 **Which statement about the two figures is true?**
   A   Angle *L* is congruent to angle *F*.
   B   $\overline{CD}$ is congruent to $\overline{KL}$.
   C   Angle *F* is congruent to angle *J*.
   D   $\overline{DE}$ is perpendicular to $\overline{JM}$.
   E   Angle *K* is similar to angle *C*.

8 **Which segment corresponds to $\overline{ML}$?**
   A   $\overline{FE}$
   B   $\overline{CD}$
   C   $\overline{DC}$
   D   $\overline{CF}$
   E   $\overline{DE}$

9 **What is the perimeter of quadrilateral *JKLM*?**
   A   9 cm
   B   11 cm
   C   13.5 cm
   D   15.5 cm
   E   Not enough information is given.

10 **Which condition is true about the two figures?**
   A   $\angle C \cong \angle H$
   B   $\overline{CB} \cong \overline{FH}$
   C   $\angle A \cong \angle H$
   D   $\overline{AC} \cong \overline{HF}$
   E   $\overline{AB} \cong \overline{FH}$

11 **What is the length of the side of Triangle 2 that corresponds to $\overline{AC}$ of Triangle 1?**
   A   0.5 in.
   B   0.75 in.
   C   1.25 in.
   D   1.5 in.
   E   Not enough information is given.

12 **What is the length of the side of Triangle 2 that corresponds to $\overline{AB}$ of Triangle 1?**
   A   0.5 in.
   B   0.625 in.
   C   0.75 in.
   D   0.976 in.
   E   1 in.

**Directions:** Questions 13 and 14 are based on the information below.

The corresponding angles of the two triangles are congruent, and the lengths of the corresponding sides are in proportion. The corresponding angles and sides of the two trapezoids are congruent.

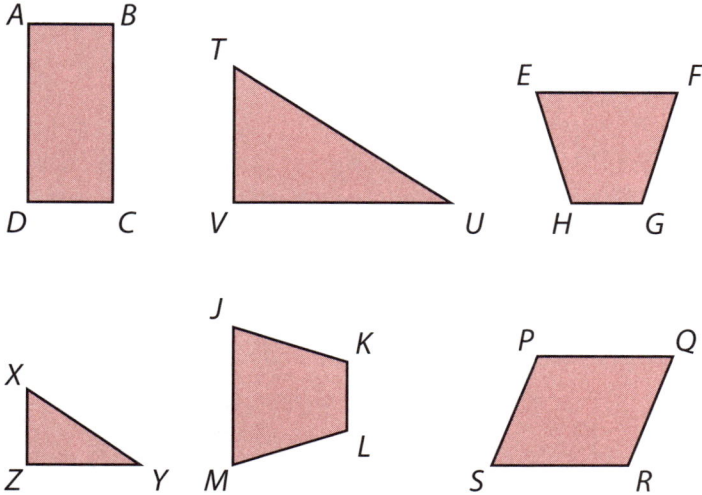

**Directions:** Questions 16 through 18 are based on the information below.

The corresponding angles of these two triangles are congruent. The corresponding sides of △GIH are twice the length of the sides of △BCD.

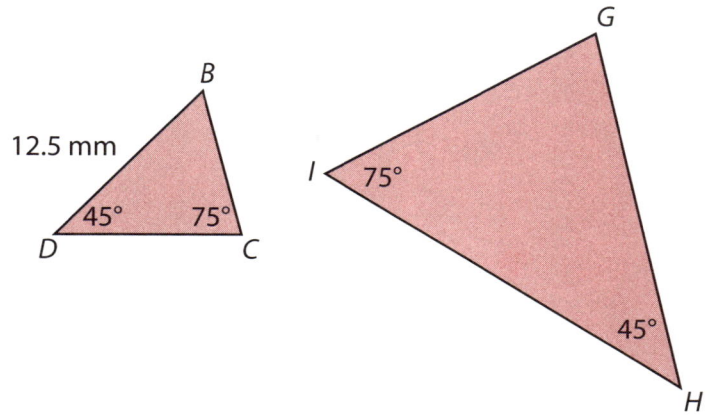

**13** Which statement about the figures is true?

A  △XYZ ≅ △TUV
B  quadrilateral EFGH ~ quadrilateral PQRS
C  △TUV ~ △XYZ
D  quadrilateral PQRS ~ quadrilateral ABCD
E  quadrilateral JKLM ~ △XYZ

**14** Which condition is true about △TUV and △XYZ?

A  ∠Y ~ ∠V
B  $\overline{TU} \cong \overline{XY}$
C  $\overline{VU} \cong \overline{XZ}$
D  ∠Z ≅ ∠V
E  $\overline{TV} \cong \overline{XY}$

**15** Which statement is true?

A  A right triangle can be similar to an equilateral triangle.
B  A scalene triangle cannot be congruent to a right triangle.
C  An obtuse triangle cannot be congruent to a right triangle.
D  A quadrilateral cannot be congruent to a square.
E  A triangle can be similar to a rectangle.

**16** What is the measure of ∠G in △GIH?

A  45°
B  60°
C  65°
D  70°
E  75°

**17** What is the measure of the angle in △GIH that corresponds to ∠C in △BCD?

A  45°
B  60°
C  75°
D  90°
E  Not enough information is given.

**18** What is the length of $\overline{GI}$ in △GIH?

A  6.4 mm
B  12 mm
C  12.8 mm
D  25.6 mm
E  Not enough information is given.

**Directions:** Questions 19 through 21 are based on the information below.

A sculptor's design consists of two identical steel triangles shown below. However, the plans provided by the sculptor are incomplete. The manager of the foundry at which the sculpture is being cast knows, however, that the pieces are to be identical and that the sides of the triangles are congruent.

**SCULPTURE PLANS**

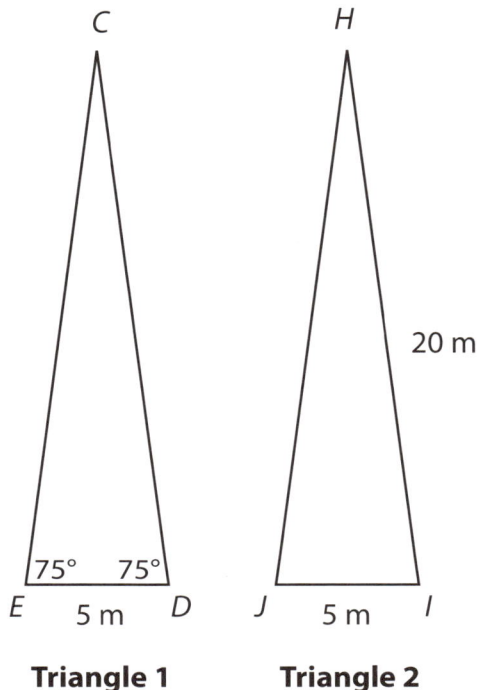

Triangle 1          Triangle 2

**19** What is the length of the side of Triangle 1 that corresponds to $\overline{HI}$ of Triangle 2?

- **A**  5 m
- **B**  10 m
- **C**  20 m
- **D**  25 m
- **E**  Not enough information is given.

**20** What is the measure of ∠H?

- **A**  15°
- **B**  30°
- **C**  75°
- **D**  105°
- **E**  Not enough information is given.

**21** What is the perimeter of each triangular sculpture?

- **A**  25 m
- **B**  40 m
- **C**  45 m
- **D**  90 m
- **E**  Not enough information is given.

**Directions:** Questions 22 and 23 are based on the information below.

A carpet company makes two sizes of a popular carpet. The sides of the smaller carpet will be two-thirds the length of the larger carpet's sides. This illustration shows the smaller carpet lying over top of the larger carpet.

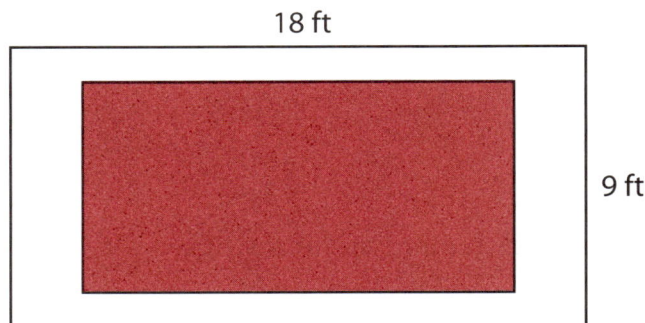

**22** The larger carpet is 18 feet long. What is the length of the smaller carpet?

- **A**  5 ft
- **B**  6 ft
- **C**  10 ft
- **D**  12 ft
- **E**  Not enough information is given.

**23** What is the difference between the areas of the two carpets?

- **A**  90 ft²
- **B**  120 ft²
- **C**  162 ft²
- **D**  234 ft²
- **E**  Not enough information is given.

# Indirect Measurement and Scale Drawings

## ① Learn the Skill

**Indirect measurement** involves the use of proportions and corresponding parts of similar figures to find a measurement that you cannot find directly. **Scale drawings**, or drawings that represent an actual object, can use indirect measurement.

## ② Practice the Skill

You will use indirect measurement to solve problems on mathematics tests. Read the example and strategies below. Use this information to answer question 1.

UNIT 2

**A** The scale provides a ratio of drawing measurements to actual measurements. In this case, one centimeter equals 200 meters.

**SCALE DRAWING**
Scale: 1 cm : 200 m

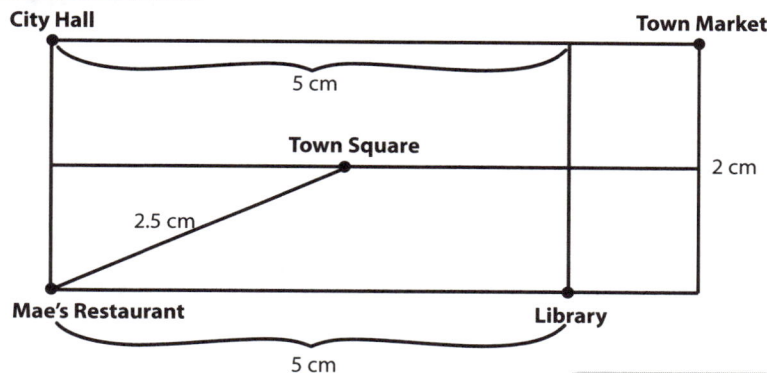

City Hall      5 cm      Town Market

Town Square

2.5 cm      2 cm

Mae's Restaurant      Library

5 cm

$$\frac{\text{Map distance}}{\text{Actual distance}} = \frac{1 \text{ cm}}{200 \text{ m}} = \frac{5 \text{ cm}}{x}$$

$$1(x) = 5(200) = 1{,}000 \text{ m}$$

**B** To find the actual distance between Mae's Restaurant and the library, write a proportion using the scale. Be sure to write the corresponding parts of the proportion in the correct order.

### ✓ TEST-TAKING TIPS

Pay close attention to the units involved in a proportion. For example, a question posed about the map above could be "How many kilometers from City Hall is the library?" To answer this question correctly, you must convert meters to kilometers.

**1** Two cities are 5 inches apart on a map. The map scale is 1 in. : 2.5 miles. What is the actual distance in miles between the two cities?

A  0.5

B  5

C  10

D  12.5

E  15

**2** A tree 14 feet tall casts a shadow that is 2.5 feet long. At the same time of day, a person casts a shadow that is 1 foot long. Which proportion can be solved to find the height of the person?

A $\frac{x}{1} = \frac{2.5}{14}$

B $\frac{1}{x} = \frac{2.5}{14}$

C $\frac{14}{1} = \frac{2.5}{x}$

D $\frac{1}{2.5} = \frac{14}{x}$

E Not enough information is given.

**3** Erika drove from Plymouth to Manchester and back again. On a map, these two cities are 2.5 cm apart. If the map scale is 1 cm : 6 km, how many kilometers did she drive?

A 2.4
B 8.5
C 15
D 22.5
E 30

**4** A map scale is 2 in. : 4.8 miles. What is the actual distance in miles between two points that are 5.5 inches apart on the map?

A 1.75
B 2.29
C 8.3
D 13.2
E 26.4

**5** A person who is 6 feet tall casts a shadow that is 8.5 feet long. At the same time of day, a person who is $4\frac{1}{4}$ feet tall would cast how long of a shadow?

A 12.0 feet
B 6.75 feet
C 6.23 feet
D 6.02 feet
E 3.0 feet

**Directions:** Questions 6 through 8 are based on the information below.

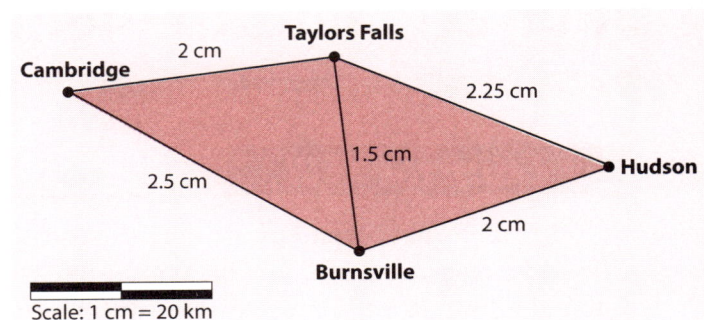

**6** What is the actual distance in kilometers between Burnsville and Taylors Falls?

A 13.3
B 15
C 20
D 23.3
E 30

**7** Jack drove from Cambridge to Burnsville. Pedro drove from Hudson to Burnsville. How much farther did Jack drive than Pedro?

A 0.5 km
B 10 km
C 19.5 km
D 40 km
E Not enough information is given.

**8** Carl drives from Cambridge to Taylors Falls and then to Hudson to go to work each day. Each night he drives the same way home to Cambridge. How many kilometers does he commute each workday?

A 170
B 85
C 70
D 45
E 40

**9** A furniture maker made a model of a table design. The model of the table is 12 inches long. The actual table will be 60 inches long. What is the scale of the model?

A 6 in. : 1 in.
B 1 in. : 4 in.
C 1 in. : 5 in.
D 1 in. : 12 in.
E 6 in. : 36 in.

UNIT 2

**Directions:** Questions 10 and 11 are based on the information below.

A scale drawing of a living room in a house is shown.

Scale : 1 inch : 5 feet

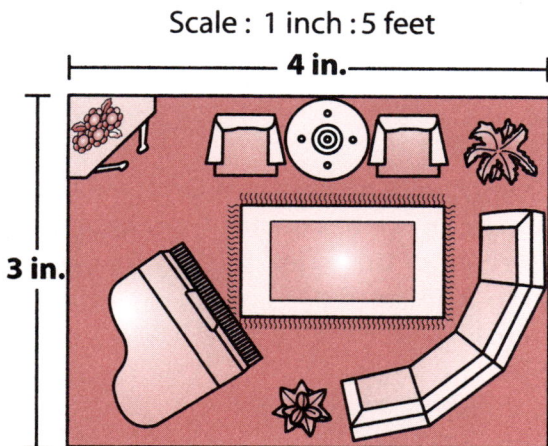

**10** The length of the throw rug in the actual living room is 8 feet. How many inches long is the throw rug in the floor plan?
A   0.625
B   0.975
C   1.6
D   5.8
E   40

**11** The homeowners want to replace the living room wall-to-wall carpet. How many square feet of carpet will they need for the living room?
A   35
B   180
C   250
D   300
E   600

**12** Two towns on a map are 3.5 inches apart. The map scale is 1 inch : 5.5 miles. What is the actual distance in miles between the two towns?
A   2.25
B   7.25
C   9.25
D   11.25
E   19.25

**13** Lake Superior is 350 miles in length. If the map scale is 1 inch : 25 miles, how many inches long is Lake Superior on the map?
A   14
B   16
C   325
D   1,050
E   10,500

**Directions:** Questions 14 and 15 are based on the information below.

On the map below, 2 centimeters represent 6.5 kilometers in actuality.

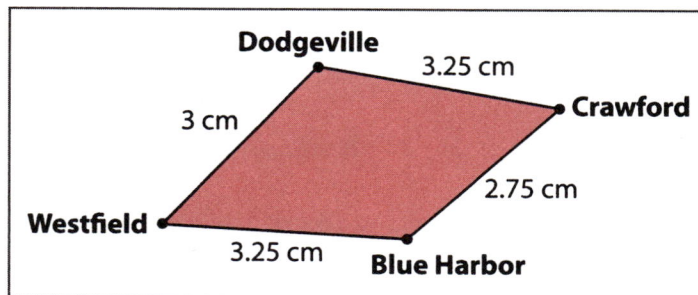

**14** What is the actual distance in kilometers between Dodgeville and Crawford?
A   1.1
B   4.0
C   9.8
D   10.6
E   12.2

**15** About how many more kilometers would you drive from Blue Harbor to Dodgeville through Westfield than from Blue Harbor to Dodgeville through Crawford?
A   0.1
B   0.3
C   0.4
D   0.6
E   0.8

**Directions:** Questions 16 through 18 are based on the information below.

In this floor plan, the scale is $\frac{1}{2}$ inch = 5 feet.

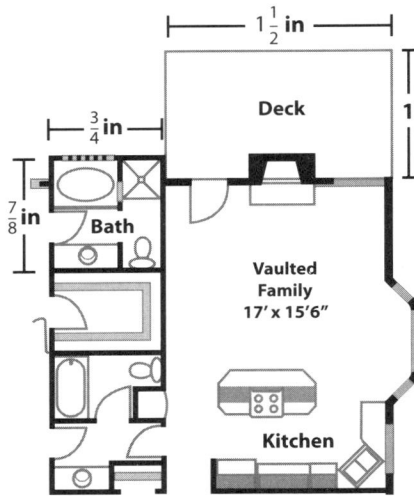

**16** What are the dimensions of the actual deck?
A   1.5 feet by 5 feet
B   5.5 feet by 10 feet
C   5 feet by 7.5 feet
D   10 feet by 15 feet
E   12 feet by 18 feet

**17** What is the actual length of the longer side of the bathroom?
A   $8\frac{3}{4}$ ft
B   $8\frac{1}{2}$ ft
C   8 ft
D   $7\frac{1}{2}$ ft
E   $4\frac{3}{8}$ ft

**18** What are the dimensions of the actual bath?
A   $7\frac{1}{2}$ feet by $8\frac{3}{4}$ feet
B   $5\frac{1}{2}$ feet by $8\frac{1}{2}$ feet
C   5 feet by 8 feet
D   5 feet by $7\frac{1}{2}$ feet
E   $3\frac{3}{4}$ feet by $4\frac{3}{8}$ feet

**19** Peri biked 19.2 miles. She drew a map of her route for a friend using a scale of 2 inches = 3.2 miles. How many inches long is Peri's route on the map she drew?
A   8
B   10
C   12
D   14
E   16

**20** Bloomington is 48 km from Orchard Point. On a map, these towns are 4 cm apart. What is the scale of this map?
A   1 cm = 16 km
B   1 cm = 12 km
C   1 cm = 8 km
D   1 cm = 4.5 km
E   1 cm = 0.5 km

**21** A 22-foot pole casts a shadow that is 31.9 feet long. At the same time of day, how many feet long would a 55-foot building's shadow be?

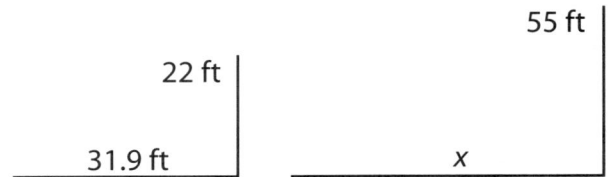

A   12.8
B   31.9
C   37.9
D   64.9
E   79.8

**22** An animal that is 4.2 feet tall casts a shadow that is 3.8 feet long. At the same time of day, a second animal casts a shadow that is 6.8 feet long. How tall is the second animal?
A   7.5 ft
B   7.2 ft
C   6.4 ft
D   6.2 ft
E   2.3 ft

**Lesson 8 | Indirect Measurement and Scale Drawings**

**23** A model car was built using the scale 1 in. : 32 in. If the length of the bumper of the actual car is 108.8 in., how many inches long is the bumper on the model car?

A   3.1 in.
B   3.3 in.
C   3.4 in.
D   3.6 in.
E   3.7 in.

**24** A rectangular floor plan is drawn to a scale of 1 in. : 6 ft. The dimensions of a living room on the plan are $2\frac{1}{4}$ in. by $3\frac{1}{8}$ in. What is the actual area of the living room?

A   $64\frac{1}{2}$ square feet

B   $213\frac{1}{8}$ square feet

C   $233\frac{1}{4}$ square feet

D   $243\frac{1}{8}$ square feet

E   $253\frac{1}{8}$ square feet

**25** A map scale is 3 cm : 18.6 km. Two cities are 10 cm apart on the map. If Stacey drives at an average speed of 90 km/hr from one city to the other, about how long will it take her?

A   20 minutes
B   40 minutes
C   1 hour
D   1 hour 10 minutes
E   1 hour 20 minutes

**26** A scale drawing of a square lot has a perimeter of 22 cm. The scale of the drawing is 2 cm : 5 yards. How long is one side of the square in actuality?

A   2.2 yd
B   5.5 yd
C   8.5 yd
D   13.75 yd
E   55 yd

**Directions:** Questions 27 and 28 are based on the information below.

Michael had to submit the scale drawing below with his application for a permit to build a new deck.

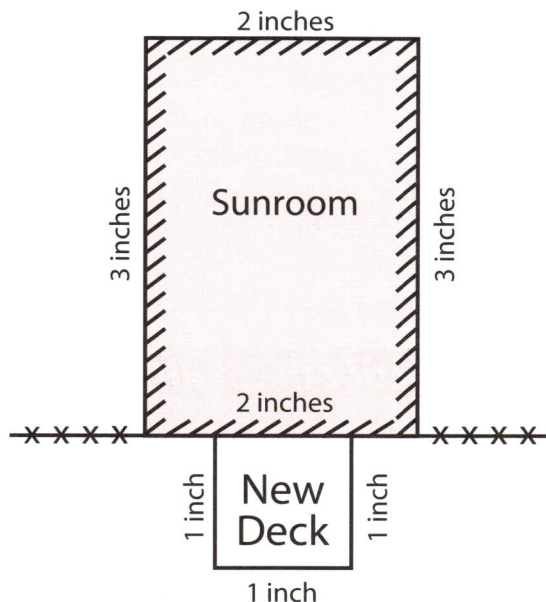

**27** If the longer side of the actual sunroom is 30 feet long, what is the scale of the drawing?

A   1 in. : 7 ft
B   1 in. : 8 ft
C   1 in. : 10 ft
D   1 in. : 12 ft
E   1 in. : 14 ft

**28** If the drawing has a scale of $\frac{1}{2}$ in. : 6.5 ft, what are the actual dimensions of the new deck?

A   6.5 feet by 6.5 feet
B   7 feet by 7 feet
C   10 feet by 10 feet
D   11.5 feet by 11.5 feet
E   13 feet by 13 feet

# Prisms and Cylinders

## ① Learn the Skill

A **solid figure** is a 3-dimensional figure. Solid figures include cubes, prisms, pyramids, cylinders, and cones. The **volume** of a solid figure is the amount of space it takes up, as measured in cubic units. The volume of a prism or cylinder is the product of the area of its base and its height. The **surface area** is the sum of the areas of its two bases and the area of its lateral surfaces.

If you know the volume of a prism and either its base area or its height, you can calculate the other quantity. Similarly, if you know the volume of a cylinder and either its radius or height, you can find the other dimension.

## ② Practice the Skill

You will compute the surface area and volume of prisms and cylinders on mathematics tests. Read the examples and strategies below. Use this information to answer question 1.

**A** A **prism** has two parallel bases. For a rectangular prism, any parallel faces can be used as bases. Prisms are named for the shape of their bases. A prism with triangular bases is called a **triangular prism**.

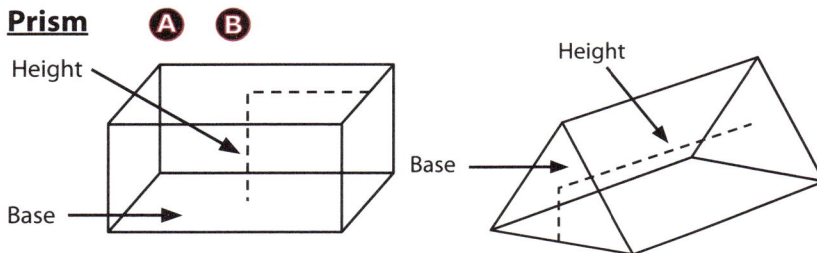

**Prism** Ⓐ Ⓑ

Height

Base

Height

Base

**B** The volume of a prism is the product of the area of its base and its height, or $V = Bh$. The **height** of a prism is the perpendicular distance between its bases. The surface area of a prism is the area of its bases and its lateral faces. To find the area of the base of a triangular prism, use the formula $A = \frac{1}{2}bh$, where the base and height of the triangle are perpendicular. To find the surface area of a triangular prism, you must find the sum of the area of its bases and its 3 lateral sides.

**Cylinder** Ⓒ

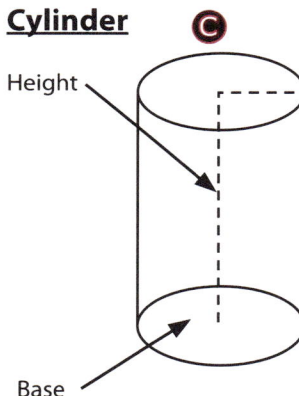

Height

Base

**C** A **cylinder** has two congruent circular bases connected by a curved surface. The volume of a cylinder is the product of the area of its circular base and its height, or $V = \pi r^2 h$. Meanwhile, the surface area of a cylinder is the area of its two circular bases, plus its lateral area. The **lateral area** is the product of the circumference and the height.

☑ **TEST-TAKING TIPS**

Questions relating to solid figures and volume may ask you to solve for diameter, radius, or area of a base, or the length, width, or height of a figure. Read the problem carefully to decide what you need to solve for.

**1** **A company sells oatmeal in a cylindrical canister. The canister has a height of 8 in., and the radius of the base is 3 in. What is the volume of the container to the nearest cubic inch?**

A   24 in.³

B   72 in.³

C   75 in.³

D   226 in.³

E   678 in.³

**2** A rectangular bale of hay has the following dimensions: length = 40 in., height = 20 in., and width = 20 in. Darla had 50 hay bales delivered to her farm. How much hay did she have delivered?

A 4,000 in.$^3$
B 8,000 in.$^3$
C 16,000 in.$^3$
D 160,000 in.$^3$
E 800,000 in.$^3$

**Directions:** Questions 3 and 4 are based on the information below.

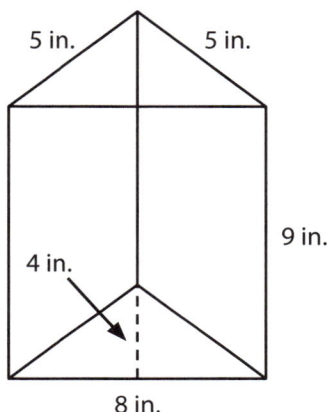

5 in.    5 in.

9 in.

4 in.

8 in.

**3** Which expression can be used to find the surface area of the triangular prism above?

A $(8 × 4) + (9 × 5) + (9 × 8)$
B $(8 × 4) + 2(9 × 5) + (9 × 8)$
C $\frac{1}{2}(8 × 4) + (9 × 5) + (9 × 8)$
D $\frac{1}{2}(8 × 4) + 2(9 × 5) + (9 × 8)$
E $2(8 × 4) + 2(9 × 5) + (9 × 8)$

**4** A prism has a triangular base with an area of 24 in.$^2$. The prism has the same volume as the triangular prism above. What is the height of the prism?

A 3 in.
B 6 in.
C 9 in.
D 12 in.
E 16 in.

**5** A plastic display case is in the shape of a rectangular prism. The prism is 8 in. long, 6 in. wide, and 10 in. high. How much plastic was used to make the display case?

A 188 in.$^2$
B 240 in.$^2$
C 256 in.$^2$
D 376 in.$^2$
E 480 in.$^2$

**6** The rain barrel has a volume of 9,156.24 cm³.

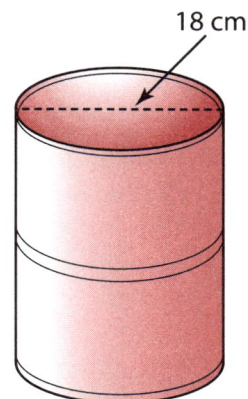

18 cm

**What is the height of the rain barrel?**
A 2.25 cm
B 9 cm
C 36 cm
D 113.04 cm
E 162 cm

**7** Morgan needs to find the circumference of a cylindrical can. He knows the height and volume. He divides the volume by the height and gets an answer of $x$ cm². Which describes the next steps Morgan should take to find the circumference?

A Divide $x$ by 3.14, then take the square root to find the radius. Multiply the radius by 3.14.
B Divide $x$ by 3.14, then take the square root to find the radius. Multiply the radius by 6.28.
C Take the square root of $x$ to find the product of the radius and 3.14. Multiply by 2.
D Take the square root of $x$ to find the product of the radius and 6.28.
E Multiply $x$ by 3.14, then multiply the square of the radius by 2.

UNIT 2

**8** A grocery store sells sausage in the cylindrical tube-shaped package shown.

10 in.

If the radius of the package is about 1.5 in., how many cubic inches is the package?

A   23.55
B   70.65
C   93.40
D   104.67
E   141.30

**9** A cylinder has a diameter of 10 m and a height of 7 m. What is the volume of the cylinder?

A   109.9 m³
B   549.5 m³
C   769.3 m³
D   1725.4 m³
E   2198 m³

**Directions:** Questions 10 and 11 are based on the information below.

A cube-shaped box has a length, width, and height of 18 inches.

**10** What is the volume of the box to the nearest cubic foot?

A   3
B   18
C   36
D   324
E   5,832

**11** What is the surface area of the box to the nearest square foot?

A   14
B   19
C   324
D   1944
E   5,832

**12** Anaya's perfume bottle is a rectangular prism. The bottle is 2.5 cm wide and 13 cm tall. If the volume of the bottle is 97.5 cm³, what is the length of the perfume bottle to the nearest tenth of a centimeter?

A   2.5 cm
B   3.0 cm
C   3.3 cm
D   4.3 cm
E   7.5 cm

**13** A shipping package has the shape of the triangular prism shown.

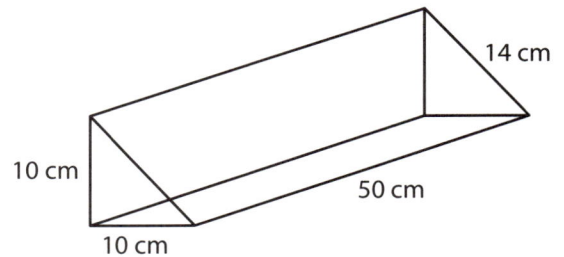

Assuming the package has no gaps or overlaps, what area of cardboard is needed to produce the shipping package?

A   1,300 cm²
B   1,750 cm²
C   1,800 cm²
D   2,500 cm²
E   5,000 cm²

**14** Calvin says that the surface area of a cube is always greater than its volume. Which side length of a cube shows that Calvin is incorrect?

A   0.5 ft
B   1 ft
C   2 ft
D   5 ft
E   8 ft

**15** What is the diameter of a cylinder with a volume of 235.5 in.³ and a height of 3 in.?

A   2.5 in.
B   5 in.
C   10 in.
D   25 in.
E   78.5 in.

**Directions:** Questions 16 and 17 are based on the information below.

An open-top flour canister has a circumference of 31.4 cm. The bottom and outside are made of cardboard and the lid is made of plastic.

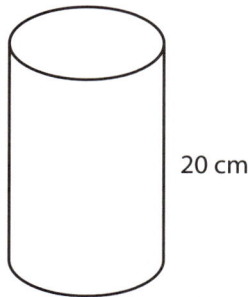

20 cm

**16** To the nearest square centimeter and assuming there is no overlap, what is the area of cardboard needed to make the canister?
A   157
B   314
C   628
D   707
E   785

**17** A second canister holds the same volume of flour as the canister above but is 15 cm high. To the nearest tenth of a centimeter, what is the circumference of the second canister?
A   33.3
B   36.2
C   72.5
D   103.4
E   133.3

**18** The two prisms have the same volume.

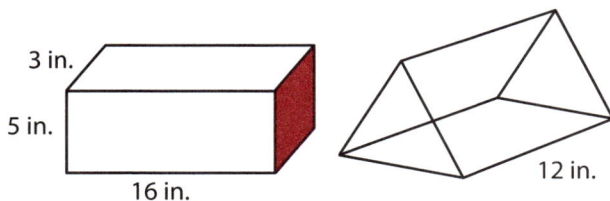

3 in.
5 in.
16 in.
12 in.

What is the area of the base of the triangular prism, in square inches?
A   10
B   11
C   15
D   20
E   24

**Directions:** Questions 19 and 20 are based on the information below.

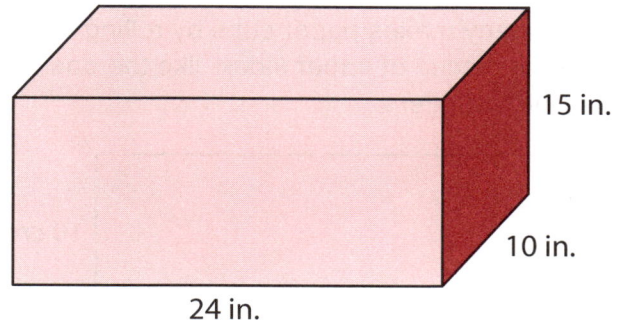

15 in.
10 in.
24 in.

**19** Alyssa has the fish tank above. If she pours water into the tank until it is half full, how many cubic inches of water will be in the tank?
A   7,200
B   3,600
C   1,800
D   1,200
E   450

**20** Alyssa pours water into the tank until the water reaches one inch below the top of the tank. How many cubic inches of water are in the tank?
A   3,599
B   3,360
C   3,240
D   240
E   Not enough information is given.

**21** A can of soda is 3 in. in diameter and holds 28.26 in.³. What is the height of the can?
A   1 in.
B   2 in.
C   3 in.
D   4 in.
E   6 in.

**Directions:** Questions 22 and 23 are based on the information below.

A company makes paper cups by rolling rectangular pieces of paper stock, like the one shown, into cylinders.

10 cm

**22** If a cup has a radius of 3.5 cm, what is the volume of the cup?
- **A** 384.7 cm³
- **B** 219.8 cm³
- **C** 192.3 cm³
- **D** 128.2 cm³
- **E** 109.9 cm³

**23** If a cup has a radius of 3.5 cm, what is the lateral area of the cup?
- **A** 35 cm²
- **B** 70 cm²
- **C** 110 cm²
- **D** 122.5 cm²
- **E** 220 cm²

**24** A food service company ships boxes that contain 40 cans of shortening. Each can has a radius of 2 in. and a height of 5 in. After the cans are placed in a box like the one shown, how much empty space is in the box?

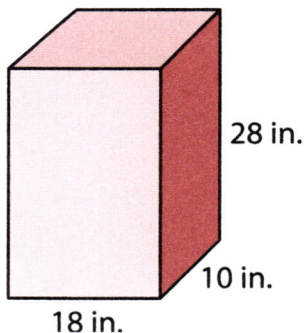

28 in.
10 in.
18 in.
- **A** 2,512 cubic inches
- **B** 2,528 cubic inches
- **C** 3,784 cubic inches
- **D** 4,202 cubic inches
- **E** 4,973 cubic inches

**25** Kaya is planting flowers in the cylindrical-shaped flowerpot shown.

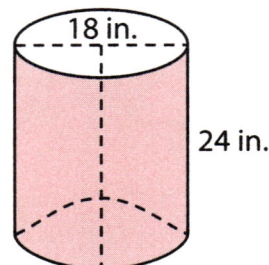

18 in.
24 in.

She begins by filling the bottom 3 inches of the flowerpot with rocks. Then she fills the rest of the pot with potting soil. About how many cubic inches of potting soil does she use?
- **A** 760
- **B** 1,190
- **C** 5,340
- **D** 6,100
- **E** 21,360

**Directions:** Questions 26 and 27 are based on the information below.

A rectangular prism with volume 4,050 in.³ has a base that is twice as long as it is wide.

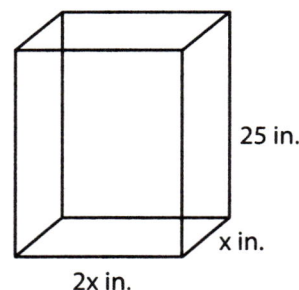

25 in.
x in.
2x in.

**26** What is the length of the prism?
- **A** 4.5 in.
- **B** 9 in.
- **C** 18 in.
- **D** 27 in.
- **E** 54 in.

**27** What is the surface area of the prism?
- **A** 1,674 in.²
- **B** 1,198 in.²
- **C** 859.5 in.²
- **D** 837 in.²
- **E** 599 in.²

# Pyramids, Cones, and Spheres

## ① Learn the Skill

A **pyramid** is a 3-dimensional figure that has a polygon as its single base and triangular faces. A **cone** has one circular base. The volume of a pyramid is $V = \frac{1}{3}Bh$. The volume of a cone is $V = \frac{1}{3}\pi r^2 h$.

The **surface area** of a solid figure is the sum of the areas of surfaces. The surface area of a pyramid is the sum of the area of its base and its triangular faces. Use the **slant height** (height of the triangle) to find the areas of the faces. The formula for surface area of a pyramid is $SA = B + \frac{1}{2}Ps$, where $B$ is the area of the base, $P$ is the perimeter of the base, and $s$ is the slant height. The surface area of a cone is the sum of its circular base and its curved surface. The formula for surface area is $SA = \pi r^2 + \pi rs$.

A **sphere** is shaped like a ball and has no bases or faces. The formula for volume of a sphere is $\frac{4}{3}\pi r^3$. The formula for surface area of a sphere is $4\pi r^2$.

## ② Practice the Skill

You will see problems about surface area and volume of pyramids, cones, and spheres on mathematics tests. Read the examples and strategies below. Use this information to answer question 1.

**A** A **cone** has a circular base and one vertex. The two are connected by a curved surface, which, when unwrapped, forms part of a circle. The length of the curved edge of the part of the circle is equal to the circumference of the base. The radius of the part of the circle is equal to the slant height, $s$, of the cone.

**B** A **square pyramid** has a square base and four congruent triangular faces. The faces all connect to a single point called a **vertex**. The height of a square pyramid forms a right angle with its base. The slant height, $s$, is not perpendicular to the base. It extends from the base of the triangular face to the vertex.

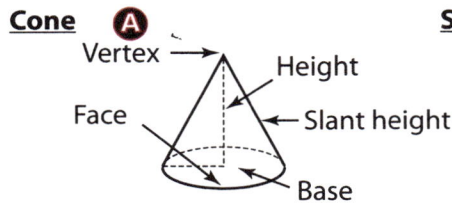

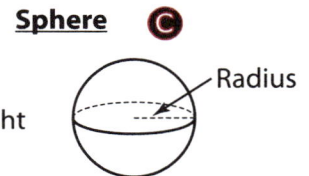

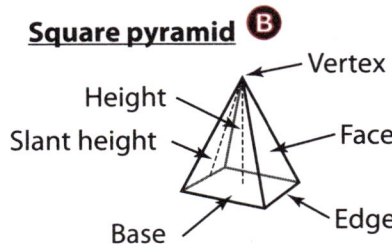

**Cone** Ⓐ — Vertex, Height, Face, Slant height, Base

**Sphere** Ⓒ — Radius

**Square pyramid** Ⓑ — Vertex, Height, Slant height, Face, Base, Edge

**C** Half of a sphere is called a **hemisphere**. The volume of a hemisphere is half the volume of the sphere. The surface area of a hemisphere is the area of half the surface of the sphere, plus the area of the circular base. The radius of the base is equal to the radius of the sphere.

### ✓ TEST-TAKING TIPS

Multiplying by $\frac{1}{3}$ is the same as dividing by 3. For example, to find the volume of a square pyramid, you can find the product of (base edge)$^2$ × height and then divide by 3.

1  A factory manufactures solid spherical rubber balls. To the nearest cubic inch, what volume of rubber is required to manufacture one ball?

A  6 in.$^3$
B  14 in.$^3$
C  28 in.$^3$
D  36 in.$^3$
E  42 in.$^3$

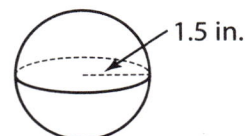

1.5 in.

**Directions:** Questions 2 and 3 are based on the information below.

A paper cup has the shape of the cone below.

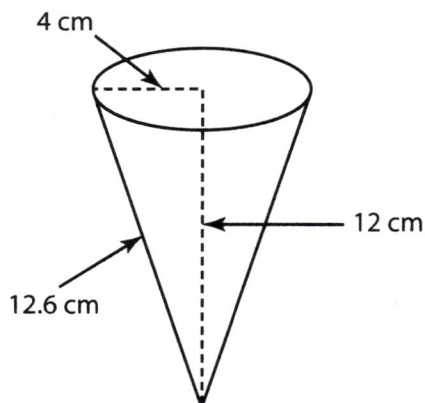

4 cm

12 cm

12.6 cm

**2  Which expression represents the volume of water the cup can hold?**

A  $\frac{1}{3} \times 3.14 \times 4^2 \times 12$

B  $\frac{1}{3} \times 3.14 \times 8^2 \times 12$

C  $3.14 \times 4^2 \times 12$

D  $3 \times 3.14 \times 4^2 \times 12$

E  $\frac{1}{3} \times 4^2 \times 12$

**3  To the nearest square centimeter, what area of paper is required to make the cone? Assume there is no overlap of paper.**

A  151 cm²
B  158 cm²
C  201 cm²
D  208 cm²
E  475 cm²

**4  A cone and a hemisphere each have a radius of 6 in. What is the height of the cone if the two figures have the same volume?**

A  4 in.
B  12 in.
C  24 in.
D  36 in.
E  72 in.

**Directions:** Questions 5 and 6 are based on the information below.

A chocolate shop makes specialty shapes and sizes of chocolate. Lia ordered the two chocolate figures shown.

**5  The pyramid has a square base with side length 2 cm and height 3 cm. What is the volume of the chocolate to the nearest tenth?**

A  12 cm³
B  6 cm³
C  4 cm³
D  2 cm³
E  1.3 cm³

**6  The cone has the same height as and twice the volume of the pyramid. What is the radius of the cone to the nearest tenth?**

A  2.6 cm
B  1.6 cm
C  1.3 cm
D  1.1 cm
E  0.8 cm

**7  A tent has the shape of the square pyramid.**

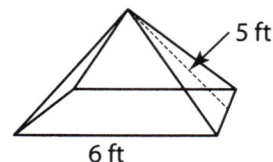

5 ft

6 ft

**What area of fabric is needed to make the tent? Assume no overlap.**

A  36 square feet
B  51 square feet
C  60 square feet
D  66 square feet
E  96 square feet

**8** Walt inflates a beach ball with a diameter of 15 in. About how many cubic inches of air does he put into the ball?

  **A** 710
  **B** 1,770
  **C** 2,820
  **D** 10,600
  **E** 14,130

**9** A square pyramid has a height of 9 cm and a base area of 36 cm². What is the volume of the pyramid to the nearest cm³?

  **A** 108
  **B** 162
  **C** 324
  **D** 432
  **E** 1,017

**10** A sphere has a surface area of about 28.26 in.³. What is the radius of the sphere?

  **A** 1.5 cm
  **B** 1.9 cm
  **C** 2.1 cm
  **D** 3.0 cm
  **E** 6.0 cm

**Directions:** Questions 11 and 12 are based on the information below.

The square pyramid has a volume of 64 ft³ and a surface area of 144 ft².

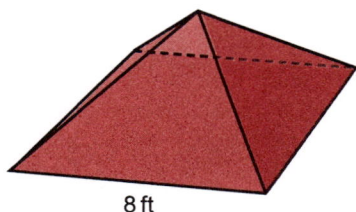

8 ft

**11** What is the height of the pyramid?

  **A** 2.25 ft
  **B** 3 ft
  **C** 5 ft
  **D** 6.75 ft
  **E** 12 ft

**12** What is the slant height of the pyramid?

  **A** 2.25 ft
  **B** 4 ft
  **C** 5 ft
  **D** 6.75 ft
  **E** 13 ft

**Directions:** Questions 13 through 15 are based on the information below.

The Sno-Cone Hut sells sno-cones in three sizes.

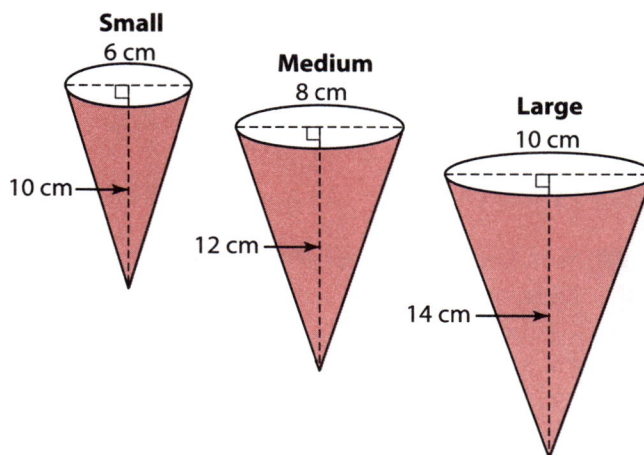

**13** What is the volume of the smallest cone to the nearest cubic centimeter?

  **A** 62
  **B** 94
  **C** 141
  **D** 283
  **E** 376

**14** The smallest cone has slant height 10.4 cm. To the nearest whole number, what area of paper is needed to make this cone? Assume no overlap.

  **A** 94 cm²
  **B** 98 cm²
  **C** 126 cm²
  **D** 188 cm²
  **E** 196 cm²

**15** About how much more does the large size hold compared to the medium?

  **A** 44 cm³
  **B** 117 cm³
  **C** 165 cm³
  **D** 274 cm³
  **E** 662 cm³

**Directions:** Questions 16 through 18 are based on the information below.

An ice sculpture company is experimenting with freezing blocks in the shapes of square pyramids as shown in the diagram.

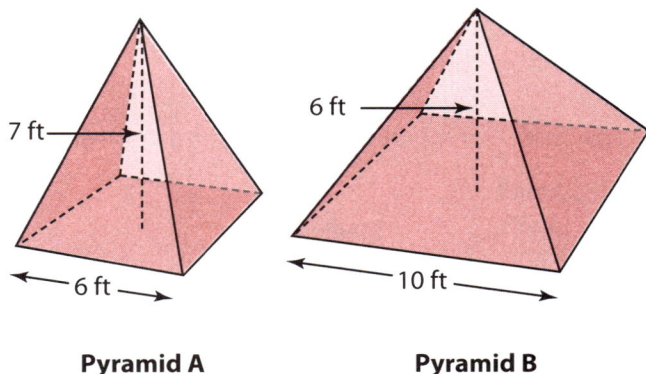

Pyramid A                    Pyramid B

**16 What is the volume of Pyramid A?**
  A  28 cubic feet
  B  42 cubic feet
  C  84 cubic feet
  D  98 cubic feet
  E  104 cubic feet

**17 What is the volume of Pyramid B?**
  A  20 cubic feet
  B  60 cubic feet
  C  120 cubic feet
  D  200 cubic feet
  E  600 cubic feet

**18 Pyramid C has the same volume as Pyramid A and the same base edge length as Pyramid B. About how tall is Pyramid C?**
  A  2.5 ft
  B  5 ft
  C  6.5 ft
  D  7 ft
  E  16.7 ft

**19 A spherical lantern has a circular hole cut out of its top. About what area of paper is needed to construct the lantern?**

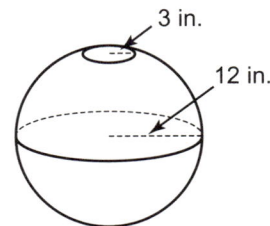

  A  1,017 square inches
  B  1,780 square inches
  C  1,810 square inches
  D  7,200 square inches
  E  7,230 square inches

**20 To the nearest cubic inch, what is the volume of a cone 8 in. tall with diameter 10 in.?**
  A  204 in.$^3$
  B  209 in.$^3$
  C  335 in.$^3$
  D  523 in.$^3$
  E  837 in.$^3$

**21 What is the volume of the square pyramid?**

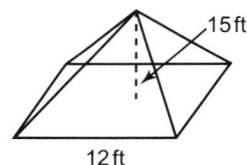

  A  2,260.8 ft$^3$
  B  920 ft$^3$
  C  720 ft$^3$
  D  504 ft$^3$
  E  240 ft$^3$

**22 A team of sand sculptors is building a dinosaur with spikes. The shape of the spikes is a cone. If they build spikes that contain about 550 in.$^3$ of sand with diameter 10 in., how tall are the cones?**
  A  5 in.
  B  7 in.
  C  10 in.
  D  12 in.
  E  21 in.

**23** What is the volume, to the nearest cubic foot, of a square pyramid with side length 2 ft 6 in. and height 3 ft 3 in.?

   A   6
   B   7
   C   9
   D   12
   E   21

**Directions:** Questions 24 and 25 are based on the information below.

The space inside the teepee shown below measures 803.84 ft³.

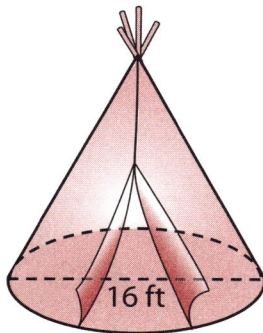

16 ft

**24** What is the height of the teepee to the nearest foot?

   A   3
   B   4
   C   12
   D   24
   E   36

**25** If 563 ft² of material is used to make the teepee and its floor, what is the slant height of the teepee to the nearest inch?

   A   7 feet 2 inches
   B   8 feet 5 inches
   C   14 feet 5 inches
   D   40 feet 10 inches
   E   42 feet 3 inches

**Directions:** Questions 26 through 28 are based on the information below.

An architect is designing a glass greenhouse in the shape of a square pyramid. The triangular faces will be constructed from glass and the base from wood. The two designs have the same volume.

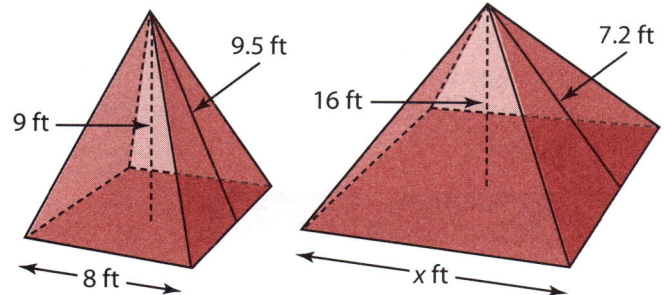

9.5 ft

9 ft

8 ft

**Greenhouse A**

7.2 ft

16 ft

x ft

**Greenhouse B**

**26** What is the volume of each greenhouse?

   A   48 cubic feet
   B   72 cubic feet
   C   96 cubic feet
   D   192 cubic feet
   E   576 cubic feet

**27** What area of wood will be needed for Greenhouse B?

   A   6 square feet
   B   12 square feet
   C   16 square feet
   D   36 square feet
   E   80 square feet

**28** The architect wants to minimize the amount of glass that will be used to build the structure. Which explains the design she should choose?

   A   Greenhouse A; the surface area of A is less than the surface area of B.
   B   Greenhouse A; the total area of the triangular faces of A is less than the total area of the triangular faces of B.
   C   Greenhouse B; the surface area of B is less than the surface area of A.
   D   Greenhouse B; the total area of the triangular faces of B is less than the total area of the triangular faces of A.
   E   Either; the amount of glass is the same for both.

# Composite Figures and Solids

## 1 Learn the Skill

Most shapes in the real world are not simple figures or solids. Many are **composite**. They are made up of several plane figures or solid shapes. To find the area or volume of a composite figure or solid, you must find the area or volume of each part and add them together. You also can find the perimeter or surface area of composite figures and solids.

## 2 Practice the Skill

Understanding how to solve problems involving composite figures and solids will improve your ability to successfully solve real-world area, volume, and perimeter problems on mathematics tests. Read the example and strategies below. Use this information to answer question 1.

**A** To find the area of an irregular figure, first divide the figure into simple shapes. This figure can be divided into three rectangles. The dimensions of the two outer rectangles are given. One side of the middle rectangle is given (6 cm). To find the other side, use the measurements of the sides you know. The length of the entire figure is 15 cm. By subtracting 4 cm and then 6 cm, you can find the length of the middle rectangle (5 cm).

**B** To find the perimeter of a composite figure, add the lengths of the sides. If no number is given for a side, find the side length by addition or subtraction. In the figure at the right, the missing sides are 15 − 4 − 6 = 5 cm long and 10 − 6 = 4 cm long. The perimeter is 10 + 4 + 4 + 5 + 4 + 6 + 10 + 15 = 58 cm.

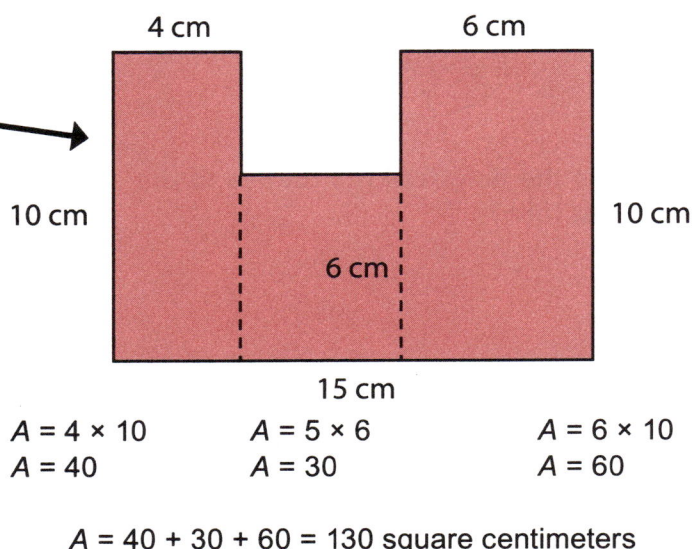

$A = 4 \times 10$     $A = 5 \times 6$     $A = 6 \times 10$
$A = 40$         $A = 30$        $A = 60$

$A = 40 + 30 + 60 = 130$ square centimeters

### ✓ TEST-TAKING TIPS

To find the volume of a composite solid, separate it into simple solids. Then find the volume of each simple solid and add to find the total volume.

**1** **What is the area in square feet of the figure?**

A 40
B 64
C 76
D 88
E 112

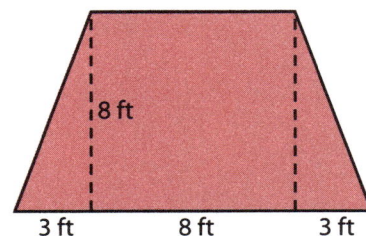

**2** Kirsten sewed a tablecloth in the shape shown. What is the area of her tablecloth to the nearest tenth of a square foot?

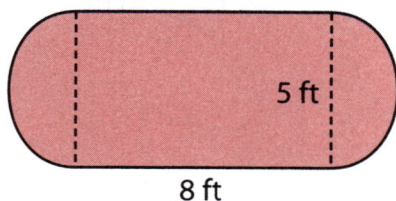

5 ft

8 ft

**A** 40.0
**B** 47.9
**C** 55.7
**D** 56.0
**E** 59.6

**Directions:** Questions 3 and 4 are based on the figure below.

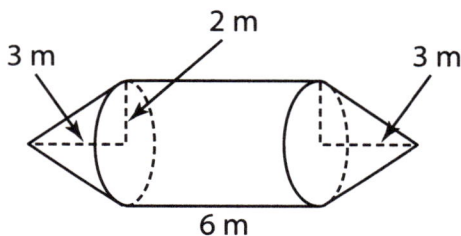

2 m
3 m          3 m
6 m

**3** What is the combined volume of the cones?
**A** 75.36 cubic meters
**B** 37.68 cubic meters
**C** 25.12 cubic meters
**D** 18.84 cubic meters
**E** 12.56 cubic meters

**4** What is the volume of the figure to the nearest cubic meter?
**A** 50
**B** 75
**C** 100
**D** 125
**E** 151

**Directions:** Questions 5 through 7 are based on the information below.

Karen and Bill had cement poured to make the patio shown in the diagram.

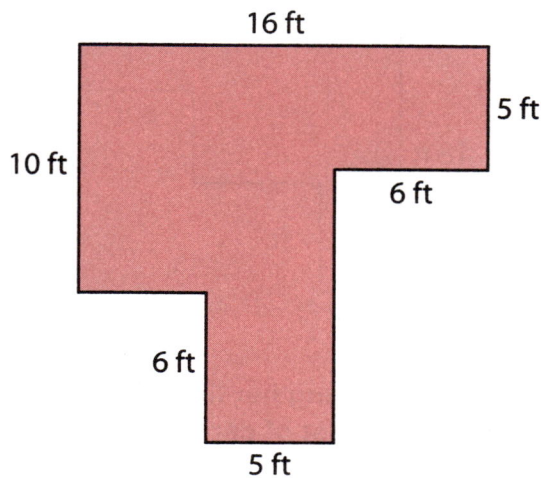

16 ft
5 ft
10 ft
6 ft
6 ft
5 ft

**5** What is the area of Karen and Bill's patio in square feet?
**A** 40
**B** 70
**C** 100
**D** 130
**E** 160

**6** Karen and Bill are laying a decorative tile border around the outside of the patio. What is the perimeter of the patio in feet?
**A** 8
**B** 64
**C** 96
**D** 192
**E** 768

**7** If Karen and Bill have the cement for the patio poured 3 inches deep, how many cubic feet of cement will they use?
**A** 32
**B** 40
**C** 90
**D** 160
**E** 480

**Directions:** Questions 8 and 9 are based on the information below.

The diagram shows Laura's living room space.

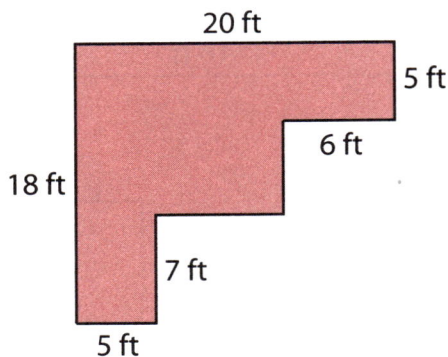

20 ft
5 ft
6 ft
18 ft
7 ft
5 ft

**8** Laura is carpeting her living room. How many square feet of carpet will she need?
   A   76
   B   189
   C   219
   D   317
   E   360

**9** Laura needs tacking strips to go on the floor around the outer edge beneath the carpet. How many feet of tacking strips will she need?
   A   72
   B   76
   C   78
   D   85
   E   90

**10** Corinne ordered a pizza with a diameter of 16 in. The pizza was cut into 8 equal slices. She ate 3 slices. What is the area of the remaining pizza?
   A   75.36 sq in.
   B   100.48 sq in.
   C   125.6 sq in.
   D   200.96 sq in.
   E   502.8 sq in.

**11** For the back-to-school season, a store is selling a container of paper clips in the shape of a pencil as shown below.

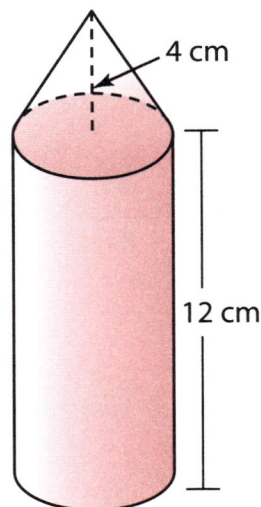

4 cm
12 cm

The radius of the pencil container is 3 cm. What is the volume of the container to the nearest cubic centimeter?
   A   126
   B   144
   C   226
   D   377
   E   452

**12** Adam is tiling the bathroom floor shown.

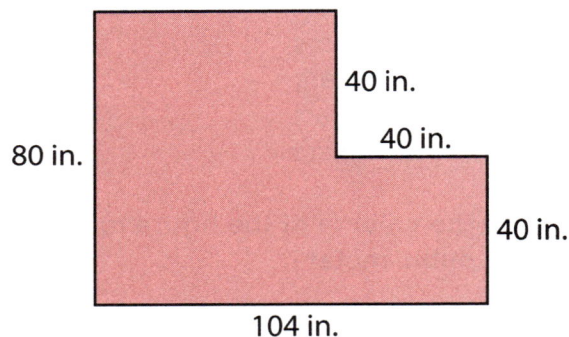

40 in.
40 in.
80 in.
40 in.
104 in.

If each tile is 8 inches by 8 inches, how many tiles does he need to cover the floor?
   A   64
   B   105
   C   116
   D   124
   E   840

**13** A theater company built the stage shown below. What is the volume of the figure?

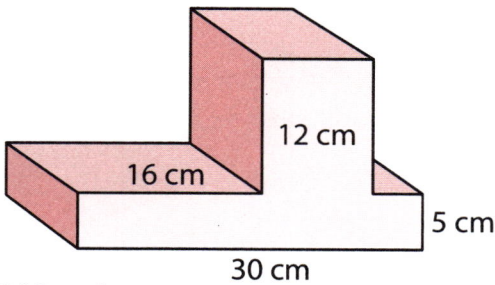

16 cm
12 cm
5 cm
30 cm

**A** 960 cm³
**B** 1,800 cm³
**C** 5,640 cm³
**D** 5,760 cm³
**E** Not enough information is given.

**Directions:** Questions 14 and 15 are based on the information below.

The diagram shows the setup of the main garden bed in a rose garden. The two larger circles are congruent.

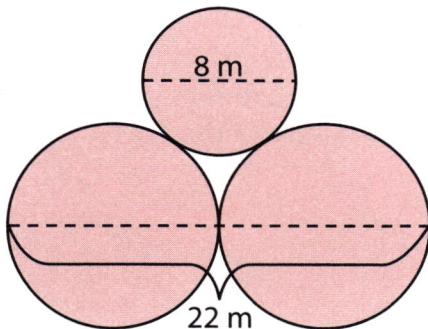

8 m
22 m

**14** What is the area of one of the larger circles?
**A** 34.5 m²
**B** 95.0 m²
**C** 379.9 m²
**D** 759.8 m²
**E** 1,519.8 m²

**15** About how many square meters is the garden bed?
**A** 145
**B** 190
**C** 240
**D** 290
**E** 380

**Directions:** Questions 16 through 18 are based on the information below.

Stew drew the figure below. The diameter of the semicircle is also the height of the rectangle.

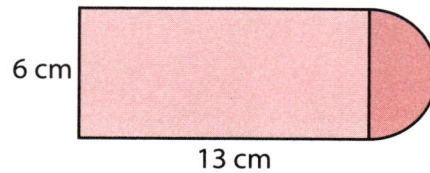

6 cm
13 cm

**16** What is the area of the figure?
**A** 87.42 cm²
**B** 92.13 cm²
**C** 96.84 cm²
**D** 106.26 cm²
**E** 134.52 cm²

**17** What is the perimeter of the figure?
**A** 50.84 cm
**B** 47.42 cm
**C** 41.42 cm
**D** 38.00 cm
**E** 36.50 cm

**18** If the figure were twice as tall, what would be the area?
**A** 156.00 cm²
**B** 174.84 cm²
**C** 184.26 cm²
**D** 193.68 cm²
**E** 212.52 cm²

**19** Which formula represents the volume occupied by the house shown below?

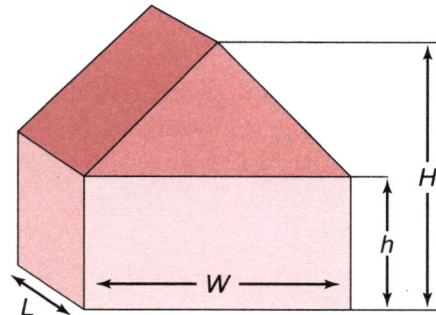

H
h
W
L

**A** $V = LW(H + h)$
**B** $V = LW(H - h)$
**C** $V = \frac{1}{2}LW(H + h)$
**D** $V = \frac{1}{2}LW(H - h)$
**E** $V = LWh + \frac{1}{2}WH - \frac{1}{2}Wh$

**Directions:** Questions 20 through 23 are based on the information below.

The figure shows a triangle set on top of a rectangle.

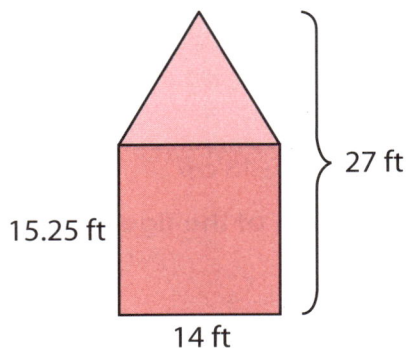

20 **What is the height of the triangle?**
   A  11.75 ft
   B  13.5 ft
   C  14 ft
   D  15.25 ft
   E  27 ft

21 **What is the area of the figure?**
   A  56.3 square feet
   B  75.0 square feet
   C  295.75 square feet
   D  378.0 square feet
   E  402.5 square feet

22 **What is the perimeter of the figure?**
   A  56.3 feet
   B  68.0 feet
   C  70.3 feet
   D  71.9 ft
   E  75.0 feet

23 **Suppose the two shapes were pulled apart, and that the triangle were equilateral. Fencing is needed to enclose each shape. How much fencing is needed?**
   A  100.5 ft
   B  93.75 ft
   C  86.5 ft
   D  71.25 ft
   E  56.25 ft

**Directions:** Questions 24 through 26 are based on the information below.

The revolving restaurant on top of a downtown skyscraper is in the shape of a cylinder with a vaulted ceiling shaped like a cone.

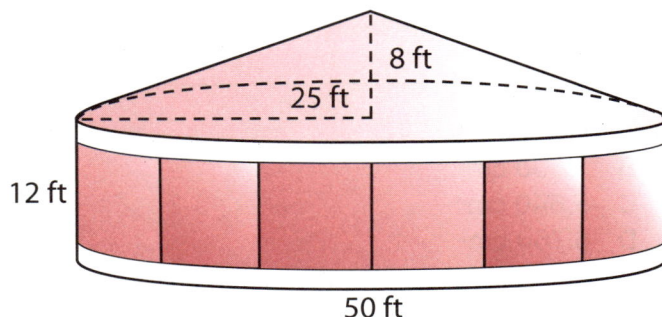

24 **What is the volume of the cone-shaped section of the restaurant in cubic feet?**
   A  209.3
   B  5,233.3
   C  7,850.0
   D  13,083.3
   E  15,700.0

25 **What is the approximate volume, in cubic feet, of the inside of the restaurant?**
   A  5,233
   B  18,317
   C  23,550
   D  28,783
   E  39,249

26 **If $r$ is the radius of the restaurant, $h$ is the height of the wall of the cylindrical portion, and $s$ is the slant height of the conical ceiling, what is the combined surface area of the wall and ceiling?**
   A  $A = \pi r(2h + s)$
   B  $A = \pi r(2h - s)$
   C  $A = \pi r(h + s)$
   D  $A = \pi r(h - s)$
   E  $A = \pi r(rh + s)$

# Unit 2 Review

The Unit Review is structured to resemble mathematics high school equivalency tests. Be sure to read each question and all possible answers very carefully before choosing your answer.

To record your answers, fill in the lettered circle that corresponds to the answer you select for each question in the Unit Review.

Do not rest your pencil on the answer area while considering your answer. Make no stray or unnecessary marks. If you change an answer, erase your first mark completely.

Mark only one answer choice for each question; multiple answers will be scored as incorrect.

UNIT 2

---

**Directions:** Questions 1 and 2 are based on the information below.

The figure below shows several rays with a common vertex.

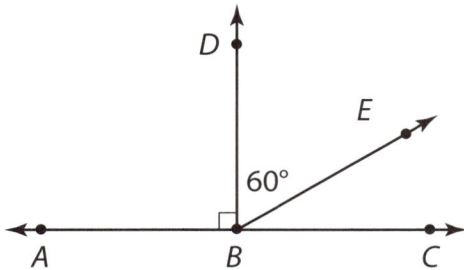

**1   Which angles are complementary?**
A   ∠ABD and ∠DBE
B   ∠DBE and ∠DBC
C   ∠ABC and ∠EBC
D   ∠ABD and ∠EBC
E   ∠DBE and ∠EBC

Ⓐ Ⓑ Ⓒ Ⓓ Ⓔ

**2   Which angle has a measure of 150°?**
A   ∠ABE
B   ∠ABD
C   ∠DBE
D   ∠CBD
E   ∠ABC

Ⓐ Ⓑ Ⓒ Ⓓ Ⓔ

**3   The distance between two cities on a map is 8.5 cm. If the map scale is 1 cm : 15 km, what is the actual distance in kilometers between the two cities?**
A   127.5
B   63.8
C   31.9
D   3.5
E   1.7

Ⓐ Ⓑ Ⓒ Ⓓ Ⓔ

**4** A shoelace has a mass of 1 gram. A textbook has a mass of about 1 kilogram. How many shoelaces would you need to gather to equal the mass of two textbooks?

| Mass |
| --- |
| 1 kilogram (kg) = 1,000 grams (g) |
| 1 gram (g) = 100 centigrams (cg) |
| 1 centigram = 10 milligrams (mg) |

**A** 100
**B** 200
**C** 1,000
**D** 2,000
**E** Not enough information is given.

Ⓐ Ⓑ Ⓒ Ⓓ Ⓔ

**5** The flight distance between Boston and Chicago is approximately 850 miles. A commercial airliner leaves Boston at 11:30 A.M. The average speed of the plane is 500 mph. The time in Boston is one hour ahead of the time in Chicago. What time will it be in Chicago when the plane lands?

**A** 11:12 A.M.
**B** 12:12 P.M.
**C** 1:07 P.M.
**D** 1:42 P.M.
**E** Not enough information is given.

Ⓐ Ⓑ Ⓒ Ⓓ Ⓔ

**6** A chocolate company sells its specialty hot cocoa in cylindrical canisters. The canister holds 3,740 cubic centimeters of cocoa.

17.5 cm

What is the diameter of the canister in centimeters?

**A** 8.25
**B** 16.5
**C** 21.65
**D** 33.0
**E** 68.0

Ⓐ Ⓑ Ⓒ Ⓓ Ⓔ

**7** The figure shows two parallel lines intersected by a transversal.

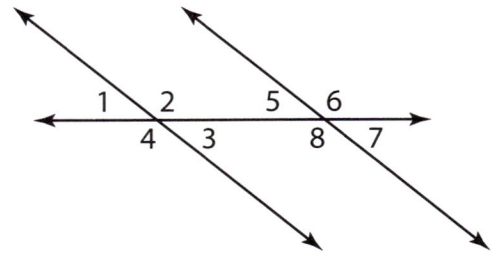

Which two angles have the same measure?
**A** ∠4 and ∠8
**B** ∠3 and ∠6
**C** ∠2 and ∠5
**D** ∠7 and ∠4
**E** Not enough information is given.

Ⓐ Ⓑ Ⓒ Ⓓ Ⓔ

**8** Instead of following the sidewalk around the outside of a park to her car, Wanda cut through the park as shown.

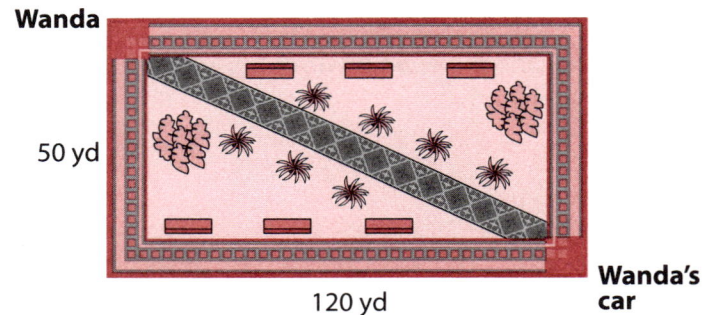

Wanda

50 yd

120 yd

Wanda's car

How many fewer yards did Wanda walk than if she had taken the sidewalk back to her car?

**A** 130
**B** 120
**C** 80
**D** 40
**E** 10

Ⓐ Ⓑ Ⓒ Ⓓ Ⓔ

UNIT 2

**Directions:** Questions 9 and 10 are based on the information below.

A human-made backyard pond is shown in the diagram.

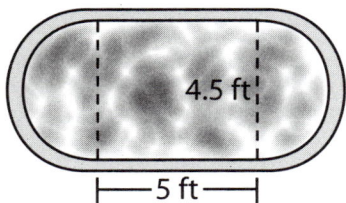

4.5 ft

5 ft

**9  What is the perimeter of the pond?**
- A  28.26 ft
- B  24.13 ft
- C  19.13 ft
- D  14.13 ft
- E  13.5 ft

Ⓐ Ⓑ Ⓒ Ⓓ Ⓔ

**10  What is the area of the pond in square feet?**
- A  30.45
- B  36.63
- C  38.40
- D  54.29
- E  86.09

Ⓐ Ⓑ Ⓒ Ⓓ Ⓔ

**Directions:** Questions 11 and 12 are based on the figure below.

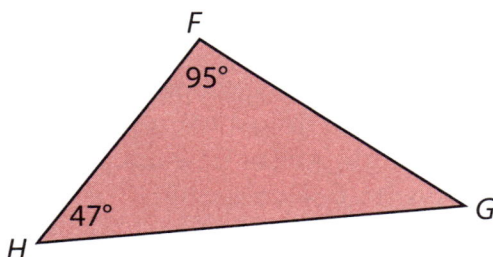

F
95°
47°
H
G

**11  What is the measure of ∠G?**
- A  48°
- B  38°
- C  28°
- D  18°
- E  Not enough information is given.

Ⓐ Ⓑ Ⓒ Ⓓ Ⓔ

**12  Which term best describes △HFG?**
- A  acute
- B  isosceles
- C  equilateral
- D  obtuse
- E  right

Ⓐ Ⓑ Ⓒ Ⓓ Ⓔ

**13  The pyramid and cone below have the same volume.**

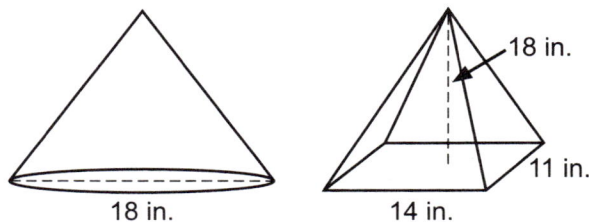

18 in.
18 in.
14 in.
11 in.

**To the nearest tenth, what is the height of the cone?**
- A  10.9 in.
- B  11.4 in.
- C  13.2 in.
- D  18.0 in.
- E  32.7 in.

Ⓐ Ⓑ Ⓒ Ⓓ Ⓔ

**14  An architect used a hemisphere and a rectangular prism to build a model of a new planetarium below.**

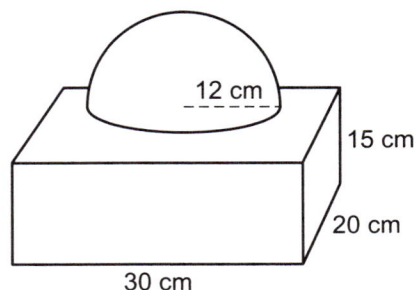

12 cm
15 cm
20 cm
30 cm

**About how many cubic centimeters is the total volume of the model?**
- A  4,500
- B  9,000
- C  12,600
- D  16,200
- E  30,700

Ⓐ Ⓑ Ⓒ Ⓓ Ⓔ

UNIT 2

**15** A scale drawing of one part of a house is shown.

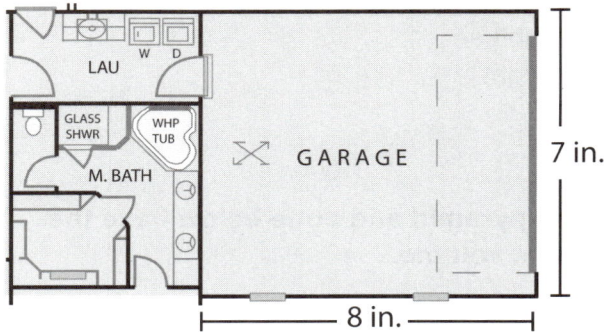

LAU

GLASS SHWR

M. BATH

WHP TUB

GARAGE

7 in.

8 in.

If the scale is 1 inch : 3 feet, what are the actual dimensions of the garage?

**A**  7 feet by 8 feet
**B**  14 feet by 16 feet
**C**  21 feet by 24 feet
**D**  24 feet by 25 feet
**E**  27 feet by 28 feet

Ⓐ Ⓑ Ⓒ Ⓓ Ⓔ

**16** Angle *C* in this parallelogram measures 125°.

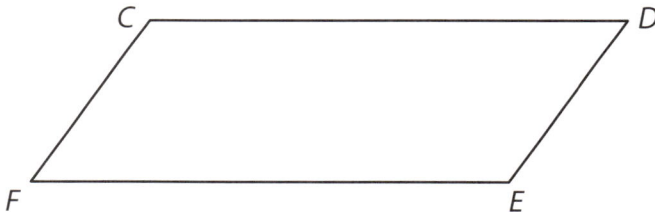

What is the measure of ∠*D*?

**A**  45°
**B**  55°
**C**  65°
**D**  125°
**E**  Not enough information is given.

Ⓐ Ⓑ Ⓒ Ⓓ Ⓔ

**17** The intersecting lines form a right triangle.

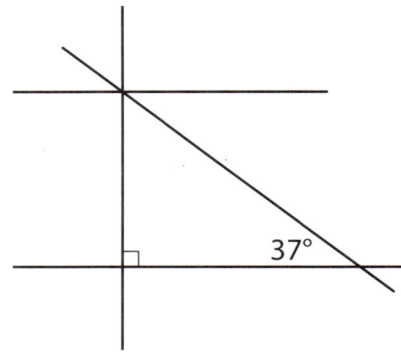

37°

What is the measure of the unknown angle of the triangle?

**A**  37°
**B**  45°
**C**  53°
**D**  74°
**E**  Not enough information is given.

Ⓐ Ⓑ Ⓒ Ⓓ Ⓔ

**18** A snack shop sells frozen cookie dough in a cone as shown in the diagram.

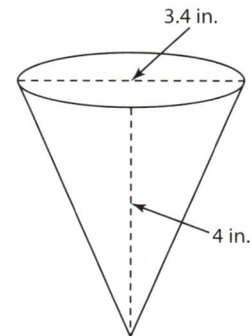

3.4 in.

4 in.

The snack shop buys cookie dough in containers that hold 480 cubic inches of cookie dough. About how many cones can be filled with one container?

**A**  about 40
**B**  about 36
**C**  about 24
**D**  about 13
**E**  about 10

Ⓐ Ⓑ Ⓒ Ⓓ Ⓔ

**19** The two pyramids in the figure have the same volume.

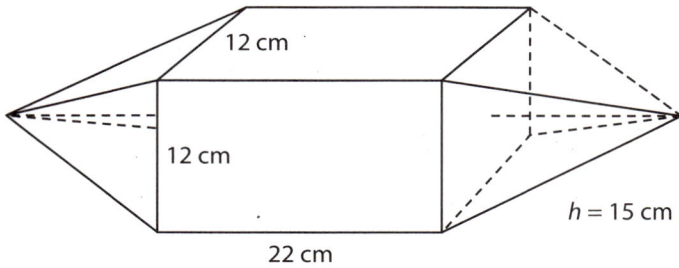

What is the volume of the two square pyramids combined?

A  120 cubic centimeters

B  720 cubic centimeters

C  1,440 cubic centimeters

D  2,160 cubic centimeters

E  4,320 cubic centimeters

Ⓐ Ⓑ Ⓒ Ⓓ Ⓔ

**20** Three points are plotted on the coordinate grid.

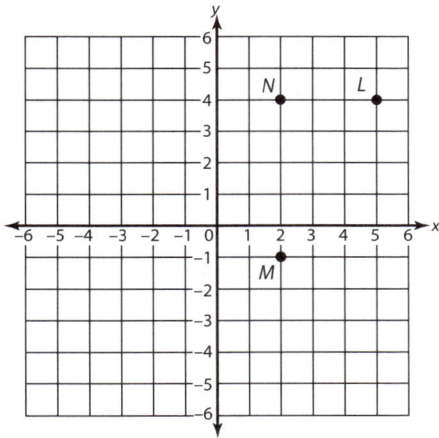

What is the distance to the nearest unit between point *L* and point *M*?

A  3

B  4

C  5

D  6

E  7

Ⓐ Ⓑ Ⓒ Ⓓ Ⓔ

**21** If the scale of the map is 2 cm : 29.8 km, what is the actual distance in kilometers between Fourth Pass and Karlton?

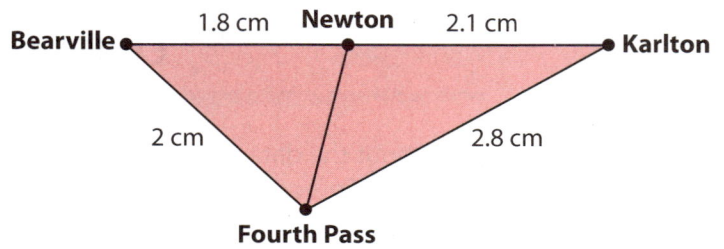

A  5.6

B  10.64

C  21.29

D  41.72

E  83.44

Ⓐ Ⓑ Ⓒ Ⓓ Ⓔ

**22** The pool of a fountain in a park is shaped like a crescent moon, as shown in the red part of the diagram.

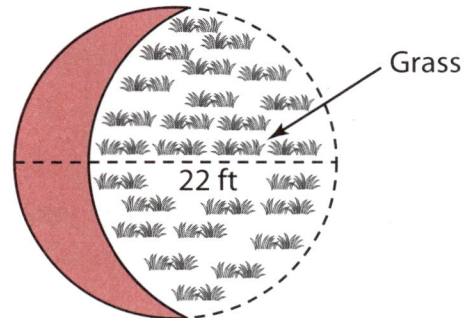

If the grass covers 253.3 square feet, what is the area of the fountain pool?

A  126.64 square feet

B  379.94 square feet

C  759.88 square feet

D  1,013.17 square feet

E  1,266.46 square feet

Ⓐ Ⓑ Ⓒ Ⓓ Ⓔ

# Unit 3

## Unit Overview

Think about the data provided in a weather report. The forecaster describes the likelihood of rain or snow. Forecasts of expected high and low temperatures over a day or week are described. Often these highs and low are compared with average highs and lows for the time of year. Charts and graphs may be used to visually represent the data. You may be surprised to learn that data analysis and probability are at the heart of your local forecast! The techniques that mathematicians use to collect, classify, analyze, and interpret data are collectively known as statistics.

In Unit 3, you will study they ways in which sets of data are analyzed, how probabilities are determined, and the ways in which data can be presented. Understanding these skills is important to your everyday life as well as to your success on mathematics high school equivalency tests. As with other general areas of math, data analysis, probability, and statistics comprise about 25 percent of questions on mathematics high school equivalency tests.

## Table of Contents

# Data Analysis/Probability/Statistics

## Key Terms

**bar graph:** a visual display that uses vertical or horizontal bars, aligned side-by-side for easy comparison, to represent data quantities

**box plot:** a visual display of a data set that shows and allows comparison of the median, the upper and lower quartile values, and the maximum and minimum values

**census:** a collection of data from an entire population

**circle graph:** a visual display that shows how parts compare to a whole

**convenience sample:** a non-random sample selected from a population because of accessibility

**correlation:** the relationship between data sets

**data table:** a tool for organizing and analyzing information in which data are arranged in rows and columns

**data:** a collection of factual values or measurement

**dimension:** the number of rows by the number of columns in a matrix

**dot plot:** an expanded number line that displays a data set by showing each occurrence of a value as a dot

**event:** a result of an experiment or trial

**experimental probability:** a ratio that expresses an actual number of outcomes as a part of a total number of trials

**frequency table:** a table that displays data as numbers of occurrences

**frequency:** the number of values for a given interval in a data set

**fundamental counting principle:** the basic rule used to count possible outcomes; if one choice has $m$ options and another choice has $n$ options, then the number of possible outcomes for the two choices is $m \cdot n$

**histogram:** a visual display of bars that correspond to an associated scale to show frequency in ranges

**line graph:** a visual display used to show how a data set changes over time

**matrix:** a tool for organizing and analyzing information in which data are arranged in rows and columns in order to make the information easier to identify and compare; no units of measure are included

**mean:** the average value of a data set

**measures of variation:** measures that describes "how scattered" a set of data is; range, variance, and standard deviation

**median:** the middle number in a set of data when the values are ordered from least to greatest

**mode:** the value that occurs most frequently in a set of data

**normal distribution:** an arrangement of data values that produce a mean, median, and mode that coincide

**observational study:** a method of collecting data in which a researcher observes behavior or outcomes in a systematic manner without influencing the behavior or outcome

**outcome:** the single result of a trial or probability experiment

**permutation:** an arrangement of a group of objects, or the same options for one choice, in a particular order

**population:** the entire group of individuals in a study

**probability:** the likelihood that an event will happen

**proportion:** an equation with equal ratios on each side

**qualitative:** descriptive details that cannot be measured

**quantitative:** characterized by measurable numbers

**random sample:** a sample, selected from a population, in which every individual in the population has an equal chance of being chosen

**range:** the spread of a data set, expressed by the difference between the greatest value and the least value

**ratio:** a comparison of two numbers; written as a fraction, by using the word to, or with a colon (:)

**sample population:** a portion of a population selected for study when data cannot be collected for the entire group

**scatter plot:** a line graph that shows how one set of data affects another

**skewed distribution:** a "lopsided" arrangement of data values, for which the mean is not equal to the mode

**theoretical probability:** a ratio that expresses the likelihood of an outcome as a part of the number of ways an event can occur

# Counting Outcomes

## ① Learn the Skill

The **fundamental counting principle** is the basic rule used to count possible outcomes. If one choice has *m* options and another choice has *n* options, then the number of possible outcomes for the two choices is *m • n*.

A **permutation** is an arrangement of a group of objects, or the same options for one choice, in a particular order. In a phone number, the digits 234 are different from the digits 342. The options can be either repeated or not repeated after an option is chosen for each choice.

A **combination** is a selection of objects or options, in which the order of the options does not matter. When playing a card game, holding cards with 2, 3, and 4 is the same as holding cards with 3, 4, and 2. Often combinations involve counting the different ways a subgroup can be made from a larger group.

## ② Practice the Skill

Several types of questions may ask you to find the number of different ways items can be arranged or combined. Understanding the question types will help you use the correct calculation to find the total possible outcomes. Read the examples and strategies below. Use this information to answer question 1.

**Ⓐ Counting principle:** There are 4 options for the first choice, 2 options for the second choice, and 3 options for the third choice. The order of choosing shirts or pants does not matter.

**Ⓒ Permutation without repetition:** Order of the choices (3 digits) matters and options (1, 2, 3, 4) cannot be repeated once they are chosen. Use *n*! (read "*n* factorial") to *x* places, for the *x* choices: $n! = n(n-1)(n-2)\ldots • 2 • 1$.

| an outfit from 4 shirts, 2 pants, and 3 jackets<br><br>$4 • 2 • 3 = 24$<br><br>24 different outfits | 3-digit numbers using 1, 2, 3, and 4 with no repetition 4! to 3 places:<br><br>$4 • 3 • 2 = 24$<br><br>24 possible numbers |
| --- | --- |
| 3-digit numbers using 1, 2, 3, and 4<br><br>$4 • 4 • 4 = 4^3 = 64$<br><br>64 possible numbers | groups of 3 from a team of 10 people<br><br>$\dfrac{10 • 9 • 8}{3 • 2 • 1} = 120$<br><br>120 different groups of 3 |

**Ⓑ Permutation with repetition:** Order of the choices (3 digits) matters and options (1, 2, 3, 4) can be repeated. Use $n^x$ when there are *n* options for *x* choices.

**Ⓓ Combination:** Order of the 3 people in the subgroup does not matter. Compute 10! to 3 places for all possible subgroups. Then divide by the number of ways to arrange the 3 people (3!) to remove the duplicate groups of 3. The formula for a subgroup of *x* from a group of *n* is $\dfrac{n! \text{ to } x \text{ places}}{x!}$.

### ✓ TEST-TAKING TIPS

Use blank lines or blank boxes to represent how many choices are needed for the combination or permutation. Then, fill in each blank with the correct number of options for that choice.

**1** A capital building has five flagpoles set up in a semicircle. There are six different flags. How many different ways can the flags be arranged on the flagpoles?

  **A** 30
  **B** 120
  **C** 720
  **D** 7,776
  **E** 15,625

**Directions:** Questions 2 through 5 are based on the information below.

Kyrah has a jewelry kit for making bracelets. The kit contains 2 kinds of bands, 3 kinds of clasps, 6 different beads, and 5 different charms.

2  Which expression represents the total number of different bracelets Kyrah can make if she uses one each of the bands, clasps, beads, and charms?
A  $2 \cdot 3 \cdot 6 \cdot 5$
B  $(2 + 3) \cdot (6 + 5)$
C  $(2 \times 3) + (6 \times 5)$
D  $4 \cdot (2 + 3 + 6 + 5)$
E  $(2 + 3 + 6 + 5)$

3  How many different bracelets can Kyrah make using one band, one clasp, and one bead or one charm?
A  28
B  55
C  66
D  90
E  180

4  Which expression represents how many different bracelets can be made using the silver clasp, and three beads arranged in a row on the pink satin band?
A  $3 + 6$
B  $3 \cdot 6$
C  $6 \cdot 5 \cdot 4$
D  $6^3$
E  $3^6$

5  Which expression represents how many different bracelets can be made if each bracelet has 8 different beads or charms arranged on a gold band with a gold clasp?
A  $8 \cdot (6 + 5)$
B  $11 \cdot 10 \cdot 9 \cdot 8 \cdot 7 \cdot 6 \cdot 5 \cdot 4$
C  $11!$
D  $11^8$
E  $8^{11}$

6  How many times do you have to toss a quarter until there are 16 possible outcomes?
A  32
B  16
C  8
D  4
E  2

7  Mrs. Bell teaches 24 learners. She divides her class into three equal groups. Partway through the year she takes one learner from each group to form a new group of 3 learners. How many different ways can the new group of three be formed?
A  $24 \cdot 23 \cdot 22$
B  $8 \cdot 8 \cdot 8$
C  $8 \cdot 7 \cdot 6$
D  $3 \cdot 3 \cdot 3$
E  $3^8$

8  Javier designs skateboards. He draws five different pictures arranged top to bottom underneath the skateboard. Last year he had 12 different pictures from which to choose. This year he has 3 more pictures. How many more possible arrangements of the five pictures are there now compared to last year?
A  45
B  243
C  265,320
D  510,543
E  759,375

9  Roberta has 16 different stones. She is choosing four of them to form a row in her garden. How many stones will be left to choose from when she picks the third stone?
A  16
B  15
C  14
D  13
E  12

**10** Which of the following questions could you solve by using an expression of the form $m \cdot m \cdot m \cdot m$?

A How many possible outcomes are there if you park 4 cars in a row?

B How many possible outcomes are there if you write 4 names on 4 lockers?

C How many possible outcomes are there if you pull 4 marbles from a bag of 24 marbles, without replacing the marbles between pulls?

D How many possible outcomes are there if you choose 4 different colors out of 16 colors?

E How many possible outcomes are there if you choose 4 cards from a deck of 52 and replace each card before picking the next card?

**Directions:** Questions 11 and 12 are based on the information below.

Clyde works with 8 other people building houses and office buildings that are designed to benefit the environment.

**11** How many possible three-worker crews can be made?

A 56 possible crews
B 84 possible crews
C 336 possible crews
D 504 possible crews
E 729 possible crews

**12** Three houses each need one construction worker to finish. How many different ways could the construction workers be assigned to complete all three houses?

A 27
B 504
C 729
D 362,880
E 387,420,489

**13** How many 3-digit odd numbers can be formed using the digits 0, 1, 2, 3, and 4? Which solution uses the correct strategy and solves the problem?

A You are using the conditions on each digit to find the number of options. Compute $4 \cdot 5 \cdot 2$ to get 40 different odd numbers.

B You are picking subgroups from a single group, so order doesn't matter. Compute $\frac{5 \times 4 \times 3}{3 \times 2 \times 1}$ to get 10 different odd numbers.

C You can repeat the digits and order matters. Compute $5 \cdot 5 \cdot 5$ to get 125 different odd numbers.

D You cannot repeat a digit and order matters. Compute $5 \cdot 4 \cdot 3$ to get 60 different odd numbers.

E Not enough information is given.

**14** Which computation could be used to find the total possible outcomes for a permutation without repetition?

A $3^3$
B $6^4$
C $6 \cdot 4 \cdot 3$
D $2 \cdot 9 \cdot 16$
E $13 \cdot 12 \cdot 11$

**15** Maria has a grooming business for pets. There are 7 dogs in her waiting room. She will groom them one at a time. Which statement is true?

A There are 49 ways in which she can choose the order of grooming the dogs.

B She has $7^7$ possible ways to groom the dogs.

C She will have four dogs to choose from when she makes her fourth choice.

D There will be 823,543 different ways to choose the order of grooming the dogs.

E She has 5,040 different ways to choose the first dog to groom.

Lesson 1 | Counting Outcomes

**Directions:** Questions 16 through 20 are based on the information below.

There are 10 multiple-choice questions on a test and each question has 5 options.

**16** **How many ways can the test be answered?**
  A  50
  B  30,240
  C  100,000
  D  3,628,800
  E  9,765,625

**17** **Which expression represents the different ways three of these tests with different content can be answered?**
  A  $3 \cdot 10^5$
  B  $3 \cdot 5^{10}$
  C  $10^{15}$
  D  $5^{30}$
  E  $3 \cdot 10 \cdot 9 \cdot 8 \cdot 7 \cdot 6$

**18** **How many selections or choices does a student make to complete the test?**
  A  10
  B  50
  C  30,240
  D  100,000
  E  9,765625

**19** **How many options does the test display?**
  A  5
  B  50
  C  30,240
  D  3,628,800
  E  9,765,625

**20** **Which expression represents how many options four of these tests with different content display?**
  A  $4 \cdot 50$
  B  $5^4$
  C  $4 \cdot 30,240$
  D  $4 \cdot 3,628,800$
  E  9,765,625

**Directions:** Questions 21 through 23 are based on the information below.

An online gamers club has 112 members.

**21** **Which expression represents how many different ways a president, a vice president, a treasurer, and a secretary can be elected from the members?**
  A  $112 \div 4$
  B  $4 \times 112$
  C  $112 \times 111 \times 110 \times 109$
  D  $112!$
  E  $112^4$

**22** **Which expression represents how many different committees of four people could be formed with the 112 members?**
  A  $\dfrac{112 \cdot 111 \cdot 110 \cdot 109}{4 \cdot 3 \cdot 2 \cdot 1}$
  B  $112 \times 111 \times 110 \times 109$
  C  $112 \div 4$
  D  $4 \times 12$
  E  $112^4$

**23** **Which expression represents how many different ways the members could be assigned a number 1 through 112 so that each number is assigned once?**
  A  $112 \div 2$
  B  $112 \cdot 112$
  C  $112^{112}$
  D  $112!$
  E  $112$

**24** **Which expression represents the number of different 5-card hands that are possible from a deck of 52 cards?**
  A  $5 \cdot 52$
  B  $52!$
  C  $52^5$
  D  $52 \cdot 51 \cdot 50 \cdot 49 \cdot 48$
  E  $\dfrac{52 \cdot 51 \cdot 50 \cdot 49 \cdot 48}{5 \cdot 4 \cdot 3 \cdot 2 \cdot 1}$

25 Which scenario's total possible outcomes is represented by the expression 6 • 4?

A A company has 24 employees. The manager wants to assign each employee to a bowling team made of 6 employees.

B A company produces different beads that have one color and one shape. There are six different colors and four different shapes.

C A company designs six different pieces of pottery and packs four different pieces in a box for shipping.

D A company sells six items in six different colors.

E A company produces six different tiles and places four of them in the lobby.

26 There are 18 diplomats attending a peace conference. Each diplomat shook hands once with every other diplomat. Which expression represents how many different handshakes there were?

A 18 • 17

B 18 • 18

C 9 • 9

D $\frac{18 \cdot 17}{2 \cdot 1}$

E Not enough information is given.

27 Lindsay has 6 different places to go mountain biking and 6 friends that can ride with her today. How many different bike rides can Lindsay possibly go on today if she chooses one place to ride and one friend to go with her?

A 6

B 15

C 30

D 36

E 64

28 Which scenario's total possible outcomes equals 243?

A Charlie can have three out of the five different snacks throughout the day.

B A store displays three types of snacks from five types of snacks that can be displayed.

C Shana is allowed to have five snacks throughout the day and there are three different types of snacks.

D A company packs boxes with five snacks and has five different types of snacks.

E A cafeteria lunch includes one snack and there are three different snacks available.

29 Why are different counting strategies used in these two scenarios?

• **the number of different items when three items each have seven options**

• **the number of different groups of three people from a team of seven people**

A The order matters when finding items, and the order does not matter when choosing groups.

B The order matters when choosing groups, and the order does not matter when finding items.

C One situation has three separate types of choices and the other situation has one type of choice without repetition.

D One situation is items and the other situation is people.

E The same counting strategy is used for both situations.

**Directions:** Questions 30 and 31 are based on the information below.

A noodle bar offers 5 different noodles, 6 different sauces, and 12 different vegetables. A $12.99 noodle bowl includes one choice of noodle, one choice of sauce, and two different choices of vegetables.

30 Which expression represents how many different types of noodle bowls there are?

A 5 • 6 • 12

B 5 • 6 • 12 • 11

C 5 • 6 • 12 • 12

D $5 \cdot 6 \cdot \frac{12 \cdot 11}{2 \cdot 1}$

E $5 \cdot 6 \cdot \frac{12 \cdot 12}{2 \cdot 1}$

31 The noodle bar sold one of each type of noodle bowl and gave the proceeds to charity. How much money did the restaurant donate?

A $4,676.40

B $9,352.80

C $25,720.20

D $28,058.40

E $51,440.40

# Probability

## ① Learn the Skill

When you flip a quarter, you have an equal chance of flipping heads or tails. The chances of heads can be expressed as *1:2*, where *1* represents the number of favored outcomes (flipping heads) and *2* represents the number of possible outcomes. This ratio expresses the **theoretical probability** of the event. In theory, each time you flip a coin, you have a 50% chance of flipping heads.

Probability based on the results of an experiment is called **experimental probability**. As with theoretical probability, you can express experimental probability as a ratio, fraction, or percent. If you toss a quarter ten times and get heads six times, the experimental probability is $\frac{6}{10}$, which simplifies to $\frac{3}{5}$.

## ② Practice the Skill

By mastering the skill of probability, you will improve your study and test-taking skills. Read the example and strategies below. Use this information to answer question 1.

**A** In question 1, by choosing a striped marble from the bag during the first event and not replacing it, Marc affected the outcome of the second event. The two events are said to be **dependent**. When events are dependent, the number of outcomes changes.

If Marc had replaced the marble after the first event, the first event would not have affected the outcome of the second event. In this case, the first event and the second event would have been **independent**.

A bag of 10 marbles contains 7 striped marbles and 3 black marbles.

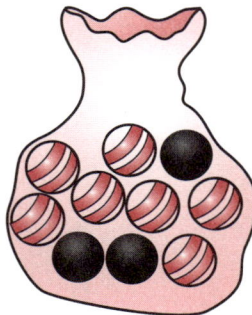

**B** Probability can be expressed as a ratio. The probability of drawing a black marble during the first event would be 3:10. There are 3 black marbles and 10 possible outcomes. The same probability can be expressed as a fraction $\left(\frac{3}{10}\right)$, a decimal (0.3), and a percentage (30%).

**C** If an event has a probability of 0%, it is said to be **impossible**. If the probability is less than 50%, the event is **unlikely**. If the chances are greater than 50%, it is **likely**. If the probability of an event is 100%, it is **certain**.

### ✔ TEST-TAKING TIPS

When answering a probability problem, always check whether the events are independent or dependent. Then determine the probability in the form that is easiest for you.

**1** In the first event, Marc draws a striped marble. He does not replace it. In the next three events, Marc draws 2 striped marbles and 1 black marble. He does not replace the marbles. What is the probability that he will select a black marble on the fifth event? **A**

   **A** 1:10
   **B** 1:3
   **C** 2:7
   **D** 2:3
   **E** 2:2

## ③ Apply the Skill

**Directions:** Questions 2 through 4 are based on the information below.

Maude uses this spinner to conduct a probability experiment.

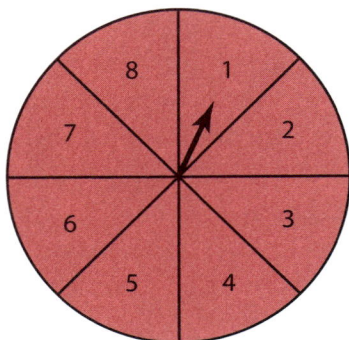

**2** On the first spin, what is the probability that the spinner will land on 6?

A   1:8
B   1:7
C   1:6
D   6:8
E   Not enough information is given.

**3** On the second spin, what is the probability that the spinner will land on 4 or 8?

A   0.48
B   0.28
C   0.25
D   0.16
E   0.13

**4** Maude spins the spinner twice. She lands on 4 and 6. So far, what is her experimental probability of spinning an odd number?

A   $\frac{1}{6}$
B   $\frac{1}{8}$
C   $\frac{1}{2}$
D   $\frac{1}{1}$
E   $\frac{0}{2}$

**Directions:** Questions 5 and 6 are based on the information below.

A large chain store keeps track of its daily customer complaint calls.

| COMPLAINT CALLS | |
|---|---|
| **Department** | **Number of Complaints** |
| Electronics | 6 |
| Housewares | 4 |
| Automotive | 2 |
| Clothing | 3 |

**5** What is the experimental probability that the next complaint call to the store will concern the clothing department?

A   20%
B   25%
C   30%
D   50%
E   Not enough information is given.

**6** What is the experimental probability that the next complaint call will concern the electronics department or the housewares department?

A   $\frac{4}{15}$
B   $\frac{1}{2}$
C   $\frac{3}{5}$
D   $\frac{2}{3}$
E   $\frac{4}{5}$

**7** Ian read in the newspaper that there is a 40% chance of rain tomorrow. What is the probability that it will not rain tomorrow?

A   $\frac{1}{25}$
B   $\frac{3}{50}$
C   $\frac{3}{5}$
D   $\frac{2}{3}$
E   $\frac{3}{2}$

**Directions:** Questions 8 through 10 are based on the spinner below.

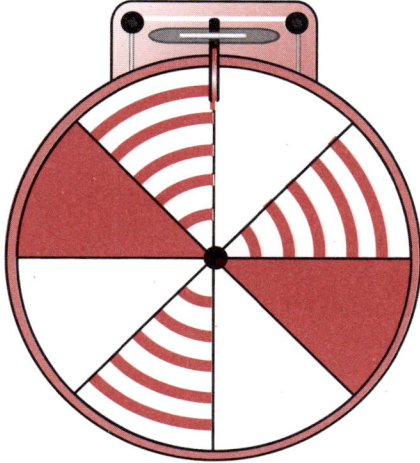

8  What is the probability that the spinner will land on a striped or white wedge?

A  2:8

B  3:8

C  3:4

D  6:6

E  Not enough information is given.

9  What is the probability that the spinner will land on a red or striped wedge?

A  $\frac{1}{4}$

B  $\frac{3}{8}$

C  $\frac{1}{2}$

D  $\frac{5}{8}$

E  $\frac{1}{1}$

10  Which word best describes the chances of the spinner landing on a green wedge?

A  impossible

B  unlikely

C  likely

D  certain

E  Not enough information is given.

**Directions:** Questions 11 through 13 are based on the information below.

Jenna has a bag of marbles that contains 7 striped marbles and 5 black marbles.

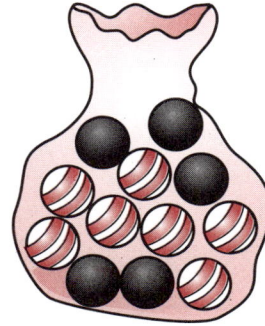

11  What is the probability that Jenna will pick a black marble?

A  1:6

B  5:12

C  1:2

D  7:12

E  1:1

12  Jenna picks a striped marble and does not replace it. Then Jenna picks a black marble and does not replace it. What is the probability that she will pick a striped marble in the third event?

A  41%

B  50%

C  60%

D  100%

E  Not enough information is given.

13  Jenna experiments to find the probability of picking a striped marble. In the first event, she picks a black marble and places it back in the bag. In the second and third events, she picks a striped marble, replacing the marble after each pick. What is the experimental probability of picking a striped marble?

A  $\frac{1}{3}$

B  $\frac{2}{3}$

C  $\frac{1}{2}$

D  $\frac{3}{4}$

E  $\frac{1}{1}$

**Directions:** Questions 14 through 17 are based on the information below.

Marta uses this spinner to conduct probability experiments.

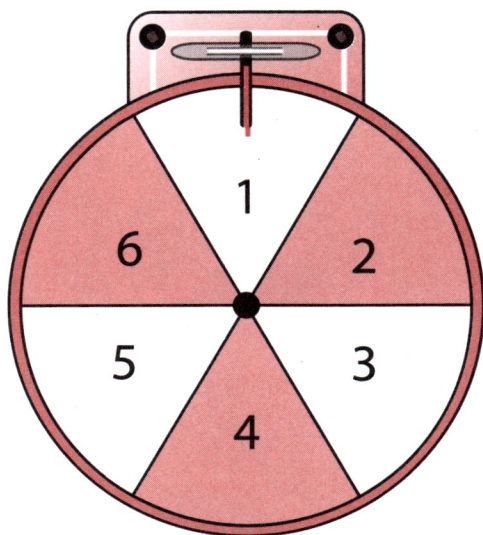

**14** What is the probability that Marta will land on either a red wedge or an odd number?

- **A** 20%
- **B** 40%
- **C** 50%
- **D** 100%
- **E** Not enough information is given.

**15** Marta spins and lands on 4. What is the probability that on her second spin, she will land on 4?

- **A** 1:2
- **B** 1:4
- **C** 1:6
- **D** 2:5
- **E** 2:6

**16** What is the probability that Marta will land on 6, 2, or one of the white wedges?

- **A** 1:6
- **B** 1:2
- **C** 2:3
- **D** 5:6
- **E** 6:6

**17** Which word best describes the chances of the spinner landing on a factor of 12?

- **A** impossible
- **B** unlikely
- **C** likely
- **D** certain
- **E** Not enough information is given.

**Directions:** Questions 18 through 20 are based on the information below.

Chuck is conducting probability experiments using a single die.

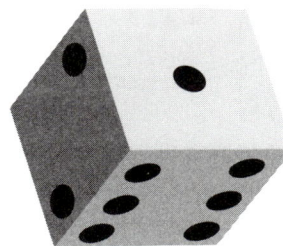

**18** Chuck rolls the die once, and it lands on 2. What is the probability that Chuck will roll a 2 on his second turn?

- **A** 1:6
- **B** 1:3
- **C** 1:2
- **D** 2:3
- **E** Not enough information is given.

**19** Chuck rolls the die once, and it lands on 3. He rolls it again, and it lands on a 5. What is the experimental probability so far for landing on an odd number?

- **A** 0:2
- **B** 1:6
- **C** 1:3
- **D** 2:3
- **E** 1:1

**20** What is the probability that Chuck will roll an even number?

- **A** 16.7%
- **B** 33.3%
- **C** 50.0%
- **D** 66.7%
- **E** 83.3%

**Directions:** Questions 21 through 23 are based on the information below.

Ryan took a random survey of 100 cars. The following table displays the results.

| SURVEY OF CARS | |
|---|---|
| **Color of Car** | **Number** |
| Black | 32 |
| Blue | 15 |
| Red | 25 |
| White | 18 |
| Other | 10 |

**21** **Based on Ryan's survey, what is the probability that the next car he spots will be blue or red?**
A  0.15
B  0.40
C  0.60
D  0.75
E  0.85

**22** **What is the probability that the next car Ryan sees will be a color other than black, blue, red, or white?**
A  0%
B  10%
C  30%
D  50%
E  100%

**23** **The probability is greatest for seeing which color of car next?**
A  blue
B  red
C  black
D  white
E  other

**Directions:** Questions 24 and 25 are based on the information below.

Julian has a bag of marbles. He knows that the bag contains 10 marbles, some black and some red. He conducts experiments to predict how many of each color marble is in the bag.

**24** **In the first event, Julian picks a black marble. He replaces the marble. In the second and third events, he picks a red marble. He replaces the marble after each event. What is the experimental probability that he will pick a red marble next?**
A  $\frac{1}{5}$
B  $\frac{3}{10}$
C  $\frac{1}{3}$
D  $\frac{2}{3}$
E  $\frac{3}{4}$

**25** **In the first event, Julian picks a black marble out of the bag. He does not replace it. In the second event, he picks another black marble out of the bag and does not replace it. What is the probability that the next marble he picks out of the bag will be black?**
A  $\frac{1}{10}$
B  $\frac{1}{5}$
C  $\frac{4}{5}$
D  $\frac{1}{1}$
E  Not enough information is given.

# Mean, Median, Mode, and Range

## 1 Learn the Skill

The mean, median, and mode are **measures of central tendency**, which are ways to describe the "center" of a set of data. The **mean** is the average value of a data set. The **median** is the middle number in a set of data when the values are ordered from least to greatest. In the number set 23, 24, 27, 29, 75, the median is 27. Notice that the median was not affected by the value of 75, which is much larger than the other values. As a result, the median more accurately describes the set than the mean (35.6) does. The **mode** is the value that occurs most frequently in a set of data. *Range* is a **measure of variation**, which describes the spread of a data set. To find the **range**, subtract the least value from the greatest value.

## 2 Practice the Skill

By mastering the skills of finding mean, median, mode, and range, you will study data in a meaningful way. Read the example and strategies below. Use this information to answer question 1.

**A** To find the median of a data set, list the values in order from least to greatest. The number 65 is listed three times in the table. When ordering numbers, be sure to list 65 three times.

**B** When a number set consists of an odd number of values, the middle number is the median, when the numbers are listed in order. When the set consists of an even number of data points, find the mean of the two middle numbers. Note that the median may not be a number in the set of data.

Felipe measured and recorded the heights of the runners participating in a neighborhood relay race.

| HEIGHTS OF RELAY RACE RUNNERS | |
| --- | --- |
| Runner | Height (inches) |
| Carol | 63 |
| Steven | 68 |
| Pedro | 65 **A** |
| Julia | 65 |
| Chantell | 67 |
| Camille | 64 |
| Frank | 72 |
| William | 71 |
| Jane | 65 **A** |
| Jake | 72 |

### ☑ TEST-TAKING TIPS

There are ten values in the table. When listing values from least to greatest, check that you listed a total of ten values.

**1** **What is the median height of the runners?**
A 65 in.
B 65.2 in.
C 66 in.
D 67 in.
E 67.2 in.

**Directions:** Questions 2 through 4 are based on the information below.

The running times of a charity 100-meter race are shown below.

| TIMES FOR THE 100-METER RACE | |
|---|---|
| Runner | Time (seconds) |
| David | 13.5 |
| Sanya | 16.0 |
| Jeremy | 12.6 |
| Erica | 15.2 |
| Chen | 12.8 |
| Yusuf | 11.8 |
| Matt | 17.2 |
| Sarah | 12.1 |

2   What is the range of the runners' times in the 100-meter race?
   A   4.2 seconds
   B   5.4 seconds
   C   6.4 seconds
   D   11.8 seconds
   E   13.9 seconds

3   What is the median time in the race?
   A   5.4 seconds
   B   12 seconds
   C   13.15 seconds
   D   13.9 seconds
   E   16 seconds

4   What is the difference between Sarah's time and the mean time of the runners?
   A   1.35 seconds
   B   1.8 seconds
   C   13.9 seconds
   D   13.15 seconds
   E   Not enough information is given.

5   The owner of Ice Cream Palace listed the number of milkshakes sold each day for one week.

| DAILY MILKSHAKE SALES | | | | | | |
|---|---|---|---|---|---|---|
| Day | Mon. | Tues. | Wed. | Thurs. | Fri. | Sat. |
| Number Sold | 22 | 16 | 20 | 26 | 24 | 85 |

Which value best describes the number of milkshakes sold at Ice Cream Palace on a typical day?
   A   20
   B   21.6
   C   23
   D   32.16
   E   69

6   The sneaker sales at Sneaker World were recorded each day for six days.

| SNEAKER WORLD SALES | |
|---|---|
| Day | Total Sales |
| Monday | $5,229 |
| Tuesday | $3,598 |
| Wednesday | $6,055 |
| Thursday | $3,110 |
| Friday | $3,765 |
| Saturday | ? |

The mean sales for this one week were $3,743.14. The manager misplaced her records for Saturday. What were the sales on Saturday?
   A   $634.99
   B   $701.84
   C   $852.19
   D   $975.92
   E   $1,681.15

**Directions:** Questions 7 through 9 are based on the information below.

| FINAL EXAM SCORES | |
|---|---|
| **Student** | **Grades** |
| David | 87 |
| Marla | 72 |
| Elena | 75 |
| Jeff | 85 |
| Tyrell | 89 |
| Jasmine | 93 |
| Kim | 68 |
| Chris | 97 |
| Jessica | 85 |
| Mel | 70 |
| Jean | 91 |

**7** What is the range of grades for the exam?
  A  29
  B  68
  C  72
  D  85
  E  97

**8** What is the mode for the set of exam grades?
  A  29.0
  B  82.9
  C  85.0
  D  93.0
  E  97.0

**9** What is the difference between Elena's grade and the median grade on the exam?
  A  7
  B  8
  C  10
  D  22
  E  24

**Directions:** Questions 10 through 12 are based on the information below.

Fred's Bike Shop kept track of its sales for one year. Its monthly sales (rounded to the nearest ten dollars) are shown in the table below.

| BICYCLE SALES | |
|---|---|
| **Month** | **Sales** |
| Jan. | $8,320 |
| Feb. | $7,200 |
| March | $11,820 |
| April | $18,560 |
| May | $23,630 |
| June | $26,890 |
| July | $24,450 |
| Aug. | $22,110 |
| Sept. | $23,450 |
| Oct. | $19,300 |
| Nov. | $15,340 |
| Dec. | $16,980 |

**10** What is the median for January through June?
  A  $15,190
  B  $16,070
  C  $18,560
  D  $18,930
  E  $19,690

**11** What is the mean sale for July through December? Round to the nearest dollar.
  A  $9,110
  B  $18,171
  C  $20,272
  D  $20,705
  E  $22,110

**12** What is the range of sales throughout the year?
  A  $15,310
  B  $17,250
  C  $18,570
  D  $19,690
  E  $26,890

**Directions:** Questions 13 through 18 are based on the information below.

Jessica is conducting a random survey to find out how much time people spend on the Internet. The following table shows the results in hours per day.

| HOURS PER DAY SPENT ON INTERNET | | | | | | | |
|---|---|---|---|---|---|---|---|
| | Mon | Tues | Wed | Thurs | Fri | Sat | Sun |
| Jen | 3 | 4.25 | 5 | 1 | 1.5 | 5.5 | 6 |
| Mila | 4 | 3.5 | 5.5 | 2.5 | 4 | 6 | 5.5 |
| Trang | 2 | 1.5 | 0.5 | 0.25 | 0.75 | 3 | 3 |
| Ron | 3.5 | 4.25 | 5 | 5.5 | 2.25 | 6.25 | 7 |
| Yusef | 2.5 | 1 | 1 | 1.5 | 0.5 | 5.25 | 7.5 |

**13** What is the range for the data set showing the number of hours spent per day on the Internet?
   A   0.0
   B   3.4
   C   5.5
   D   7.0
   E   7.25

**14** What is the mode of the data set?
   A   1.0
   B   1.5
   C   3.5
   D   5.5
   E   7.5

**15** What is the median number of hours spent on the Internet over a weekend (Sat. and Sun.)?
   A   4.50
   B   5.50
   C   5.75
   D   6.00
   E   6.25

**16** What is the range for the number of hours per day that Ron spends on the Internet?
   A   7
   B   5
   C   4.75
   D   3.5
   E   2.25

**17** What is the median number of hours Trang spends on the Internet in a day?
   A   1.1
   B   1.5
   C   1.6
   D   3.0
   E   2.8

**18** What is the difference between the mean of Wednesday's data and the mean of Sunday's data?
   A   1.25
   B   1.5
   C   2.0
   D   2.4
   E   2.5

**Directions:** Questions 19 and 20 are based on the information below.

The table shows the number of points scored by the Pirates football team for the first six games of the season.

| FOOTBALL SCORES | | |
|---|---|---|
| Game | Points Scored by Pirates | Points Scored by Opponents |
| 1 | 24 | 0 |
| 2 | 7 | 14 |
| 3 | 13 | 21 |
| 4 | 12 | 6 |
| 5 | 0 | 12 |
| 6 | 36 | 30 |

**19** What is the average number of points scored by the Pirates?
   A   12.50
   B   13.83
   C   15.33
   D   17.67
   E   18.40

**20** What is the median number of points scored by the Pirates' opponents?
   A   12.5
   B   13
   C   13.8
   D   14
   E   16.5

**Directions:** Questions 21 and 22 are based on the information below.

The table shows the results of a cross-country team's times in a recent 3-mile race.

| CROSS-COUNTRY RUNNING TIMES | |
|---|---|
| Name | Times (minutes:seconds) |
| Holly | 25:21 |
| Karen | 21:07 |
| Ana | 20:58 |
| Jessie | 26:10 |
| Sonya | 23:27 |

**21** Which teammate ran the fastest?
- **A** Holly
- **B** Karen
- **C** Ana
- **D** Jessie
- **E** Sonya

**22** What is the median time of the team's results?
- **A** 4:37
- **B** 20:58
- **C** 23:27
- **D** 23:41
- **E** Not enough information is given.

**23** Which data set has a range of $6\frac{1}{2}$?
- **A** $3, 0, 6\frac{1}{2}, 4\frac{1}{2}$
- **B** $6, \frac{1}{2}, 3, 4\frac{1}{2}$
- **C** $4, 6\frac{1}{2}, 3, \frac{1}{2}$
- **D** $3\frac{1}{2}, 6\frac{1}{2}, 6, \frac{1}{2}$
- **E** Not enough information is given.

**Directions:** Questions 24 through 26 are based on the information below.

Kyle wants to know about how many runs his favorite baseball team scores per game. To do so, he sets up a table and records the number of runs the team scores each game.

| RUNS SCORED BY THE PATRIOTS | |
|---|---|
| Game | Runs Scored |
| 1 | 3 |
| 2 | 9 |
| 3 | 3 |
| 4 | 4 |
| 5 | 7 |
| 6 | 5 |
| 7 | 6 |
| 8 | 3 |
| 9 | 1 |
| 10 | 5 |

**24** What is the range of the data set that Kyle collected?
- **A** 1
- **B** 3
- **C** 5
- **D** 7
- **E** 8

**25** Kyle wants to determine the average number of runs scored by the Patriots. What is the mean of Kyle's data?
- **A** 3
- **B** 4.5
- **C** 4.6
- **D** 5.5
- **E** 9

**26** What is the mode of the data?
- **A** 3
- **B** 4.5
- **C** 4.6
- **D** 5
- **E** 8

UNIT 3

# Data Tables and Matrices

## 1 Learn the Skill

A **data table** is a tool for organizing and analyzing information. Data in tables are arranged in rows and columns in order to make the information easier to identify and compare.

A **matrix** is another way to display data values. By showing only numerical values, a matrix simplifies the display and can be used more easily when operations are performed on the data. Both data tables and matrices are arrays, with entries in rows and columns. The size, or **dimension**, of a matrix is the number of rows by the number of columns. A matrix that has 4 rows and 5 columns is a $4 \times 5$ matrix (read as 4 by 5 matrix).

## 2 Practice the Skill

Tables often contain more information than you need to solve a problem. When taking mathematics tests, carefully read questions involving tables to precisely identify the information you need. Read the examples and strategies below. Use this information to answer question 1.

**A** The category of the column on the far left is "Week." The four weeks appear below this label. All the values that appear in the other columns relate to the category that appears at the top of each column.

**B** A matrix lists the data values in the same rows and columns as in the table. The values within a matrix do not include the units of measure and are enclosed by large brackets. In some matrices, row and column headers are added outside of the brackets.

Sophia made a data table to analyze her first month of training for a 5k race.

| WEEK | DAY 1 | DAY 2 | DAY 3 | DAY 4 | DAY 5 |
|------|-------|-------|-------|-------|-------|
| **SOPHIA'S RUNNING DISTANCES** | | | | | |
| 1 | 1 mi | 0 mi | 1 mi | 0 mi | 1 mi |
| 2 | 1 mi | 0 mi | 2 mi | 1 mi | 1 mi |
| 3 | 2 mi | 1 mi | 2 mi | 1 mi | 1 mi |
| 4 | 2 mi | 1 mi | 2 mi | 1 mi | 2 mi |

Sophia also made a matrix of the same data.

$$\begin{bmatrix} 1 & 0 & 1 & 0 & 1 \\ 1 & 0 & 2 & 1 & 1 \\ 2 & 1 & 2 & 1 & 1 \\ 2 & 1 & 2 & 1 & 2 \end{bmatrix}$$

### ✓ TEST-TAKING TIPS

Before answering questions about tables, be sure to read the title and the headings for the rows and columns. They contain important information that will help you understand what the data present.

**1** **On which day did Sophia run two miles for the first time?**
  A  day 1, week 3
  B  day 3, week 3
  C  day 3, week 1
  D  day 3, week 2
  E  day 5, week 4

**UNIT 3**

**Directions:** Questions 2 through 5 are based on the information below.

Steve is training for a triathlon. He tracks his speeds in the table, where SP is steady pace, AP is average pace of alternating between fast and slow, and RP is race pace.

|  | Swim Speeds | Run Speeds | Bike Speeds |
|---|---|---|---|
| **SP** | 1.3 mph | 6 mph | 12 mph |
| **AP** | 1.35 mph | 6.5 mph | 14 mph |
| **RP** | 1.5 mph | 8 mph | 18 mph |

**2** **How much faster is Steve's alternating pace than his steady pace for swimming?**
   **A**  5 mph
   **B**  1 mph
   **C**  0.5 mph
   **D**  0.1 mph
   **E**  0.05 mph

**3** **For what pace does Steve travel six and a half miles in one hour?**
   **A**  Biking at a steady pace
   **B**  Running at an alternating pace
   **C**  Swimming at a race pace
   **D**  Running at a steady pace
   **E**  Biking at an alternating pace

**4** **How many minutes would it take Steve to finish a three-mile triathlon consisting of one mile in each sport at his race pace? Round to the nearest whole number.**
   **A**  1
   **B**  28
   **C**  51
   **D**  58
   **E**  61

**5** **If Steve made a matrix to show the difference between his run speeds and swim speeds, what would 5.15 mean?**
   **A**  Steve runs 5.15 mph slower than swimming at his race pace.
   **B**  Steve runs 5.15 mph faster than swimming at his steady pace.
   **C**  Steve runs 5.15 mph slower than swimming at his alternating pace.
   **D**  Steve runs 5.15 mph faster than swimming at his alternating pace.
   **E**  Steve runs 5.15 mph faster than swimming and biking for one of the paces.

**Directions:** Questions 6 and 7 are based on the information below.

The data table represents the amount of two different charges for three hotels.

|  | Rent for Banquet Hall | Price Per Guest for Food |
|---|---|---|
| **Hotel A** | $200 | $12.95 |
| **Hotel B** | $175 | $13.25 |
| **Hotel C** | $150 | $13.40 |

**6** **How much more would it cost for food to have a party with 90 guests at Hotel B than at Hotel A?**
   **A**  $2.00
   **B**  $13.50
   **C**  $27.00
   **D**  $40.50
   **E**  $2,250.00

**7** **What would be the cost to have a family reunion at Hotel C for 250 family members?**
   **A**  $3,350.00
   **B**  $3,437.50
   **C**  $3,487.50
   **D**  $3,500.00
   **E**  $40,850.00

**Directions:** Questions 8 through 10 are based on the information below.

The amounts in three brands of vitamin and mineral supplements are listed in the data tables.

| VITAMIN C | |
|---|---|
| Brand 1 | 75 mg |
| Brand 2 | 90 mg |
| Brand 3 | 100 mg |

| VITAMIN B$_6$ | |
|---|---|
| Brand 1 | 2 mg |
| Brand 2 | 3 mg |
| Brand 3 | 2 mg |

| CALCIUM | |
|---|---|
| Brand 1 | 500 mg |
| Brand 2 | 200 mg |
| Brand 3 | 250 mg |

| IRON | |
|---|---|
| Brand 1 | 18 mg |
| Brand 2 | 12 mg |
| Brand 3 | 0 mg |

**8** **Which vitamin or mineral is provided in the largest amount?**
A Brand 3's vitamin C
B Brand 1's calcium
C Brand 2's calcium
D Brand 1's iron
E Brand 2's vitamin B$_6$

**9** **Which matrix represents the data for iron?**

A $\begin{bmatrix} 75 \\ 90 \\ 100 \end{bmatrix}$

B $\begin{bmatrix} 2 \\ 3 \\ 2 \end{bmatrix}$

C $\begin{bmatrix} 18 \\ 12 \\ 0 \end{bmatrix}$

D $\begin{bmatrix} 18 \\ 12 \end{bmatrix}$

E $\begin{bmatrix} 2 \\ 3 \end{bmatrix}$

**10** **If the daily recommended dose of calcium is 1,000 mg, what percent of the dose would you get by taking a Brand 3 tablet?**
A 25%
B 50%
C 75%
D 100%
E 400%

**Directions:** Questions 11 through 14 are based on the information below.

Five stores sell T-shirts. The third column is the number of T-shirts that need to be purchased to receive the discount listed in the fourth column.

| STORE | PRICE, 1 SHIRT | SHIRTS TO BUY | DISCOUNT |
|---|---|---|---|
| Wear This | $5.50 | 50 | 30% |
| Look at Me | $5.00 | 50 | 25% |
| Got Shirt? | $4.50 | 75 | 20% |
| Off My Back | $4.20 | 100 | 10% |
| Not Clear? | $4.00 | 100 | 5% |

**11** **How much do 100 T-shirts cost at Wear This?**
A $550
B $385
C $330
D $220
E $165

**12** **Which company has the least expensive price for 51 T-shirts?**
A Wear This
B Look at Me
C Got Shirt?
D Off My Back
E Not Clear?

**13** **Which company's price per shirt is the lowest for 100 T-shirts?**
A Wear This
B Look at Me
C Got Shirt?
D Off My Back
E Not Clear?

**14** **Which company's discount is the least amount of money per shirt?**
A Wear This
B Look at Me
C Got Shirt?
D Off My Back
E Not Clear?

**Directions:** Questions 15 through 17 are based on the information below.

Arthur has to bring party favors to a New Year's Eve celebration. He made a table to help him decide how many to buy.

| NEW YEAR'S PARTY FAVORS | | |
|---|---|---|
| Item | Items per Pack | Cost per Pack |
| Noisemakers | 6 | $3.99 |
| Year Eyeglasses | 2 | $12.99 |
| Mardi Gras Beads | 8 | $4.50 |
| Blinking Hats | 1 | $5.25 |
| Confetti Poppers | 5 | $2.99 |

**15** Which statement is true?
- **A** The cost of buying 8 Mardi Gras beads is $36.00.
- **B** The dimensions of a matrix displaying the data in the table is 2 × 5.
- **C** The total cost of purchasing one pack of each item is $29.72.
- **D** Arthur can get eight times as many blinking hats as Mardi Gras beads for about the same price.
- **E** For $10, Arthur can buy 15 noisemakers.

**16** Which item has the lowest unit rate?
- **A** noisemakers
- **B** year eyeglasses
- **C** Mardi Gras beads
- **D** blinking hats
- **E** confetti poppers

**17** How much would Arthur spend if he has to buy one of each item for 24 people?
- **A** $1.24
- **B** $118.88
- **C** $325.69
- **D** $326.29
- **E** $713.28

**Directions:** Questions 18 and 19 are based on the information below.

A newspaper reporter keeps track of the average price of gasoline in his county for eight weeks.

| AVERAGE COST OF GASOLINE IN MAY AND JUNE | |
|---|---|
| Week | Average Cost per Gallon |
| 1 | $3.98 |
| 2 | $3.89 |
| 3 | $4.24 |
| 4 | $4.09 |
| 5 | $4.45 |
| 6 | $4.62 |
| 7 | $4.04 |
| 8 | $3.99 |

**18** Between which two weeks did the average price of a gallon of gasoline increase by the greatest amount?
- **A** Week 1–Week 2
- **B** Week 2–Week 3
- **C** Week 3–Week 4
- **D** Week 4–Week 5
- **E** Week 5–Week 6

**19** Which statement is true about the average price of gasoline per gallon during Week 2?
- **A** It represents the greatest decrease in price from one week to the next.
- **B** It is the last time the average price fell from one week to another.
- **C** It is the lowest average price per gallon during the eight-week period.
- **D** The price increase from Week 2 to Week 3 is the sharpest increase shown.
- **E** It is the last time in the eight-week period that the average price was under four dollars.

Jada runs a small online bookstore. The table displays her sales and expenses for one year.

| BOOKSTORE FINANCES | | |
|---|---|---|
| Months | Sales | Expenses |
| Jan–Feb | $502.10 | $57.12 |
| Mar–Apr | $628.23 | $91.85 |
| May–June | $614.15 | $81.14 |
| July–Aug | $609.79 | $84.06 |
| Sept–Oct | $872.49 | $124.43 |
| Nov–Dec | $1,512.81 | $159.47 |

**20** **How much more did Jada sell in November and December than she did between January and April?**
  A  $10.50
  B  $382.48
  C  $884.58
  D  $1,010.71
  E  $1,130.33

**21** **What was the percent change for Jada's expenses from July and August to September and October?**
  A  68% increase
  B  68% decrease
  C  48% increase
  D  48% decrease
  E  40% increase

**22** **What trend can you infer from the table?**
  A  Sales increased throughout the year.
  B  Sales decreased throughout the year.
  C  Sales remained the same throughout the year.
  D  Sales were highest toward the end of the year.
  E  Sales were highest in summer months, but dipped in the fall and winter.

Ernesto works for a moving company based in Portland. He uses the following mileage chart to determine the amount of gasoline for each trip.

| MILEAGE CHART | | | | | |
|---|---|---|---|---|---|
| City | Eureka | Redding | Sacra. | San Fran. | Portland |
| Portland | 414 | 422 | 581 | 636 | — |
| Eureka | — | 148 | 289 | 272 | 414 |
| Redding | 148 | — | 165 | 217 | 422 |
| Sacramento | 289 | 165 | — | 91 | 581 |
| San Fran. | 272 | 217 | 91 | — | 636 |

**23** **Ernesto drives from Portland to San Francisco, where he spends the night. The following day, he drives to Sacramento. How many miles has he driven altogether?**
  A  91
  B  545
  C  636
  D  727
  E  818

**24** **Ernesto plans to move a business from Redding to Eureka. It will take two trips to complete the move. He first will have to drive from Sacramento to Redding and from Redding to Eureka and back, and then make a final trip to Eureka. How many miles will he have to drive in all?**
  A  165
  B  313
  C  461
  D  609
  E  757

UNIT 3

# Collecting Data

## 1 Learn the Skill

**Data** (singular datum) are factual details collected for analysis. Data can be qualitative (descriptive details that cannot be measured) or **quantitative** (measurable numbers). Crucial decision making comes from the analysis of data, so it is important that data be accurate and valid. The way in which data are collected affects the reliability. Data can be collected through observation, surveys, and experiments. Researchers and statisticians seek to minimize factors that lead to unreliable data, misinformation, or faulty conclusions.

Researchers and statisticians study populations. A **population** is the *entire group* of individuals being studied. A **census** is a study that collects information from an entire population. A **sample population** is a part of the population, studied when data cannot be collected for the entire group. For data from a sample population to be considered valid, the sample must be made up of individuals chosen at random. Individuals who are selected out of convenience or because they volunteer do not produce a true **random sample**.

## 2 Practice the Skill

Understanding the collection and display of data can help you improve your test-taking skills. Read the example below. Use this information to answer question 1.

**A** Having chosen employee preference as the factor they will use to make their decision about mall hours, the management needs to collect data that accurately reveals the employees' preferences.

**B** Whether to extend hours is essentially a "yes or no" survey question. It produces quantitative data as a result—measurable numbers of "yes" and "no" responses that can be compared. The total number of responses equals 100. The scenario does not reveal whether the 100 employees polled constitute the entire employee population.

**A** The management of a shopping mall took a poll of 100 employees to find out whether the mall should extend its hours of operation. The decision about whether to extend hours is based on a simple majority of the employees polled. The results of the poll are shown in the table below.

| SHOPPING MALL EMPLOYEE POLL | | |
|---|---|---|
| | **Favor Extended Hours** | **Oppose Extended Hours** |
| **Women** | 15 | 30 |
| **Men** | 25 | 30 |

### ☑ TEST-TAKING TIPS

Determine how data are collected. If data are from a sample population, determine whether the sample was chosen at random. These factors will give you clues about whether a set of data might be biased.

1 **If the mall employs 400 people, what factor of the survey would best validate the management's decision?**

A if the employees surveyed were sampled randomly

B if the employees volunteered for the survey

C if an equal number of men and women participated in the survey

D if the employees surveyed were selected based on seniority

E if the employees surveyed had experience working overtime

**Directions:** Questions 2 through 8 are based on the information below.

The police used a video camera located on one block to tape the traffic on Elm Street all day from Monday through Friday of one week. Inez and a group of volunteers kept track of traffic from 7 A.M. to 10 P.M. for seven consecutive days on a different part of Elm Street. The police and Inez are collecting data to determine whether or not a traffic light should be installed.

| TRAFFIC ON ELM STREET | | |
|---|---|---|
| | Video Camera | Inez |
| Bicycles | 53 | 39 |
| Motorcycles | 116 | 78 |
| Cars | 691 | 433 |
| Trucks | 393 | 97 |
| Total | 1253 | 647 |

**2   What type of study was used by the police?**
A   survey
B   qualitative
C   experiment
D   observation
E   random sample

**3   What type of study was used by Inez?**
A   survey
B   qualitative
C   experiment
D   observation
E   random sample

**4   Which statement best describes the population for the police study.**
A   all traffic on Elm Street
B   a random sample of the traffic on Elm Street
C   a voluntary sample of the traffic on Elm Street
D   a convenience sample of the traffic on Elm Street
E   a biased sample of the traffic on Elm Street

**5   What information is missing in Inez's study?**
A   data on traffic during the weekend
B   the number of trucks that Inez saw on Elm Street for seven days
C   traffic between 10 P.M. and 7 A.M.
D   the total number of vehicles she saw on Elm Street for seven days
E   data on traffic during weekday mornings

**6   How could the police change their study to make it a more accurate representation of the traffic on Elm Street?**
A   study for seven days to include weekend traffic
B   focus data collection on times of the day with peak traffic
C   combine their data with Inez's
D   ask Inez to track traffic for one more week
E   survey Elm Street residents about their driving habits

**7   If Inez saw one vehicle on Elm Street while she was walking where the police camera was set up, what is the probability that she would see a truck?**
A   77%
B   48%
C   37%
D   31%
E   3%

**8   Which factor might explain why the police observed a much higher proportion of trucks than Inez?**
A   The police camera was set up on a busier block of Elm Street.
B   Inez's data represented a voluntary sample.
C   Inez's population was smaller.
D   Truck traffic is higher on a weekend.
E   Truck traffic is higher at night when Inez did not gather data.

**Directions:** Questions 9 through 13 are based on the information below.

A national advertising agency visited a few locations near its own office and asked people there the following question: "Do you watch television shows more on traditional cable or via online streaming?" The responses to the survey are in the data table below.

| | Movie Theater | Exercise Club | Family Restaurant |
|---|---|---|---|
| Streaming | 77 | 38 | 43 |
| Cable | 123 | 12 | 157 |
| | Electronics Store | Diner | Barber Shop |
| Streaming | 233 | 12 | 2 |
| Cable | 67 | 88 | 28 |

**9** **Which statement best describes the population in this study?**
  A   people who stream television shows online
  B   people who watch television shows on traditional cable
  C   people who attend six businesses near the ad agency
  D   people who watch television shows via different media
  E   people who participated in a random study

**10** **What type of study was used by the advertising agency?**
  A   observation
  B   experiment
  C   sample survey
  D   census
  E   multistage sampling

**11** **Which statement best describes the groups of people in this study?**
  A   simple sample
  B   random sample
  C   volunteer sample
  D   convenience sample
  E   population

**12** **Which question would yield more useful data than the one the ad agency asked?**
  A   Do you watch television shows both online and on cable?
  B   How much time do you spend watching television shows streaming online and on traditional cable?
  C   How much time do you want to spend watching television shows streaming online?
  D   How much time do you spend watching television shows on traditional cable?
  E   How much time do you spend not watching television shows?

**13** **Which statement best describes this study.**
  A   This study is not a good representation of people who watch television shows because the sample is too small.
  B   This study includes a sample that is not randomly selected from the population of all cable and online streaming television viewers.
  C   This study provides a random sample of people who watch television because the sample includes people from six different locations.
  D   This study is a good representation of people who like to watch television shows on a computer because the ad agency polled an electronics store.
  E   This study is a good representation of people who eat out because they polled the diner and family restaurant.

**14** **Which sample method is a simple random sample?**
  A   a computer program that picks names in consecutive order from an alphabetical list
  B   a computer program that picks names at random without a repeating rule
  C   a computer program that picks every 9th entry in a list of names
  D   a computer program that picks names from a list based on a common characteristic
  E   a person choosing employees to participate in a study

Lesson 5 | Collecting Data

**Directions:** Questions 15 through 17 are based on the information below.

A supervisor puts all 42 names of her employees on paper and throws them in a bowl. Without looking, she mixes the names around and picks a name. She does not replace the name and mixes again before picking another name. She repeats until five names are picked. She pulls the five employees' files to see how long they have worked for her.

**15** **What is the probability of a particular employee being selected at random on the third drawing?**
  A  0%
  B  2.38%
  C  2.44%
  D  2.5%
  E  2.56%

**16** **Which statement describes the method the supervisor used to collect her data?**
  A  an observational study and a simple random sample of her employees
  B  an observational study and a voluntary sample of her employees
  C  an experimental study and a random sample of her employees
  D  an experimental study of her entire population of employees
  E  an experimental study of a convenience sample of employees

**17** **Approximately what percent of the population in the study does her sample represent?**
  A  10%
  B  12%
  C  15%
  D  20%
  E  42%

**18** **Which of the following studies could be based on a census of an entire population in its research?**
  A  an observational study that tracks participation in school sports for the Red Valley High School
  B  an experimental study that compares average practice time for high school sports in the United States
  C  a survey that asks parents of elementary school children about their lunchtime eating habits
  D  an observational study of the number of school children who are homeless in the United States
  E  a survey about office habits of employees in multinational corporations

**19** **Which statement best explains the significance of the population in a study?**

**The population in a study is important because**
  A  it can skew the results of the data.
  B  you can use a numerical value to represent it.
  C  it could change.
  D  it explains how you collected your data.
  E  it reveals whether a sample used to represent the population provides reliable data.

**20** **Campaign workers circulate a petition to get a candidate's name added to a ballot. The petition is what type of study?**
  A  census
  B  observational study
  C  experimental study
  D  voluntary sample survey
  E  convenience population survey

Cameron wanted to start a runner's marathon club. He posted signs in athletic stores and athletic clubs for marathon runners to contact him by e-mail. He also posted his sign on local running Web sites. He developed a list of 127 names of possible members for his running club.

**21 Which terms most apply to the method Cameron used to obtain data about possible members for his running club?**
- A *experimental* and *random*
- B *population* and *questionnaire*
- C *observational* and *sample*
- D *volunteer* and *convenience*
- E *random* and *population*

**22 Was Cameron's method of collecting data appropriate for his goal?**
- A Yes, because he was able to collect data about the specific sample of the population he was interested in.
- B Yes, because all members of the population had an equal chance of being selected for the club.
- C No, because his convenience sample was too limited.
- D No, because he did not get very many voluntary responses.
- E No, because the responses did not reflect the population he wanted to study.

Dylan wants to know if his community is interested in a park that is designed to be accessible for children and adults with special needs. He uses the phone book to call the first person listed under each letter of the alphabet to ask for his or her opinion on the issue. He collects data from 26 people. He records whether each person supports accessible parks. He also records each person's reasons for supporting or opposing the project.

**23 Which method best describes the way Dylan obtained data for his study?**
- A systematic sample survey
- B random sample survey
- C observational simple random sample
- D population census
- E observational volunteer sample

**24 Can Dylan's study be characterized as qualitative?**
- A No, because the number of "yes" and "no" responses can be counted.
- B Yes, because the study is not a random sample.
- C No, because he reached 26 individuals in his targeted population.
- D Yes, because he recorded data about the reasons for each respondent's opinion.
- E No, because the study did not collect data from people who are not listed in the phone book.

# Bar and Line Graphs

## 1 Learn the Skill

Graphs are used to visually display data. **Bar graphs** use vertical or horizontal bars to show data. These graphs are typically used to compare data. **Line graphs** are best suited for showing how a data set changes over time. Graphs may include scales that provide detail about the data.

**Scatter plots** are a type of line graph that show how one set of data affects another. The relationship between data sets is known as its **correlation.** A correlation may be positive (extending upward from the origin to *x*- and *y*- points) or negative (extending downward from the *y*-axis to the x-axis), or it may not exist at all.

## 2 Practice the Skill

By mastering the skill of interpreting bar and line graphs, you will improve your study and test-taking skills, especially as they relate to mathematics high school equivalency tests. Read the example and strategies below. Use this information to answer question 1.

**A** Multiple sets of data can appear on a bar graph or a line graph. When this occurs in a line graph, such as this one, you will see two or more line patterns. The lines for each park are a different color.

**B** When using a graph, first examine its different parts. The title describes the topic of the graph. Labels along the vertical and horizontal axes describe the data. The scale of the vertical axis shows the interval being used. You will find categories along the horizontal axis. This line graph also has a key that shows the color code used for the two different parks.

This line graph shows the monthly rainfall through the spring and summer at two state parks.

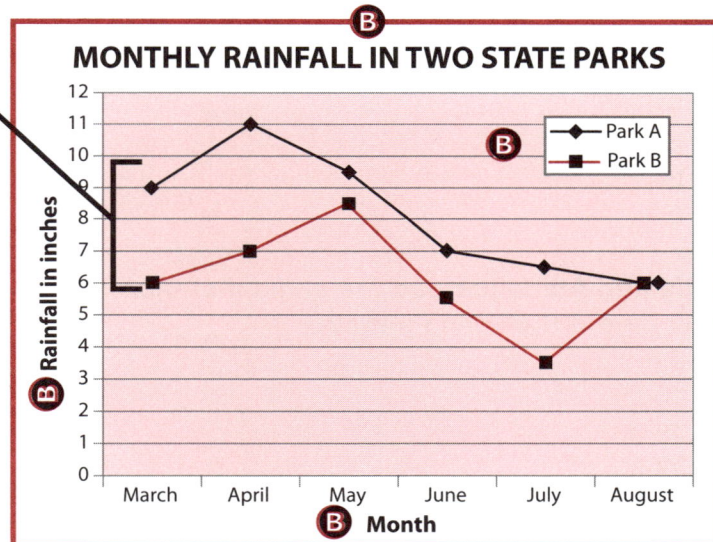

**B**

**MONTHLY RAINFALL IN TWO STATE PARKS**

**B** Rainfall in inches

**B** Month

### TEST-TAKING TIPS

Be sure you know precisely what the question is asking. For example, you can tell at a glance that May and August are not realistic answer choices.

1 **During which month was the difference in rainfall between the two parks the greatest?**
A March
B April
C May
D June
E July

UNIT 3

**Directions:** Questions 2 through 4 are based on the information below.

Fred records the long jump results in a track meet. He creates the following bar graph to show the results online.

**LONG JUMP RESULTS**

**2** **Which contestant jumped exactly half as far as the contest winner?**
A  A
B  B
C  C
D  D
E  E

**3** **Katie and Alana jumped the same distance. How far did they each jump?**
A  5 ft
B  10 ft
C  15 ft
D  17 ft
E  20 ft

**4** **What is the range of scores from the event?**
A  10
B  12.5
C  12.7
D  13
E  15

**Directions:** Questions 5 and 6 are based on the information below.

An educational services company compared student scores on a mathematics high school equivalency test with the amount of hours they prepared for it. Their findings are shown in the scatter plot.

**STUDY TIME FOR MATH TEST**

**5** **Anton hopes to earn greater than 80% on the mathematics high school equivalency test. How many hours should he study?**
A  0
B  2
C  4
D  6
E  8

**6** **Which statement does the data imply?**
A  There is no correlation between hours spent studying and scores on the mathematics test.
B  A student is more likely to score higher on the mathematics test if he/she spends more hours studying.
C  A student cannot score well on the mathematics test if he/she only studies for 4 hours.
D  A student is less likely to score higher on the mathematics test if he/she spends more hours studying.
E  Not enough information is given.

**Directions:** Questions 7 through 9 are based on the information below.

The bar graph shows the population of a town over a 50-year period.

## POPULATION OF SMITHVILLE

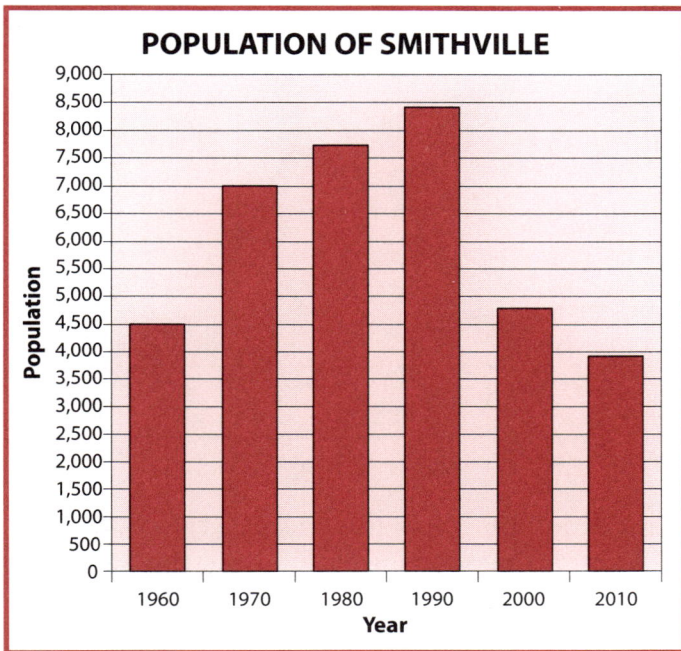

**7** The sharpest decrease in Smithville's population occurred between which years?

**A** 1960 to 1970
**B** 1970 to 1980
**C** 1980 to 1990
**D** 1990 to 2000
**E** 2000 to 2010

**8** In which two years was the population roughly the same?

**A** 1960, 1970
**B** 1960, 2000
**C** 1980, 2010
**D** 1990, 2000
**E** 1990, 2010

**9** Which sentence best describes the population trend between 1960 and 2010?

**A** The population increased.
**B** The population decreased.
**C** The population increased, then decreased.
**D** The population decreased, then increased.
**E** The population remained the same.

**Directions:** Questions 10 and 11 are based on the information below.

A marketing company conducted a study that compared annual earnings of workers with their ages. Their findings are shown on the scatter plot.

## ANNUAL EARNINGS

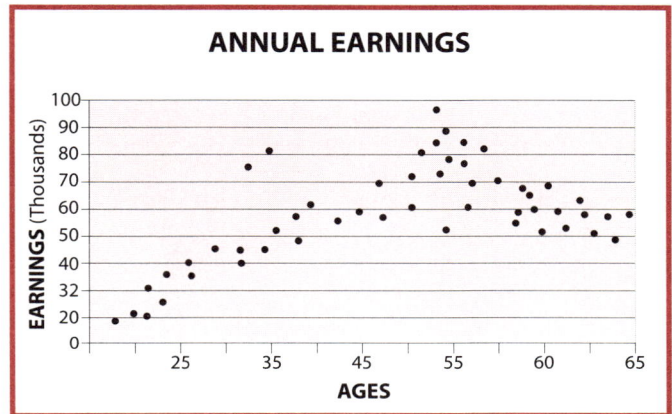

**10** At what age do workers earn the highest annual salaries?

**A** 25
**B** 35
**C** 45
**D** 55
**E** 65

**11** What correlation can you draw from the information on the scatter plot?

**A** Employees with less experience tend to earn the highest salaries.
**B** Higher levels of experience result in greater levels of pay, up to a certain point.
**C** Experience plays no role in workers' salaries.
**D** Employees with 10 years of experience earn more than employees with 20 years of experience.
**E** Workers earn their highest salaries immediately before retirement.

**Directions:** Questions 12 through 14 are based on the information below.

Every five years, ecologists record the type and number of mammals living in Pond Park. The bar graph shows their counts for 2005 and 2010.

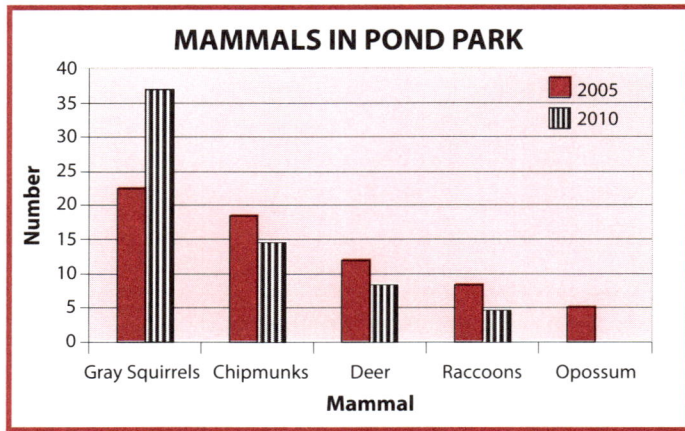

**MAMMALS IN POND PARK**

**Directions:** Questions 15 through 17 are based on the information below.

The line graph compares the average yearly rainfall for Anchorage, Alaska, to the average yearly rainfall in Honolulu, Hawaii.

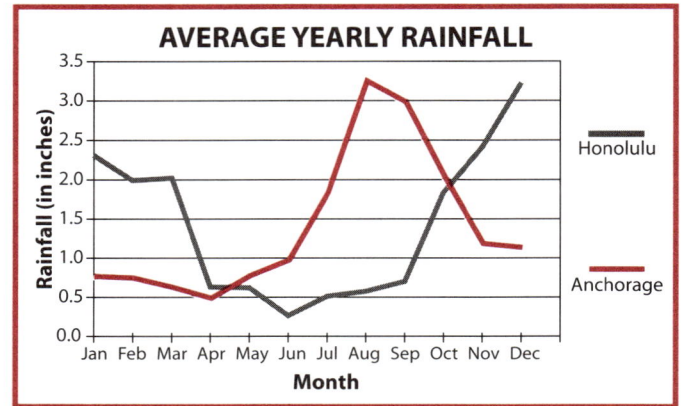

**AVERAGE YEARLY RAINFALL**

Source: www.ncdc.noaa.gov, accessed 2014

**12** Which mammal increased in numbers over the 5-year period?
  **A** gray squirrels
  **B** chipmunks
  **C** deer
  **D** raccoons
  **E** opossum

**13** Which mammal was not seen at Pond Park in 2010?
  **A** gray squirrels
  **B** chipmunks
  **C** deer
  **D** raccoons
  **E** opossum

**14** The number of deer in 2010 is equal to the number of which animal in 2005?
  **A** gray squirrels
  **B** chipmunks
  **C** deer
  **D** raccoons
  **E** opossum

**15** In which month is the average rainfall for Anchorage and Honolulu about the same?
  **A** February
  **B** April
  **C** June
  **D** September
  **E** November

**16** During which month is the difference between the two average rainfall amounts the greatest?
  **A** January
  **B** March
  **C** May
  **D** August
  **E** December

**17** What trend can you determine from the graph?
  **A** Average rainfall amounts are about the same in Anchorage as in Honolulu.
  **B** Average rainfall in Anchorage is consistently less than rainfall in Honolulu.
  **C** Average rainfall in Anchorage is consistently greater than rainfall in Honolulu.
  **D** Average rainfall in Anchorage is greatest in the summer, while, in Honolulu, it is greatest in the winter.
  **E** Both Anchorage and Honolulu have average rainfalls of less than 3 inches.

UNIT 3

**Directions:** Questions 18 through 20 are based on the information below.

The following bar graph shows the occupations of women in Centre City.

**OCCUPATIONS OF WOMEN IN CENTRE CITY**

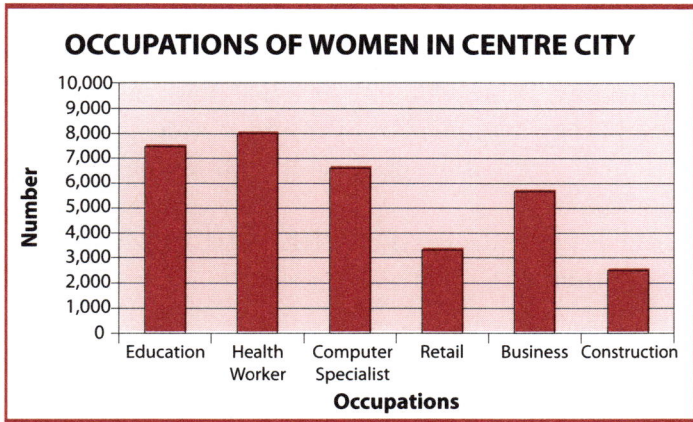

18 About how many more women work in education than in construction?
   A   3,000
   B   5,000
   C   7,000
   D   10,000
   E   11,000

19 What occupation is held by about twice the number of women who hold occupations in retail?
   A   business
   B   education
   C   health worker
   D   computer specialist
   E   construction

20 Which occupation is likely to show an increase in numbers over the next ten years?
   A   education
   B   health worker
   C   computer specialist
   D   retail
   E   Not enough information is given.

**Directions:** Questions 21 through 23 are based on the information below.

The line graph shows the average amount of daylight throughout the year in Pine Town.

**HOURS OF DAYLIGHT IN PINE TOWN**

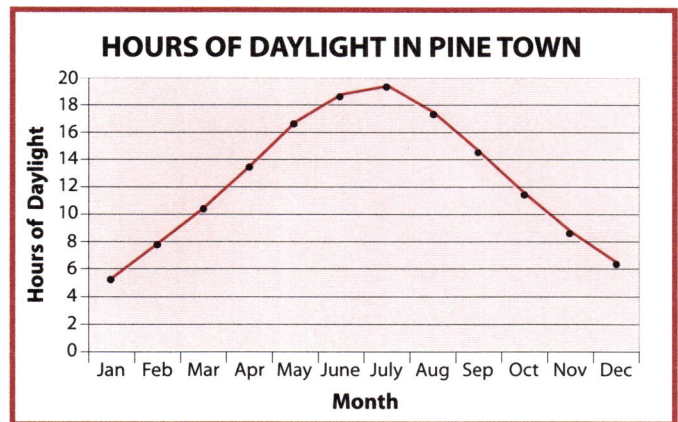

21 During which months are there more than 17 hours of sunlight?
   A   January, February, March
   B   March, April, May
   C   June, July, August
   D   September, October
   E   November, December

22 Rounded to the nearest hour, what is the approximate difference between the number of daylight hours in August and the number of daylight hours in December?
   A   8
   B   9
   C   10
   D   11
   E   12

23 Which month has about double the hours of daylight as the month of January?
   A   February
   B   March
   C   April
   D   November
   E   December

# Circle Graphs

## ① Learn the Skill

Like bar and line graphs, circle graphs show data visually. Whereas a line graph shows how data changes over time, a **circle graph** shows how parts compare to a whole. For example, a circle graph of sales from each department in a store can show at a glance the most productive department, as well as how each department's sales compares to that of the whole store.

Values of the circle graph sections may be expressed as percents, decimals, fractions, or whole numbers. You may need to convert from one form to another.

## ② Practice the Skill

By mastering the skill of interpreting circle graphs, you will improve your study and test-taking skills, as well as your understanding of percents, decimals, and fractions. Read the example and strategies below. Use this information to answer question 1.

**A** Some circle graphs, such as this one, are labeled with only categories, rather than categories and percent. In whatever manner a circle graph is labeled, the whole circle represents 1, or 100%. If a section is half of the circle, it represents 50%, 0.50, or ½. If a section is a quarter of the circle, it represents 25%, 0.25, or ¼.

**B** Use a category's size to estimate its value. Notice that car maintenance and gasoline each represent about a quarter of the whole, or 25%. This helps you estimate the percentages of other categories.

Jerry creates a circle graph showing his monthly budget.

**JERRY'S BUDGET**

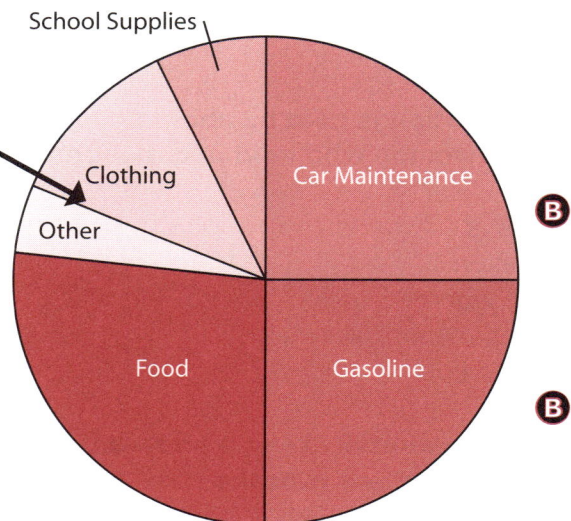

### ✓ TEST-TAKING TIPS

Since the section for food is larger than the section for gasoline, you can estimate that Jerry budgets more than 25% for food. So, answer choices A, B, and C can be eliminated right away.

**1** **Approximately what percentage per month does Jerry budget for food?**
   A   10%
   B   15%
   C   20%
   D   30%
   E   45%

## ③ Apply the Skill

**Directions:** Questions 2 and 3 are based on the information below.

The circle graph below shows the methods of transportation that employees use to get to work.

**HOW EMPLOYEES GET TO WORK**

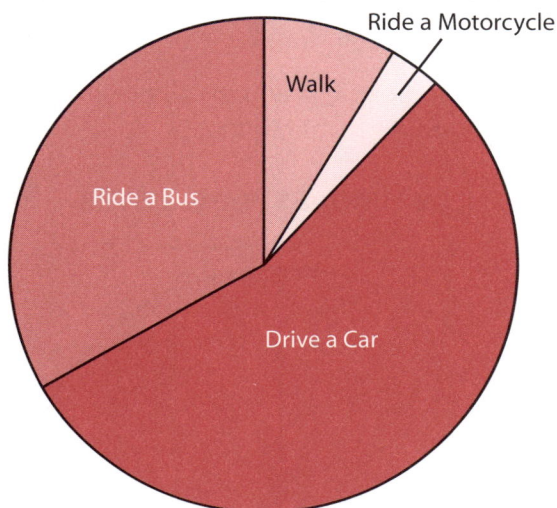

2   **About what part of the employee population drives a car to work?**
   A   20%
   B   25%
   C   30%
   D   50%
   E   60%

3   **Which of the following events would likely result in an increase in the amount of people who walk or ride a bus to work each day?**
   A   a decrease in road construction
   B   a sharp rise in gasoline prices
   C   greater affordability of hybrid vehicles
   D   lower prices from motorcycle manufacturers
   E   wearing down of infrastructure, such as sidewalks

**Directions:** Questions 4 through 6 are based on the information below.

A library creates a circle graph of the types of books checked out by readers in September.

**WHAT PEOPLE READ IN SEPTEMBER**

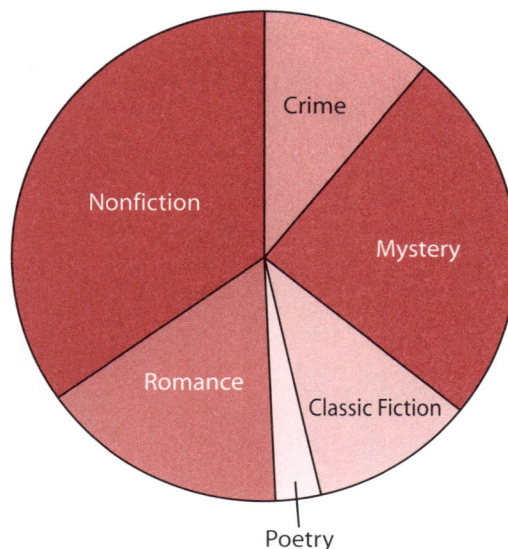

4   **Which category of books has fewer readers than crime books?**
   A   mystery
   B   classic fiction
   C   romance
   D   nonfiction
   E   poetry

5   **Which categories of books could a librarian make the best argument to order in August?**
   A   nonfiction and mystery
   B   mystery and romance
   C   nonfiction and crime
   D   romance and crime
   E   classic fiction and poetry

6   **If 30,000 books were checked out in September, approximately how many were romance?**
   A   18,000
   B   15,000
   C   7,500
   D   4,500
   E   2,000

**Directions:** Questions 7 through 9 are based on the information below.

As part of a civics project, Randall created a circle graph showing the most commonly used heating fuels in Smallsburg.

**SOURCES OF HEATING FUEL IN SMALLSBURG**

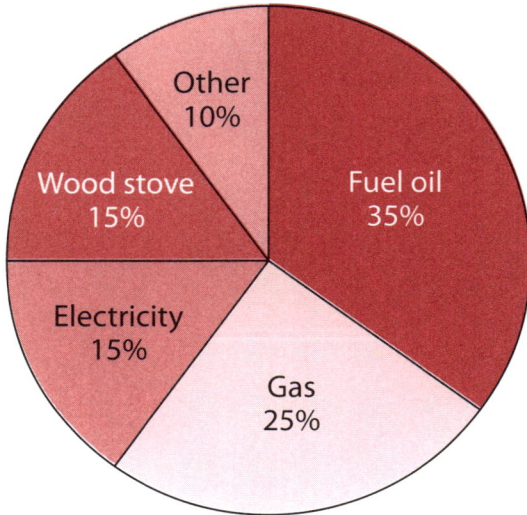

**7** Which two sources of fuel together represent more than 50% of the total commonly used heating fuels?

A   gas and wood stove
B   fuel oil and electricity
C   wood stove and fuel oil
D   fuel oil and gas
E   gas and electricity

**8** Which two sources of fuel are used by the same percentage of the population?

A   wood stove and gas
B   electricity and other
C   fuel oil and gas
D   gas and electricity
E   electricity and wood stove

**9** What percentage of the population uses a source other than gas?

A   25%
B   35%
C   50%
D   75%
E   Not enough information is given.

**Directions:** Questions 10 and 11 are based on the information below.

The following circle graph shows the voting habits of the residents of Middlesburg.

**HOW MIDDLESBURG VOTES**

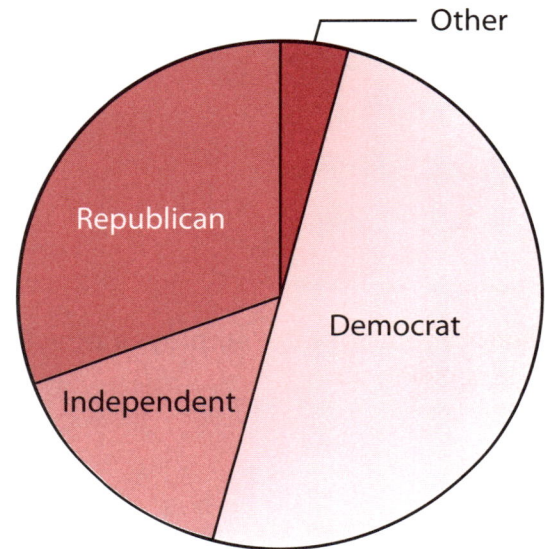

**10** For which party did half of the population vote?

A   Republican
B   Independent
C   Democrat
D   Other
E   Not enough information is given.

**11** About what percentage of the population voted for either Independent or Republican candidates?

A   15%
B   30%
C   35%
D   45%
E   50%

**Directions:** Questions 12 through 14 are based on the information below.

Tom asked 100 of his friends and family members to name their favorite form of exercise. The circle graph below displays his findings.

**FAVORITE WAY TO EXERCISE**

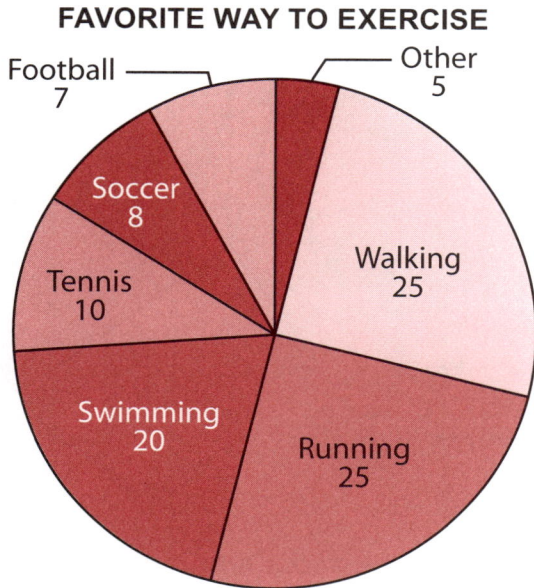

**12** **Which statement about the circle graph is accurate?**

- **A** Less than 20 percent of people prefer walking.
- **B** The same number of people prefer soccer as prefer swimming.
- **C** About half of the people prefer either running or walking.
- **D** Fewer than 10 people fall into the category "swimming."
- **E** About 70 people prefer an exercise other than swimming.

**13** **About what fraction of people prefer an exercise other than running?**

- **A** $\frac{1}{4}$
- **B** $\frac{1}{2}$
- **C** $\frac{3}{4}$
- **D** $\frac{4}{5}$
- **E** $\frac{9}{10}$

**14** **About what percentage of the people prefer skiing?**

- **A** more than 4%
- **B** more than 10%
- **C** more than 25%
- **D** more than 50%
- **E** Not enough information is given.

**Directions:** Questions 15 and 16 are based on the information below.

The circle graph shows the percentages of tree species in a state park.

**TREE SPECIES**

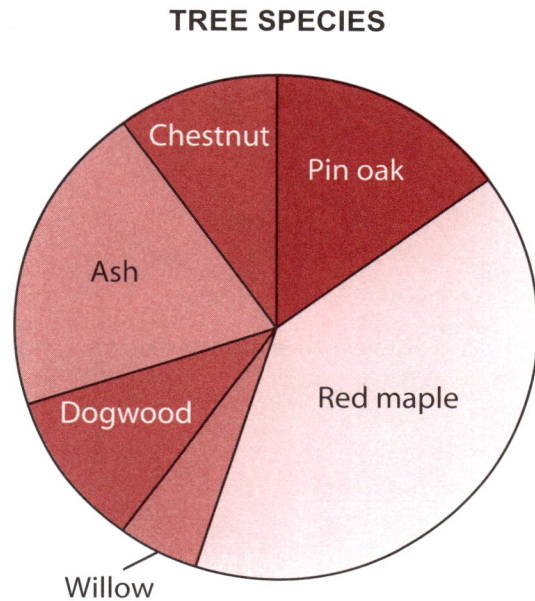

**15** **What percentage of trees are red maples?**

- **A** 5%
- **B** 10%
- **C** 20%
- **D** 30%
- **E** 40%

**16** **If there were 300 trees in the park, how many red maple trees would you expect there to be?**

- **A** 40
- **B** 80
- **C** 120
- **D** 160
- **E** 200

The circle graph shows the languages spoken by students at the Marbletown International High School.

**LANGUAGES SPOKEN BY STUDENTS**

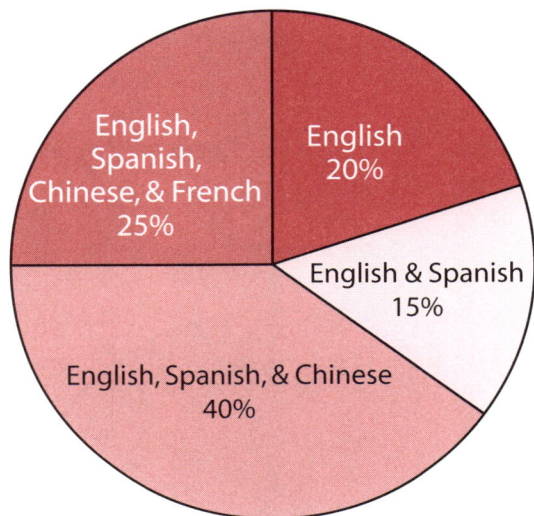

17 **What percentage of students speak more than two languages?**
   A   15%
   B   20%
   C   25%
   D   40%
   E   65%

18 **What percentage of students do not speak Chinese?**
   A   15%
   B   20%
   C   30%
   D   35%
   E   65%

19 **Which of the following statements is accurate?**
   A   Half of the students speak French.
   B   One-fourth of the students speak Spanish.
   C   All students speak English.
   D   Less than half of the students speak two or more languages.
   E   More than half of the students speak exactly two languages.

**DIEGO'S MONTHLY BUDGET**

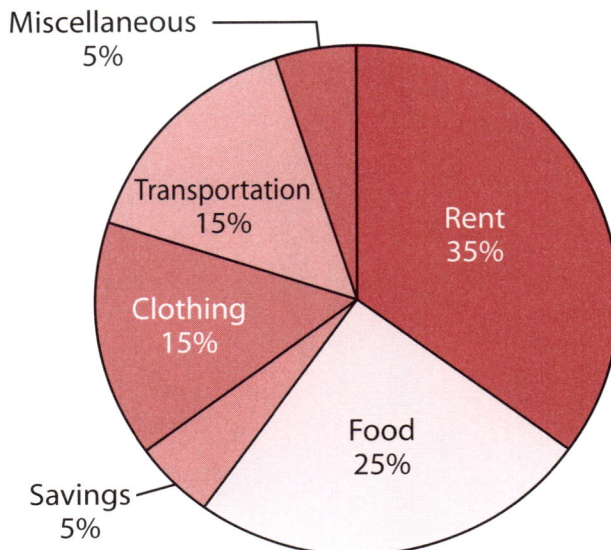

20 **For every $100, about how much should Diego spend on food?**
   A   $5
   B   $15
   C   $25
   D   $35
   E   Not enough information is given.

21 **If Diego earns $2,200 this month, how much should he put into savings?**
   A   $5
   B   $50
   C   $100
   D   $110
   E   $1,100

22 **Which of the following statements is accurate?**
   A   Diego spends most of his monthly budget on food.
   B   Diego spends more on transportation and miscellaneous expenses than on food.
   C   Diego spends more on transportation than on clothing costs.
   D   Diego saves more than 10% of his monthly income.
   E   Food and rent make up the largest percentage of Diego's monthly expenses.

UNIT 3

# Frequency Tables and Histograms

## 1 Learn the Skill

As you learned in Lesson 4, **tables** organize data so that information is easier to identify and compare. On mathematics high school equivalency tests, you will be asked to interpret and compare data from different kinds of tables. These might include **frequency tables**, which are explained below. **Histograms** are made up of adjoining bars of equal width. A histogram's bars have lengths that correspond to an associated scale. Histograms may be used with any size data set and are used to show frequency.

Tables can contain more information than you want or need in solving a problem. When taking the mathematics high school equivalency test, carefully read questions involving tables to identify precisely the information that you need in solving a problem.

## 2 Practice the Skill

By mastering the skill of using tables, you will improve your study and test-taking skills, especially as they relate to mathematics high school equivalency tests. Read the example and strategies below. Use this information to answer question 1.

**A** The number of data values for given intervals is known as the **frequency**. A frequency table presents the exact number of data values for a given interval. This frequency table displays five intervals.

**B** The table enables you to quickly eliminate certain answer choices. You can see at a glance that the data for the first and last frequencies show the least amount in the table. The sharp decline in frequency for the 46–55 age group also helps you eliminate that interval as a possible answer choice.

Advertisers are interested in knowing when people of various ages watch television. Such information helps advertisers know which products to advertise at certain times of the day.

### NIGHTTIME TELEVISION VIEWERS BY AGE GROUP

| Age Group | Frequency | |
|---|---|---|
| | 7 P.M.–9 P.M. | 9 P.M.–11 P.M. |
| 16–25 | 355 | 790 |
| 26–35 | 1,047 | 1,532 |
| 36–45 | 1,212 | 1,519 |
| 46–55 | 1,357 | 399 |
| 56–65 | 887 | 352 |

### TEST-TAKING TIPS

Before answering questions about frequency tables or histograms, be sure to read the title and column headings. They contain important information that will help you understand what the data represent.

1  **Which age group would advertisers most likely want to target between the hours of 7 P.M. and 11 P.M.?**
   A  16–25
   B  26–35
   C  36–45
   D  46–55
   E  56–65

**Directions:** Questions 2 through 5 are based on the information below.

The Northside Bowling Club recorded its nightly scores in the frequency table below. Each bowler bowled one game.

| BOWLING SCORES | |
|---|---|
| Score | Frequency |
| 100–119 | 3 |
| 120–139 | 5 |
| 140–159 | 8 |
| 160–179 | 9 |
| 180–199 | 5 |

**2** How many bowlers scored between 100 and 139?

A 3
B 5
C 8
D 13
E Not enough information is given.

**3** What score range had the highest frequency?

A 100–119
B 120–139
C 140–159
D 160–179
E 180–199

**4** How many bowlers scored between 157 and 189?

A 1
B 3
C 8
D 9
E Not enough information is given.

**5** What percentage of bowlers scored between 160 and 179?

A 10%
B 20%
C 30%
D 33%
E 40%

**Directions:** Questions 6 through 9 are based on the information below.

A survey was taken of Mr. Macon's students to find the number of pets each has owned.

| PET OWNERSHIP | |
|---|---|
| Number of Pets | Frequency |
| 1 | II |
| 2 | IIII I |
| 3 | IIII II |
| 4 | II |
| 5 | I |

**6** What is the mode for the number of pets owned?

A 1
B 2
C 3
D 4
E 5

**7** What is the median number of pets owned?

A 1
B 2
C 3
D 5
E Not enough information given.

**8** What is the mean number of pets owned?

A 2.7
B 3.0
C 3.4
D 4.0
E 9.2

**9** If one more student in the class had 5 pets, what would be the mean?

A 2.7
B 2.83
C 3.0
D 4.0
E 5.0

**Directions:** Questions 10 and 11 are based on the information below.

A new prime-time animated television comedy recently aired its first episode. Although targeted to young adults, the new show excited network officials because of its broader appeal across age groups. The histogram below illustrates the new show's ratings popularity—as measured in millions of viewers—across various age groups.

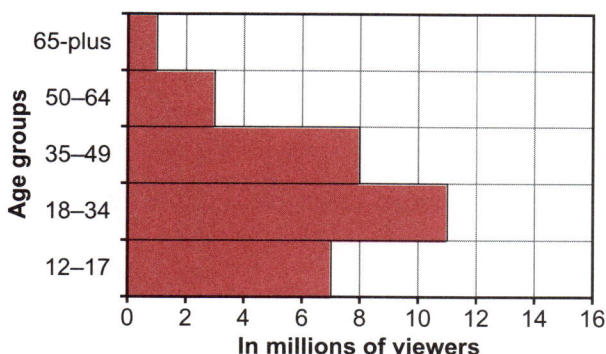

10 **What pattern can you identify based on the viewership data?**
   A   The show was equally popular among all age groups.
   B   The show was most popular among the 35–49 age group.
   C   The show enjoyed high ratings among teens and adults under 50.
   D   The show had little popularity among young viewers.
   E   The show had the highest popularity among viewers over 65.

11 **What is the median age in the age group with the most viewers of the show?**
   A   15
   B   26
   C   42
   D   57
   E   65

**Directions:** Questions 12 and 13 are based on the information below.

A swim club is open for 24 weeks from the middle of April through the middle of September. Monthly visitor totals are plotted below.

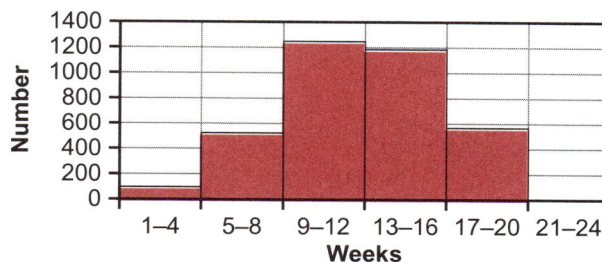

12 **In which weeks did the greatest number of people visit the swim club?**
   A   weeks 1–4 and 5–8
   B   weeks 9–12 and 13–16
   C   weeks 13–16 and 17–20
   D   weeks 17–20 and 21–24
   E   weeks 1–4 and 21–24

13 **During which weeks should the operators of the pool consider closing it?**
   A   weeks 1–4
   B   weeks 5–8
   C   weeks 9–12
   D   weeks 17–20
   E   weeks 21–24

14 **About how many weeks of the year could be classified as "very active pool season"?**
   A   2
   B   4
   C   6
   D   8
   E   12

**Directions:** Questions 15 and 16 are based on the information below.

A company that makes steel rods has one particular stock item that is supposed to be 60.0 cm long. The lengths of a random sample of 53 rods are measured, and the results are tabulated in the following histogram.

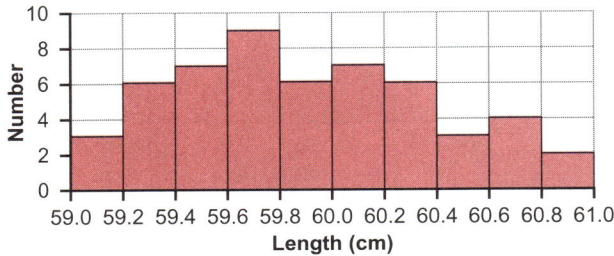

**15** The company decides to reject rods with lengths that fall outside the range of 59.4 cm to 60.6 cm. Based on this sample, roughly what percentage of rods will be rejected?
   A   9.4%
   B   17.0%
   C   28.3%
   D   47.2%
   E   60.8%

**16** How many rods in the sample are too short to meet the company's standard for acceptable length?
   A   3
   B   6
   C   9
   D   12
   E   15

**Directions:** Questions 17 and 18 are based on the information below.

The following histogram lists the average number of viewers of a new movie, grouped by age.

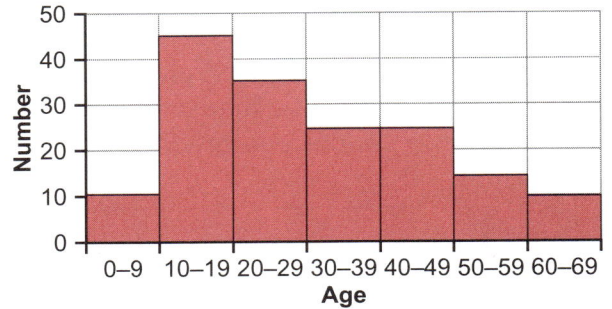

**17** What is the mode age of viewers for the new movie?
   A   10 to 19
   B   20 to 29
   C   30 to 39
   D   40 to 49
   E   0 to 69

**18** To which group is the movie most likely marketed?
   A   parents with young children
   B   teens
   C   young couples without children
   D   mature audiences
   E   senior citizens

**Directions:** Questions 19 through 21 are based on the information below.

The teacher of a keyboarding class tested all the students in the class on the number of words they could type in a minute. The teacher recorded the data in the histogram below.

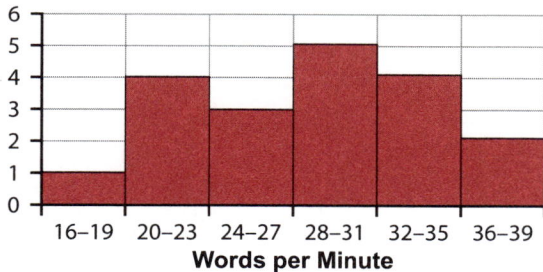

**Words per Minute**

**19 How many students are in the class?**
- A  5
- B  6
- C  20
- D  31
- E  39

**20 What is the mode of words typed per minute in the keyboarding class?**
- A  20 to 23
- B  24 to 27
- C  28 to 31
- D  32 to 35
- E  16 to 39

**21 How many students are meeting the goal of typing at least 28 words per minute?**
- A  5
- B  6
- C  7
- D  11
- E  20

**22 Which of the following statements about frequency tables and histograms is true?**
- A  A frequency table is a more useful tool than a histogram.
- B  A histogram is a more useful tool than a frequency table.
- C  A frequency table displays data while a histogram does not.
- D  A histogram displays data while a frequency table does not.
- E  A histogram visually displays data with bars while a frequency table generally reports numbers or occurrences.

**23 Which of the following statements about histograms is false?**
- A  Histograms are made up of adjoining bars of equal width.
- B  The length of a histogram's bars corresponds to a scale.
- C  Histograms are used to show frequency.
- D  Histograms can only be used to show a small data set.
- E  Histograms can be oriented vertically or horizontally.

**24 Which scenario would best be displayed in a frequency table?**
- A  total monthly sales of a bookstore
- B  daily temperature changes in a coastal city in January
- C  numbers of students who walk or ride the bus each month of the school year
- D  running times for cross-country team members in a single race
- E  gas price increases in April

# Dot Plots and Box Plots

## ① Learn the Skill

**Dot plots** provide an easy way of organizing sets of data with modest numbers of values (e.g., those less than 50). They consist of a number line on which each occurrence of a value is noted by a dot; the number of dots associated with each value indicates the frequency of that value in the data set.

**Box plots** are a convenient way of showing and comparing sets of numerical data using five characteristics of each data set: the median value, the lower (25%) and upper (75%) quartile values, and the maximum and minimum values. Both of these methods of displaying data allow you to visually identify whether the distribution of values is normal or skewed. In a normal distribution of data points, the mean, median, and mode coincide, and the display looks symmetrical. When the mean is greater than or less than the mode, the distribution is skewed and the display appears "lopsided."

## ② Practice the Skill

Read the example and strategies below. Use this information to answer question 1.

**A** A dot plot contains detailed information about a data set and allows for determination of quantities such as mean, mode, and range. For example, since there is an odd number of students (33), the median score value will be the 17th value (8.0), counting in from either end. This appears as the mid-line in the box plot.

**B** Since the median value is an actual data point, that data point is not considered to be part of the upper or lower halves of the data set. The **lower quartile** is the median of the lower half of the data set. The **upper quartile** is the median of the upper half of the data set. Since there are 16 points in each half, the quartile values will be half-way between the 8th and 9th data values. In the case of the upper quartile, both values are 9; the upper quartile value then is 9.0. This appears as the upper bound of the box in the box plot.

A class of 33 students takes a 10-point quiz. The following dot plot (top) and box plot (bottom) represent the distribution of student scores.

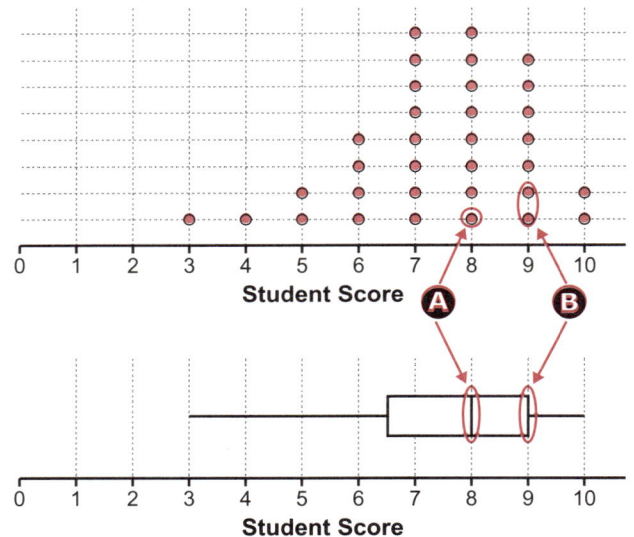

### ✓ TEST-TAKING TIPS

The circled points were found by counting dots from left to right, starting at the top of each column. You can check your work by counting from right to left, also starting at the top of each column of dots.

**1** Using the dot plot, what is the lower quartile student score for this quiz?

A  6.0
B  6.5
C  7.0
D  7.5
E  8.0

## ③ Apply the Skill

**Directions:** Questions 2 through 5 are based on the information below.

A sleep study is conducted on 40 people for one week. The average number of hours of sleep per night per subject is rounded to the nearest hour. The tabulated results are shown in the following dot plot.

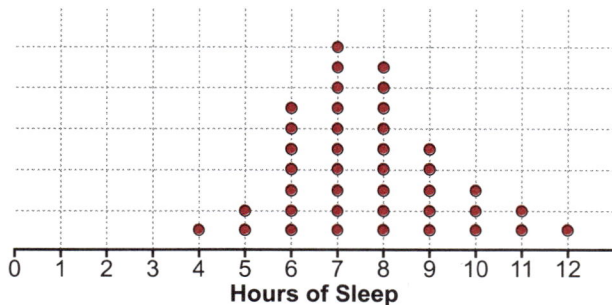

**Hours of Sleep**

**2** What is the median value of the hours of sleep reported in this study?
  A  6.5 hours
  B  7.0 hours
  C  7.5 hours
  D  8.0 hours
  E  8.5 hours

**3** What is the mode value of the distribution?
  A  6.5 hours
  B  7.0 hours
  C  7.5 hours
  D  8.0 hours
  E  8.5 hours

**4** What is the range of the distribution?
  A  4 hours
  B  5 hours
  C  7 hours
  D  8 hours
  E  9 hours

**5** How many subjects had 9 hours of sleep?
  A  5
  B  6
  C  7
  D  9
  E  10

**Directions:** Questions 6 through 8 are based on the information below.

A class of 16 students is asked to report on the number of brothers and sisters they each have. The number of siblings is plotted on the following dot plot.

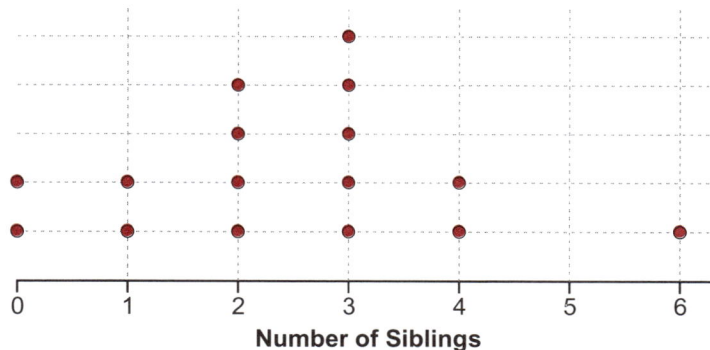

**Number of Siblings**

**6** The mode of the distribution corresponds to what number of siblings?
  A  2.0
  B  2.5
  C  3.0
  D  5.0
  E  6.0

**7** What is the median number of siblings?
  A  2.0
  B  2.5
  C  3.0
  D  5.0
  E  6.0

**8** How does the mean number of siblings relate to the mode in this distribution?
  A  It is equal to the mode.
  B  It is greater than the mode.
  C  It is less than the mode.
  D  It is twice the value of the mode.
  E  It is half the mode.

**Directions:** Questions 9 through 12 are based on the information below.

Two 4-sided dice are tossed a total of 30 times. The dot plot below shows the sum of the two dice for each toss.

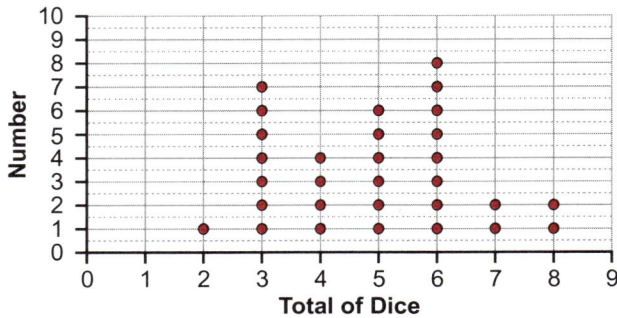

9  What is the median of the number of dice rolls?

A  4.5
B  5
C  5.5
D  6
E  6.5

10  What total of the dice represents the mode value?

A  6
B  5
C  4
D  3
E  2

11  What is the minimum value?

A  1
B  2
C  3
D  6
E  8

12  If a very large number of dice tosses were performed, what would the expected mode value be?

A  3
B  4
C  5
D  6
E  7

**Directions:** Questions 13 and 14 are based on the information below.

Four chefs' dishes are judged by customers on a scale of 0 to 10. Box plots representing the scores each chef received are presented below.

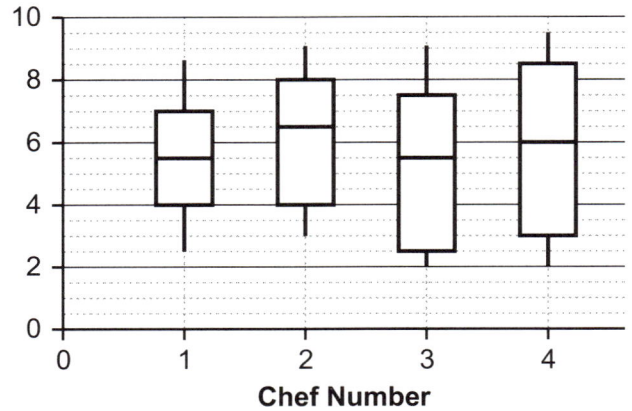

13  The chef with the highest median score is the winner. What is the number of the winning chef?

A  1
B  2
C  3
D  4
E  Chefs 2 and 4 tie.

14  If the chefs were judged based on who had the highest upper quartile, which chef would have won?

A  1
B  2
C  3
D  4
E  Chefs 2 and 4 tie.

**Directions:** Questions 15 through 17 are based on the information below.

Thomas has taken eight 10-point quizzes in his science class. His scores are as follows: 7, 8, 9, 6, 8, 7, 10, 6. The dot plot shows his scores

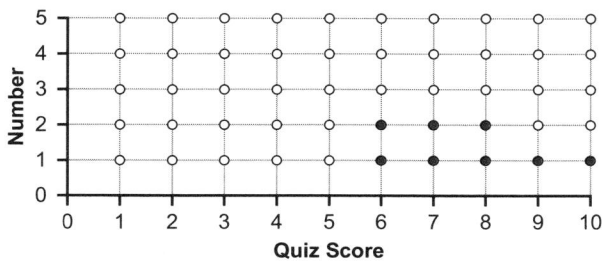

**15** What is Thomas's mean quiz score?
   A 7.025
   B 7.500
   C 7.625
   D 7.750
   E 8.225

**16** After taking the eight quizzes, Thomas has two 10-point quizzes remaining. Determine the minimum scores Thomas needs on the final two quizzes to achieve a mean, or average, score of at least 8.0.
   A 8 and 8
   B 8 and 9
   C 9 and 9
   D 9 and 10
   E 10 and 10

**17** With two additional test scores that give Thomas a total of 80 points for all ten quizzes, what is the mode of his test scores?
   A 8
   B 6, 7, and 8
   C 7, 8, and 9
   D 9 and 10
   E The mode cannot be described.

**Directions:** Questions 18 through 20 are based on the information below.

A teacher's class roster shows her students have the following ages.

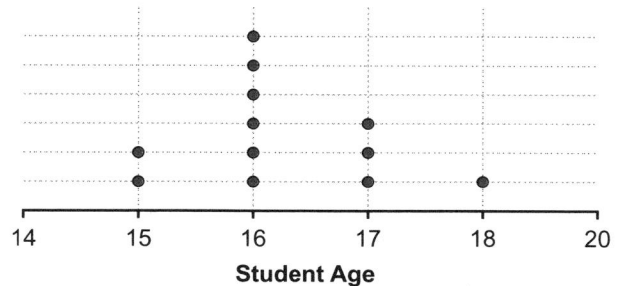

**18** What is the range of the data?
   A 3
   B 4
   C 6
   D 16
   E 18

**19** What is the mean age of the students?
   A 16 years
   B 16 years 3 months
   C 16 years 6 months
   D 16 years 9 months
   E 17 years

**20** Which statement is true about this distribution?
   A The distribution is normal.
   B The distribution is skewed because the highest value occurs with the lowest frequency.
   C The distribution is positively skewed because the mean is greater than the mode.
   D The distribution is negatively skewed because the mean is less than the mode.
   E The distribution is negatively skewed because the mode is less than the median.

**Directions:** Questions 21 through 24 are based on the information below.

Four thermostats placed in similar rooms are tested by setting the temperatures at 68°, letting the temperature settle for an hour, and then tabulating the temperature once every minute for an hour. The results are shown in the box plots below.

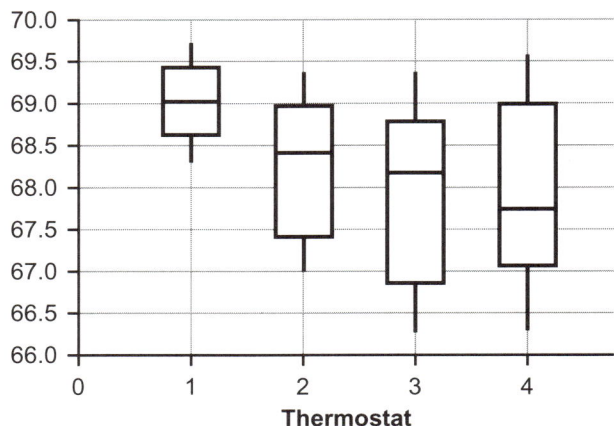

**Thermostat**

**21** Which of the four thermostats does the best job minimizing the amount that the temperature varies?

 A  1
 B  2
 C  3
 D  4
 E  Not enough information is given.

**22** Which thermostat maintains a median temperature closest to the set value?

 A  1
 B  2
 C  3
 D  4
 E  Not enough information is given.

**23** Which thermostat exhibits the greatest range in temperature?

 A  1
 B  2
 C  3
 D  4
 E  Not enough information is given.

**24** Which thermostat exhibits the smallest range in temperature?

 A  1
 B  2
 C  3
 D  4
 E  Not enough information is given.

**Directions:** Questions 25 and 26 are based on the information below.

The 19 students in a keyboarding class are tested on the number of words they can type in a minute. Their results appear in the dot plot below.

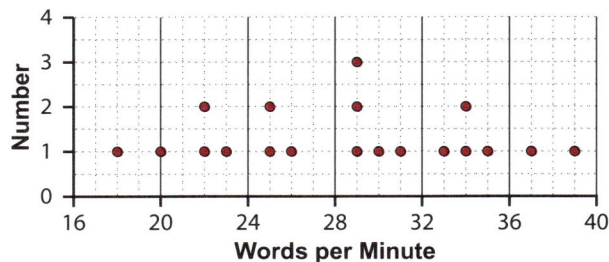

**Words per Minute**

**25** What is the median number of words per minute that students in the class can type?

 A  25
 B  26
 C  29
 D  31
 E  34

**26** What is the range of words typed per minute by students in the class?

 A  15
 B  17
 C  19
 D  21
 E  22

UNIT 3

The Unit Review is structured to resemble mathematics high school equivalency tests. Be sure to read each question and all possible answers very carefully before choosing your answer.

To record your answers, fill in the lettered circle that corresponds to the answer you select for each question in the Unit Review.

Do not rest your pencil on the answer area while considering your answer. Make no stray or unnecessary marks. If you change an answer, erase your first mark completely.

Mark only one answer choice for each question; multiple answers will be scored as incorrect.

## Sample Question

An artist needs 840 feet of ribbon for an outdoor work of art. How many yards of ribbon does the artist want to use?

A  70 yd
B  280 yd
C  300 yd
D  350 yd
E  2,520 yd

Ⓐ⬤ⒸⒹⒺ

---

**Directions:** Questions 1 through 3 are based on the information below.

A cable television company asks a family to keep track of the number of hours they spend watching television. They record weekly data for two months.

| WEEKLY TELEVISION VIEWING | |
|---|---|
| Week | Hours Watched |
| 1 | 21.5 |
| 2 | 28.0 |
| 3 | 15.5 |
| 4 | 23.0 |
| 5 | 29.0 |
| 6 | 34.0 |
| 7 | 27.0 |
| 8 | 35.0 |

1  To the nearest tenth, what is the mean number of hours watched each week by the family?

A  26.0
B  26.6
C  27.0
D  27.5
E  28.0

Ⓐ Ⓑ Ⓒ Ⓓ Ⓔ

2  Which statement best describes the mean and the median?

A  The median is slightly greater than the mean.
B  The median and mean are equal.
C  The mean is slightly greater than the median.
D  The mean is significantly greater than the median.
E  The median is significantly greater than the mean.

Ⓐ Ⓑ Ⓒ Ⓓ Ⓔ

3  What is the range of the weekly hours watched for the family?

A  15.5
B  18.5
C  19.5
D  21.5
E  35.0

Ⓐ Ⓑ Ⓒ Ⓓ Ⓔ

**Directions:** Questions 4 through 6 are based on the information below.

The die has one of the digits 1 through 6 on each side.

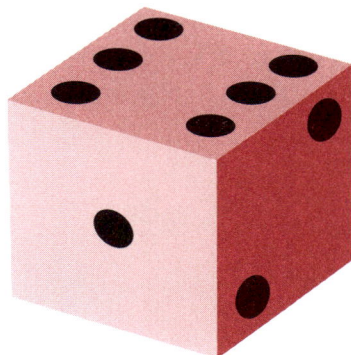

4   What is the probability of rolling an even number?

A   25%
B   33%
C   50%
D   66%
E   75%

Ⓐ Ⓑ Ⓒ Ⓓ Ⓔ

5   What is the probability of rolling a 2 or a 4?

A   25%
B   33.$\overline{3}$%
C   50%
D   66.$\overline{6}$%
E   75%

Ⓐ Ⓑ Ⓒ Ⓓ Ⓔ

6   A pair of die are rolled together. What is the probability, expressed as a fraction, of a roll that totals 2 or 4?

A   $\frac{1}{18}$
B   $\frac{1}{9}$
C   $\frac{1}{6}$
D   $\frac{1}{3}$
E   $\frac{1}{2}$

Ⓐ Ⓑ Ⓒ Ⓓ Ⓔ

**Directions:** Questions 7 and 8 are based on the information below.

A company keeps track of the bonuses its employees receive each year.

**COMPANY BONUSES: 2003–2008**

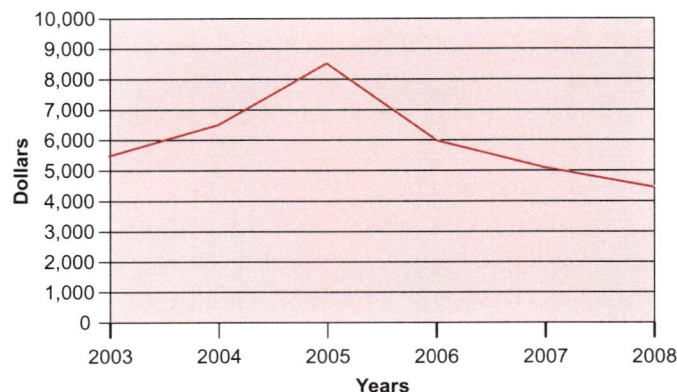

7   Between which two years did the amount of bonuses awarded show the greatest increase?

A   2003–2004
B   2004–2005
C   2005–2006
D   2006–2007
E   2007–2008

Ⓐ Ⓑ Ⓒ Ⓓ Ⓔ

8   During which year was the amount of bonuses awarded less than $5,000?

A   2004
B   2005
C   2006
D   2007
E   2008

Ⓐ Ⓑ Ⓒ Ⓓ Ⓔ

**9** A single six-sided die is rolled, and two coins are tossed at the same time. What is the probability, expressed as a fraction, of coming up with a three on the die, and two heads on the coins?

A $\frac{1}{4}$

B $\frac{1}{6}$

C $\frac{1}{10}$

D $\frac{1}{24}$

E $\frac{1}{32}$

ⒶⒷⒸⒹⒺ

**10** There are two green boxes and two red boxes. How many possible ways are there of arranging the boxes in a row?

A 4

B 6

C 12

D 24

E 36

ⒶⒷⒸⒹⒺ

**11** If three six-sided dice are thrown, what is the probability, expressed as a fraction, that they will all come up with the same number?

A $\frac{1}{6}$

B $\frac{1}{12}$

C $\frac{1}{36}$

D $\frac{1}{64}$

E $\frac{1}{216}$

ⒶⒷⒸⒹⒺ

**12** A pair of coins have been tossed two times and come up both heads each time. What is the probability, in percent, that the next time they are tossed, they will come up both heads?

A 1.56%

B 6.25%

C 12.5%

D 25%

E 50%

ⒶⒷⒸⒹⒺ

**Directions:** Questions 13 and 14 are based on the information below.

The most popular game at a carnival is *Spin for Fortune*, shown below. The "Sorry" outcome means that the player does not win a prize.

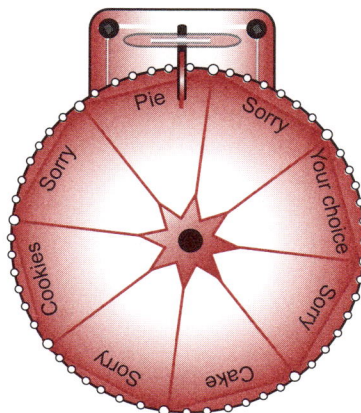

**13** What are the chances, expressed as a decimal, that a player spinning the wheel will win a prize?

A 0.5

B 0.375

C 0.25

D 0.125

E 0.05

ⒶⒷⒸⒹⒺ

**14** Suppose a player really wants to win a cake. What are the chances of that player winning a cake, expressed in percent?

A 50%

B 37.5%

C 25%

D 12.5%

E 8%

ⒶⒷⒸⒹⒺ

**Directions:** Questions 15 through 18 are based on the information below.

A public radio station is having a week-long fund drive. The results for the first five days are shown below.

| RADIO STATION FUNDRAISER | |
|---|---|
| **Fund Drive Results** | |
| Monday | $5,400 |
| Tuesday | $6,200 |
| Wednesday | $4,900 |
| Thursday | $4,400 |
| Friday | $7,600 |
| Saturday | |
| Sunday | |

**15** What is the range of the daily fund-drive results so far?

A $1,600
B $3,200
C $4,400
D $6,200
E $7,600

Ⓐ Ⓑ Ⓒ Ⓓ Ⓔ

**16** What is the daily mean of the fund-drive results for the first five days?

A $4,750
B $5,400
C $5,700
D $7,125
E $7,600

Ⓐ Ⓑ Ⓒ Ⓓ Ⓔ

**17** What is the median of the fund-drive result for the first five days?

A $4,400
B $4,900
C $5,400
D $6,200
E $7,600

Ⓐ Ⓑ Ⓒ Ⓓ Ⓔ

**18** If the goal of the fund drive is $45,000, what must be the mean of the funds raised on Saturday and on Sunday?

A $16,500
B $8,250
C $6,430
D $5,110
E $4,400

Ⓐ Ⓑ Ⓒ Ⓓ Ⓔ

**19** The following table lists what Morgan, Tom, and Dana scored on each hole of a 9-hole miniature golf course.

| SCORE FOR EACH HOLE | | | | | | | | | |
|---|---|---|---|---|---|---|---|---|---|
| **Hole** | **1** | **2** | **3** | **4** | **5** | **6** | **7** | **8** | **9** |
| Morgan | 4 | 3 | 6 | 4 | 2 | 4 | 2 | 3 | 4 |
| Tom | 3 | 2 | 4 | 3 | 5 | 4 | 2 | 5 | 3 |
| Dana | 5 | 4 | 5 | 3 | 4 | 3 | 2 | 2 | 1 |

Considering the scores of all three players together, what is the mode of the score distribution?

A 2
B 3
C 4
D 5
E 6

Ⓐ Ⓑ Ⓒ Ⓓ Ⓔ

**20** Suppose one is given four cards, numbered 1 through 4. The cards are shuffled and dealt one at a time. What is the probability, expressed as a fraction, that they will be dealt in order: 1 first, then 2, then 3, and then 4?

A $\frac{1}{4}$

B $\frac{1}{6}$

C $\frac{1}{8}$

D $\frac{1}{12}$

E $\frac{1}{24}$

Ⓐ Ⓑ Ⓒ Ⓓ Ⓔ

**Directions:** Questions 21 and 22 are based on the information below.

The heights of three people during the first 20 years of their lives are plotted in the graph below.

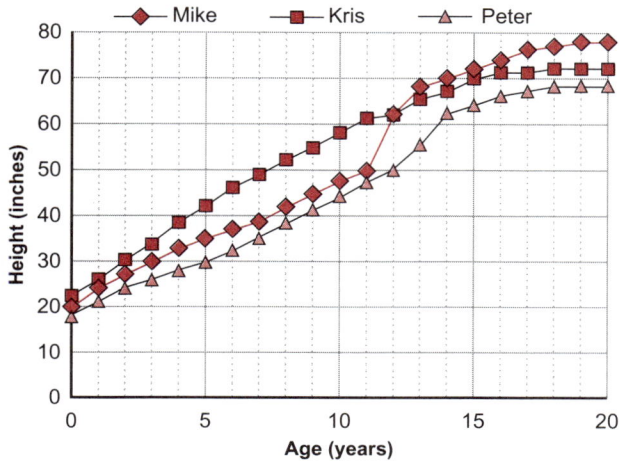

**21 What is the order of the three people at age 10, from greatest to least height?**

A   Mike, Kris, Peter
B   Mike, Peter, Kris
C   Kris, Mike, Peter
D   Kris, Peter, Mike
E   Peter, Kris, Mike

Ⓐ Ⓑ Ⓒ Ⓓ Ⓔ

**22 Which of the people experience the greatest change in height in one year, and at what age does that occur?**

A   Peter in the year following his 13th birthday
B   Mike in the year following his 11th birthday
C   Kris in the year following her 10th birthday
D   Kris in the year following her 11th birthday
E   Peter in the year following his 15th birthday.

Ⓐ Ⓑ Ⓒ Ⓓ Ⓔ

**23 Twenty-six people signed up for the Throwing Darts Tournament. Each match is between two people. How many possible combinations are there for the first round of matches?**

A   1,300
B   676
C   650
D   325
E   13

Ⓐ Ⓑ Ⓒ Ⓓ Ⓔ

**24 Which expression represents how many different ways you can lay down seven cards from a 52 card deck?**

A   $52^7$
B   $52!$
C   $\dfrac{52 \times 51 \times 50 \times 49 \times 48 \times 47 \times 46}{1 \times 2 \times 3 \times 4 \times 5 \times 6 \times 7}$
D   $7^{52}$
E   $7!$

Ⓐ Ⓑ Ⓒ Ⓓ Ⓔ

**25 The speed limits on three roads change at the county line. The speed limits are represented in the data tables.**

| FOX COUNTY | |
|---|---|
| **Street** | **Speed Limit** |
| Main | 25 mph |
| Railroad | 45 mph |
| Chestnut | 45 mph |

| BEAR COUNTY | |
|---|---|
| **Street** | **Speed Limit** |
| Main | 35 mph |
| Railroad | 35 mph |
| Chestnut | 55 mph |

**Which matrix represents the difference between the Bear County data table and the Fox County data table?**

A   $\begin{bmatrix} 10 \\ 10 \\ 10 \end{bmatrix}$   B   $\begin{bmatrix} -10 \\ 10 \\ -10 \end{bmatrix}$

C   $\begin{bmatrix} 10 \\ -10 \\ 10 \end{bmatrix}$   D   $\begin{bmatrix} -10 \\ -10 \\ 10 \end{bmatrix}$

E   $\begin{bmatrix} -10 \\ 10 \\ -10 \end{bmatrix}$

Ⓐ Ⓑ Ⓒ Ⓓ Ⓔ

# Unit 4

## Unit Overview

Algebra builds upon the core areas of mathematics, such as number sense and data measurement and analysis, by translating everyday situations into mathematical language. We use algebra to solve complex problems and to explore more sophisticated areas of mathematics. Certain jobs, such as those in high-tech fields, require strong backgrounds in algebra and other forms of higher mathematics.

As with other subject areas, algebraic concepts make up about 25 percent of questions on high school equivalency mathematics tests. In Unit 4, you will study equations, patterns, slopes and intercepts, factoring, and other skills that will help you prepare for the test.

## Table of Contents

## Key Terms

**algebraic expression:** a mathematical phrase made up of numbers, variables, and operations; variables in the expression may change in value, causing the overall expression to represent different values

**coefficient:** a number multiplied by a variable in an expression, such as the number 3 in the expression $3x + 4$

**coordinate grid:** a two-dimensional visual representation of points, or ordered pairs

**dilation:** the proportional resizing of a figure on a coordinate grid

**distributive property:** the principle that multiplying a number by a group of numbers added together is the same as doing each multiplication separately, such as $2(x + y) = 2x + 2y$

**equation:** an algebraic expression that states that two things are equal

**factors:** numbers or expressions that are multiplied together to form a product

**function:** an equation in which each input has exactly one output; can be written as the solution for one variable as in $y = 3x + 4$, or can be written in terms of the inputs as in $f(x) = 4x - 2$

**inequality:** an algebraic expression that states that one side is greater or less than the other side; similar to an algebraic equation except that $<$, $>$, $\leq$ or $\geq$ are used in place of an $=$, as in $y > 3x + 5$

**inverse operations:** operations that undo each other, such as addition and subtraction or multiplication and division; used in algebra to solve equations by grouping variable terms on one side of an equation and constant terms on the other side

**isolate a variable:** to perform operations to an equation in order get one variable alone on one side of the equation

**line segment** the line formed by connecting two points on a coordinate grid

**linear equation:** an equation that makes a straight line when it is graphed; all of the solutions to the equation lie on the line

**mathematical pattern:** an arrangement of numbers and terms created by following a specific rule

**maximum of a curve:** the point on a curve for which $y$ has the highest value

**minimum of a curve:** the point on a curve for which $y$ has the lowest value

**negative slope:** a line that falls from left to right on a coordinate grid

**ordered pair:** a pair of $x$- and $y$-values that can be represented on a coordinate grid; the $x$-value is always shown first

**origin:** the point (0, 0) where the $x$-axis and $y$-axis cross on a coordinate grid

**point-slope form:** the equation of a straight line in the form $y - y_1 = m(x - x_1)$ where $m$ is the slope of the line and $(x_1, y_1)$ are the coordinates of a given point on the line

**positive slope:** a line that rises from left to right on a coordinate grid

**quadrant:** one of the four sections on a coordinate grid created by the intersection of the $x$- and $y$-axis

**quadratic equation:** an equation in which one or more of the terms is squared but raised to no higher power; they are set in the form $ax^2 + bx + c = 0$

**reflection:** the flipping of a figure over a line, such as an axis, on a coordinate grid

**rise:** the vertical increase between two points on a line; represented as the numerator in an equation for slope

**run:** the horizontal increase between two points on a line; represented as the denominator in an equation for slope

**slope:** a number that measures the steepness of a line

**slope-intercept form:** the equation of a straight line in the form $y = mx + b$ where $m$ is the slope of the line and $b$ is its $y$-intercept

**solution:** the set of all values for the variables in an equation that make the equation true

**translation:** the sliding of a figure to a new position on a coordinate grid

**variable:** a letter used to represent a value in algebraic expressions, such as $x$ in the expression $2x + 1$

**vector:** a line segment that has magnitude and direction; one point is its starting point and the magnitude and direction describe a path to the other point

**x-intercept:** the point at which a line crosses the $x$-axis on a coordinate grid and the $y$-coordinate is 0

**y-intercept:** the point at which a line crosses the $y$-axis on a coordinate grid and the $x$-coordinate is 0

# Algebraic Expressions and Variables

## 1 Learn the Skill

A **variable** is a letter used to represent a number. Variables are used in algebraic expressions. An algebraic expression has numbers and variables, possibly connected by an operation sign. A variable may change in value, which allows the expression itself to have different values. When you **evaluate** an algebraic expression, you substitute a number for the variable and solve. For example, if $b = 3$, then $b + 12 = 15$. If $b = -1$, then $b + 12 = 11$. The first step in evaluating an expression is to **simplify** it by combining like terms.

## 2 Practice the Skill

Understanding how to use variables and how to simplify and evaluate algebraic expressions are important skills for success on mathematics high school equivalency tests. Read the example and strategies below. Use this information to answer question 1.

**A** Order is important for division and subtraction. For example, "6 less than 3" is $3 - 6$, but "the difference between 6 and 3" is $6 - 3$.

**B** To simplify an expression, add like terms. Like terms have the same variable or variables raised to the same power. For example, $2x$ and $4x$ are like terms.

If an expression has parentheses, use the distributive property to simplify.

To evaluate an expression, substitute the given values for the variables, and then follow the order of operations.

| WORDS | SYMBOLS |
|---|---|
| 4 more than a number | $x + 4$ |
| 5 less than a number | $x - 5$ |
| 3 times a number | $3x$ |
| A number times itself | $x^2$ |
| The product of 8 and a number | $8x$ |
| The product of 6 and $x$ added to the difference between 5 and $x$ | $6x + (5 - x)$ |
| The quotient of 6 and $x$ | $\frac{6}{x}$ or $6 \div x$ |
| One-third of a number increased by 5 | $\frac{1}{3}x + 5$ |

**Simplify** $4x(5x + 7) - 2x$

$(4x)(5x) + (4x)(7) - 2x$
$20x^2 + 28x - 2x$
$20x^2 + 26x$

### ☑ TEST-TAKING TIPS

Multiplication can be written in several ways. In algebraic expressions, a number next to a variable means multiplication. The expression $3y$ is the same as 3 times $y$. Parentheses and a dot (·) also indicate multiplication.

**1** Gabe's current age is 3 times his sister's current age, represented by $x$. Which expression represents Gabe's current age?

**A** $3x$

**B** $\frac{x}{3}$

**C** $x - 3$

**D** $x + 3$

**E** $3 \div x$

UNIT 4

**2** The width of Kevin's yard is twice the width of his garage, increased by 10 feet. Which expression below describes the width of Kevin's yard if *g* represents the width of his garage?

  **A** $2g(10)$

  **B** $\dfrac{2g}{10}$

  **C** $2g - 10$

  **D** $2g + 10$

  **E** $g + 10$

**3** The number of employees that work in manufacturing is 500 less than 3 times the number of employees that work in shipping. Which expression represents the number of employees who work in manufacturing?

  **A** $3s + 500$

  **B** $3s(500)$

  **C** $3s - 500$

  **D** $\dfrac{3s}{500}$

  **E** $\dfrac{500}{3s}$

**4** Michael's score on a math quiz was 8 more than one-half of his score on his science quiz. Which expression below describes Michael's score on his math quiz?

  **A** $\dfrac{s}{2} + 8$

  **B** $\dfrac{s}{8} + 2$

  **C** $\dfrac{1}{2}s - 8$

  **D** $\dfrac{1}{2}(8) + s$

  **E** $\dfrac{s}{8} - \dfrac{1}{2}$

**5** Julie left a tip that was one-sixth of her restaurant bill plus $2. If the bill was $48, how much was the tip?

  **A** $2

  **B** $4

  **C** $6

  **D** $8

  **E** $10

**6** The cost of an adult ticket to the ballet is 2 times the cost of a child's ticket then decreased by $4. If a child's ticket is $12, how much is an adult ticket?

  **A** $4

  **B** $8

  **C** $12

  **D** $20

  **E** $24

**Directions:** Questions 7 and 8 are based on the information below.

**7** Which expression represents the perimeter of the rectangle?

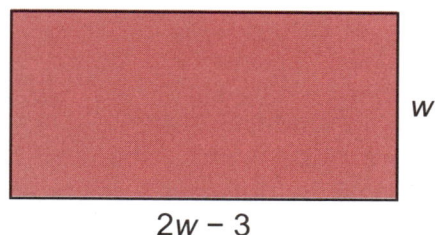

$2w - 3$

  **A** $3w - 3$

  **B** $w(2w - 3)$

  **C** $5w - 3$

  **D** $4w - 6$

  **E** $6w - 6$

**8** Which expression represents the area of the rectangle?

  **A** $w + 2w - 3$

  **B** $w(2w - 3)$

  **C** $w^2$

  **D** $w + 2w - 3 + w + 2w - 3$

  **E** Not enough information is given.

UNIT 4

9   The number of girls that registered to play basketball in a summer league is 15 fewer than twice the number of boys. Which expression describes the number of girls that registered?

A   $\frac{2b}{15}$

B   $\frac{1}{2}b + 15$

C   $15 + 2b$

D   $2b - 15$

E   $15 - 2b$

10  A school sold adult tickets and children's tickets to a football game. The number of children's tickets sold was 56 more than one-third the number of adult tickets. Which expression describes the number of children's tickets sold?

A   $\frac{1}{3}a - 56$

B   $\frac{a}{3} + 56$

C   $\frac{56a}{3}$

D   $\frac{3}{56}a$

E   $\frac{3}{a} + 56$

11  There are $p$ number of pencils in a pack. There are 50 packs in a box, and 12 boxes in a case. Julia delivers 3 cases to a store. She opens a case to remove 1 pencil to use. Which expression represents the number of pencils that are left?

A   $3 + 12 + 50 + p - 1$

B   $3(12)(50)p - 1$

C   $3(12)\left(\frac{50}{p}\right) - 1$

D   $3(12)(50)p(1)$

E   $3(12)(50)p + 1$

12  Edward drove 4 times as many miles on Tuesday as he did on Wednesday and Thursday combined. Which expression describes the number of miles he drove on Tuesday in terms of the number of miles he drove on Wednesday and Thursday?

A   $x + 4y$

B   $\frac{4}{x + y}$

C   $4x + y$

D   $\frac{x + y}{4}$

E   $4(x + y)$

13  The number of students in an incoming freshman class is 3 times the number of students in the sophomore class divided by 4. Which expression describes the number of students in the incoming freshman class?

A   $4y + 3$

B   $\frac{3y}{4}$

C   $3y + 4$

D   $\frac{4}{3y}$

E   $3y - 4$

14  Which expression represents the perimeter of the triangle?

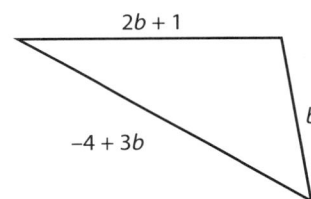

A   $5b + 3$

B   $6b + 3$

C   $6b - 3$

D   $-b + 4$

E   $3b$

**15** The age of Nick's grandfather is 5 years greater than twice the ages of his two grandchildren together. Which expression describes Nick's grandfather's age?

A  $2(x + y) + 5$

B  $\dfrac{x + y}{2} + 5$

C  $\dfrac{2(x + y)}{5}$

D  $10(x + y)$

E  $2x + 2y - 5$

**16** On Monday, a cyclist rode 20 fewer than 3 times the number of miles he rode on Sunday. If he rode 30 miles on Sunday, how many miles did he ride on Monday?

A  30

B  40

C  60

D  70

E  90

**17** The expression $3(x + 2x)$ represents the distance between two cities. What is the distance if $x = 4$?

A  48

B  36

C  20

D  18

E  Not enough information is given.

**18** Leo's age is 2 times the age of his sister decreased by 21. If his sister is 23, how old is Leo?

A  2

B  21

C  25

D  46

E  Not enough information is given.

**19** The expression $15a + 25b$ represents the amount of money a theater takes in per night if they sell $a$ number of $15 seats and $b$ number of $25 seats. How much money does the theater take in if $a = 207$ and $b = 134$?

A  $5,115

B  $5,175

C  $6,455

D  $7,185

E  $7,255

**20** A middle school has 374 male students. The number of female students is one-half the number of male students, then increased by 56. How many female students are in the middle school?

A  215

B  243

C  430

D  692

E  860

**21** A store sells women's and men's shoes. In one day, the number of women's shoes sold was 12 more than 4 times the number of men's shoes sold. How many pairs of women's shoes did the store sell?

A  48

B  72

C  96

D  108

E  Not enough information is given.

**22** The height of a triangle is 3 less than 3 times the base. If $b$ = the base of the triangle, which expression represents the area of the triangle?

A  $3b - 3$

B  $b(3b - 3)$

C  $\dfrac{1}{2}b(3b - 3)$

D  $\dfrac{1}{2}b(b)$

E  $\dfrac{3b - 3}{2}$

23 The number of minutes that Erin spent on Project A is 45 less than one-half of her time spent on Project B. Which expression best represents her time spent on Project B, if $t$ equals the number of minutes spent on Project A?

A $2t + 90$

B $2t - 45$

C $\dfrac{2t}{45}$

D $\dfrac{45}{2t}$

E $\dfrac{1}{2}t - 45$

24 A number is half the value of the sum of a second and third number. Which expression describes the first number?

A $2x + 2y$

B $\dfrac{2}{x + y}$

C $\dfrac{x + y}{2}$

D $x + y$

E $2(x + y)$

25 Sean swam 8 fewer than twice as many laps as Antonio. If Antonio swam 15 laps, how many laps did Sean swim?

A 11

B 12

C 15

D 22

E 23

26 If $-3 = x$, then what does $4y - 8(3 - 2x)$ equal?

A $-72$

B $-68y$

C $4y - 24$

D $4y + 24$

E $4y - 72$

27 The number of students who scored above average on an exam was 34 fewer than twice the number of students who scored at an average level. If 45 students scored at an average level, how many scored above average?

A 34

B 56

C 66

D 78

E 90

28 Jada wrote a check to pay for gas. The amount of the check was $5 less than one-half the amount that she had deposited into her account that day. If she deposited $84, how much did she pay for gas?

A $79

B $74

C $42

D $37

E Not enough information is given.

29 A number is 3 times the value of the quotient of a second and third number. Which expression describes the first number?

A $\dfrac{xy}{3}$

B $\dfrac{3}{x + y}$

C $\dfrac{3}{x} + y$

D $3(x + y)$

E $3\left(\dfrac{x}{y}\right)$

30 The length of a rectangle is 6 more than two-thirds the width. Which expression represents the perimeter of the rectangle?

A $\dfrac{10}{3}w + 12$

B $w\left(6 + \dfrac{2}{3}\right)w$

C $\dfrac{4}{3}w + 12$

D $\dfrac{10}{3}w + 6$

E $\dfrac{5}{3}w + 6$

# Equations

## ① Learn the Skill

An **equation** states that two things are equal. A **one-variable equation** consists of expressions involving only number values and a single variable, for example $2x + 6 = 12$. The solution of a one-variable equation is the value of the variable that makes the equation true.

To solve an equation to find the variable, use inverse operations to group variable terms on one side of the equation and constant terms on the other side of the equation. In the above example, subtract 6 from each side of the equals (=) sign so that $2x = 6$.

Next, use **inverse operations** to isolate the variable. Inverse operations are operations that undo each other. Addition and subtraction are inverse operations, as are multiplication and division. In the example above, both $2x$ and 6 can be divided by 2, isolating the $x$ so that $x = 3$.

## ② Practice the Skill

By practicing the skill of solving one-variable equations, you will improve your study and test-taking abilities, especially as they relate to mathematics high school equivalency tests. Study the information below. Use this information to answer question 1.

**A** The variable term in a one-variable equation may appear on one or both sides of the equals sign. To solve the equation, first group all of the variable terms on one side of the equals sign.

**B** Next, undo addition and subtraction. Then undo multiplication and division.

**C** You can check your solution by substituting it into the original equation. If the solution makes the equation true, then it is correct.

### Solve the equation

$$5x + 7 = 19 - 3x$$
$$5x + 3x + 7 = 19 - 3x + 3x \quad \leftarrow \quad \text{Add } 3x \text{ to both sides.}$$
$$8x + 7 = 19 \quad \leftarrow \quad \text{Group like terms.}$$
$$8x + 7 - 7 = 19 - 7 \quad \leftarrow \quad \text{Subtract 7 from both sides.}$$
$$8x = 12 \quad \leftarrow \quad \text{Group like terms.}$$
$$\frac{8x}{8} = \frac{12}{8} \quad \leftarrow \quad \text{Divide both sides by 8.}$$
$$x = 1.5 \quad \leftarrow \quad \text{Simplify.}$$

### Check:

$$5(1.5) + 7 \stackrel{?}{=} 19 - 3(1.5)$$
$$7.5 + 7 \stackrel{?}{=} 19 - 4.5$$
$$14.5 = 14.5$$

**1** What value of $x$ makes the equation $3x + 9 = 6$ true?
 A −1
 B −3
 C 5
 D 15
 E 19

UNIT 4

**2** Solve the equation for *x*.

$$0.5x - 4 = 12$$
**A** 4
**B** 8
**C** 16
**D** 32
**E** 48

**3** What value of *y* makes the equation true?

$$5y + 6 = 3y - 14$$
**A** −1
**B** −2.5
**C** −4
**D** −6
**E** −10

**4** Solve the equation for *t*.

$$\frac{1}{2}t + 8 = \frac{5}{2}t - 10$$
**A** 9
**B** 3
**C** −1
**D** −6
**E** −9

**5** Each month, Cameron earns $1,200 in salary plus an 8% commission on sales. The equation $T = 1{,}200 + 0.08s$ represents Cameron's total earnings each month. In July, Cameron earned a total *T* of $2,800. What was the value of Cameron's sales in July?
**A** $15,000
**B** $16,000
**C** $20,000
**D** $50,000
**E** $60,000

**6** A rectangular yard is *x* feet wide. The yard is 4 feet longer than it is wide. The perimeter *P* of the yard is given by the equation $P = 4x + 8$. If the perimeter of the yard is 84 feet, how long is the yard?
**A** 19 feet
**B** 23 feet
**C** 24 feet
**D** 28 feet
**E** 48 feet

**7** All of the following are used to solve for *x* in the equation below except one. Which is it?

$$9x - 2 = 4x + 8$$
**A** addition
**B** division
**C** multiplication
**D** subtraction
**E** isolation of the variable

**8** Lucas solved the equation below and got an answer of *x* = 1.25.

$$3x = (8 - 0.25x) - (3 - 0.75x)$$

Which describes Lucas' solution?
**A** The solution is correct because *x* = 1.25 makes the equation true.
**B** The solution is incorrect because Lucas divided both sides of the equation by 2.5.
**C** The solution is incorrect because Lucas added 0.75*x* to both sides of the equation.
**D** The solution is incorrect because Lucas subtracted 0.5*x* from both sides of the equation.
**E** The solution is incorrect because Lucas did not isolate the variable.

**9** Solve the equation for $x$.

$-3x + 11 = x - 5$

A $-1.5$
B $-4$
C $2$
D $4$
E $6$

**10** Solve the equation for $y$.

$0.6y + 1.2 = 0.3y - 0.9 + 0.8y$

A $1.6$
B $2.4$
C $2.6$
D $4.2$
E $19$

**11** Find the value of $n$ that makes the equation true.

$\dfrac{n}{4} - \dfrac{1}{2} = \dfrac{3n}{2} + \dfrac{3}{4}$

A $-1$
B $0$
C $1$
D $2$
E $3$

**12** Find the value of $x$ that makes the equation true.

$0.5(4x - 8) = 6$

A $-1$
B $1$
C $5$
D $7$
E $8$

**13** Solve the equation for $x$.

$3\left(\dfrac{2}{3}x + 4\right) = -5$

A $-8.5$
B $-4.5$
C $-3.5$
D $-0.5$
E $-0.4$

**14** Drew's age is 3 more than one-half Tyler's age. The equation $d = \dfrac{1}{2}t + 3$ represents the relationship between their ages. If Drew is 17, how old is Tyler?

A $7$
B $10$
C $28$
D $40$
E $42$

**15** What is the value of $x$ if
$0.25(3x - 8) = 2(0.5x + 4)$?

A $-64$
B $-40$
C $-24$
D $24$
E $40$

**16** Find the value of $w$ that makes the equation true.

$10w + 8 - 2(3w - 4) = -12$

A $-7$
B $-1$
C $-3$
D $1$
E $3$

**17** Solve the equation for $b$.

$12b - 2(b - 1) = 6 - 4b$

A $\dfrac{2}{7}$
B $\dfrac{4}{7}$
C $\dfrac{2}{3}$
D $\dfrac{4}{3}$
E $\dfrac{7}{3}$

**18** If $(x - 4)$ is 5 more than $3(2x + 1)$, what is the value of $x$?

A  $-3$
B  $-2.4$
C  $-2$
D  $-0.4$
E  $-0.2$

**19** Amelia and Brandon solve the following equation.

$$-2(4x - 5) = 3x - (6 - x)$$

Amelia gets an answer of $\frac{4}{3}$. Brandon gets an answer of $\frac{8}{5}$. Which of the following describes their answers?

A  Amelia is correct; Brandon made an error expanding the parentheses.
B  Amelia is correct; Brandon used the wrong operation to group the constant terms on one side of the equation.
C  Brandon is correct; Amelia made an error expanding the parentheses.
D  Brandon is correct; Amelia used the wrong operation to group the constant terms on one side of the equation.
E  Neither is correct.

**20** If $0.1q - 0.2(3q + 4) = -0.3(5q + 3)$, what is $q$?

A  .01
B  .05
C  .85
D  $-0.5$
E  $-0.1$

**21** The congruent sides of an isosceles triangle are 4 feet less than 3 times as long as the third side. The equation $P = x + 2(3x - 4)$ gives the perimeter of the triangle. If the perimeter is 9.5 feet, what is the length of the third side?

A  9 feet
B  4.5 feet
C  3.5 feet
D  3 feet
E  2.5 feet

**22** The perimeter $P$ of the rectangle is equal to $2x + 2(4x - 2)$. If the rectangle has a perimeter of 35 inches, what is $x$?

$$4x - 2$$

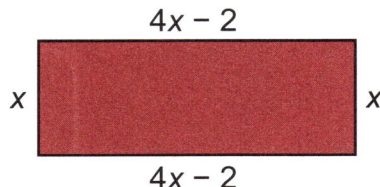

$x$         $x$

$$4x - 2$$

A  3.1
B  3.9
C  4.0
D  4.2
E  4.9

**23** All of the following are used to solve for $y$ in the equation below except one. Which is it?

$$10y - 12 = 5(2y + 3) - y$$

A  addition
B  isolation of the variable
C  multiplication
D  subtraction
E  division

**24** Solve the following equation for $t$.

$$4(\tfrac{1}{8}t + 2) = 2(t - 8) - \tfrac{1}{2}t$$

A  8
B  12
C  16
D  20
E  24

**25** The length of fencing around a garden with width $x$ feet and length $x + 7.5$ feet is given by $f = 2x + 2(x + 7.5)$. If a total of 65 feet of fencing is used, what is the value of $x$?

$$x + 7.5$$

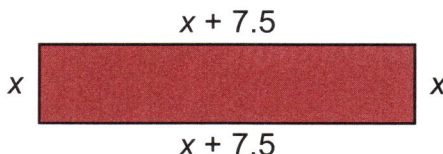

$x$         $x$

$$x + 7.5$$

A  12.5
B  14.375
C  18.125
D  20.0
E  22.375

**26** Keira is solving the equation $4(3x - 5) = -2(x + 7)$. She applies the distributive property and finds that $12x - 20 = -2x - 14$. Next, she groups together the variable terms and finds that $ax - 20 = -14$. What is the value of $a$?

A  6
B  10
C  14
D  26
E  28

**27** Quentin solved the equation $0.5(7y - 6) = 2(2y + 1)$. His solution is shown in the table.

| Step 1 | $3.5y - 3 = 4y + 2$ |
|--------|---------------------|
| Step 2 | $-0.5y - 3 = 2$     |
| Step 3 | $-0.5y = -1$        |
| Step 4 | $y = -10$           |

In which step did Quentin make an error?

A  Step 1
B  Step 2
C  Step 3
D  Step 4
E  His answer is correct.

**28** Solve the following equation for $b$.

$$10 - 3(2b + 4) = -4b - 2(5 - b)$$

A  −2
B  1
C  2
D  6
E  10

**29** Lisa solved the equation $\frac{1}{4}(2x + 12) = 3(\frac{1}{4}x - 2)$. Her solution is shown in the table.

| Step 1 | $\frac{2}{4}x + 3 = \frac{3}{4}x - 2$ |
|--------|----------------------------------------|
| Step 2 | $\frac{1}{4}x + 3 = -2$                |
| Step 3 | $\frac{1}{4}x = -5$                    |
| Step 4 | $x = 20$                               |

In which step did Lisa make an error?

A  Step 1
B  Step 2
C  Step 3
D  Step 4
E  Her answer is correct.

**30** What value of $n$ makes the equation true?

$$0.2(6n + 5) = 0.5(2n - 8) + 3$$

A  −10
B  −6
C  −4
D  0
E  4.9

**31** If $3w - 2(0.5w + 1) = -4(2 + w)$, what is $w$?

A  −6
B  −1.5
C  −1
D  3
E  4.9

# Patterns and Functions

## ① Learn the Skill

A **mathematical pattern** is an arrangement of numbers and terms created by following a specific rule. You can identify the rule used to make a pattern and apply it to find other terms in the pattern. An algebraic rule is often called a **function**. A function contains $x$- and $y$-values. There is only one $y$-value for every $x$-value. Think of a function as a machine. For every $x$-value you put into the machine, only one $y$-value will come out. A **relation** is a connection between the commonalities of two patterns.

## ② Practice the Skill

You will be asked to solve problems relating to patterns and functions on the mathematics test. Read the examples and strategies below. Use this information to answer question 1.

**A** To identify the rule used to make a pattern, study the sequence of numbers or terms. Ask yourself how each term is related to the next. In this example, the rule is "add 5."

**B** A pattern also may be geometric. In this example, three triangles are added to the first figure to form the second figure. Five triangles are added to the second figure. Seven triangles are added to the third figure. The number of triangles to be added increases by two each time.

**C** Functions are written as equations. They can show f(x) instead of $y$. Substitute each $x$-value (input) in the equation to find the value of f(x) (output). Remember, there is only one output for each input.

**Number Pattern**
−10, −5, 0, 5, 10, 15, . . .

**Geometric Pattern**

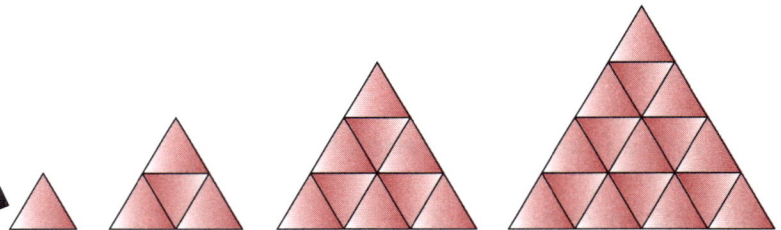

**Function**
$f(x) = 4x - 2$

| $x$ | −2 | −1 | 0 | 1 | 2 |
|------|-----|-----|-----|-----|-----|
| $f(x)$ | −10 | −6 | −2 | 2 | 6 |

### ✓ TEST-TAKING TIPS

You may need to test several rules to find the rule for a pattern. Look first for simple rules involving addition and subtraction. Then try multiplication and division. Some rules may involve more than one operation.

**1** What is the value of **f(x)** if **x = 4** in the function **f(x) = x² − 5**?

A   −3
B   −1
C   11
D   16
E   21

**Directions:** Questions 2 and 3 are based on the information below.

$$2, 4, 16, 256, 65,536, \ldots$$

**2** What is the rule for the pattern?
  A   multiply the previous term by 2
  B   add twice the previous term
  C   square the previous term
  D   divide by one-half
  E   multiply by 4

**3** What is the next term in the sequence?
  A   4,294,967,296
  B   268,435,456
  C   16,777,216
  D   1,048,576
  E   262,144

**4** For the function $f(x) = \frac{x}{5}$, which of the following $x$-values has a whole number output?
  A   19
  B   21
  C   22
  D   25
  E   26

**5** The distance an airplane travels in $t$ hours is given by the function $d = 230t$. How long does it take the airplane to travel 1,035 miles?
  A   3.5 hours
  B   4.5 hours
  C   5.5 hours
  D   6.5 hours
  E   7.5 hours

**6** Susan made a table to show the amount of sales tax due on typical purchase amounts. The function she used is $y = 0.08x$, where $x$ is the cost of the purchase and $y$ is the sales tax.

| x | 5 | 10 | 15 | 20 | 25 |
|---|---|----|----|----|----|
| y | $0.40 | $0.80 | $1.20 | $1.60 | |

How much sales tax does Susan owe if her purchases total $25?
  A   $3.20
  B   $2.80
  C   $2.40
  D   $2.00
  E   $1.80

**7** What is the sixth term in the sequence below?

**192, 96, 48, 24, . . .**
  A   1
  B   3
  C   6
  D   12
  E   15

**8** What is the $y$-value that is missing from the table?

| x | 0 | 3 | 4 | 6 | 10 |
|---|---|---|---|---|----|
| y | −1 | 14 | | 29 | 49 |

  A   3
  B   −3
  C   19
  D   20
  E   29

**9** If $f(x) = 2 - \frac{2}{3}x$, then what is $x$ when $f(x) = 4$?
  A   −3
  B   $-\frac{4}{9}$
  C   $\frac{4}{9}$
  D   3
  E   9

**Directions:** Questions 10 and 11 are based on the information below.

| x | −2 | 0 | 2 | 4 | 6 |
|---|----|----|----|----|----|
| y | −8 | −2 | 4 | 10 | |

10 Which of the following equations expresses the relationship between *x* and *y*?

A $y = 3x - 2$

B $y = \frac{1}{4}x$

C $y = 2x + 3$

D $y = \frac{2}{3}x$

E $y = 4x$

11 What number is missing from the table?

A 12
B 14
C 16
D 18
E 20

12 Which of the following rules can be used to extend the following sequence?

2, 4, 8, 16, 32, . . .

A Add 4.
B Subtract 8.
C Multiply by 2.
D Divide by 2.
E Multiply by 8.

13 What is the sixth term in the sequence?

1, 3, 9, 27, . . .

A 729
B 243
C 81
D 9
E 3

14 Solomon is following a pattern as he stacks the blocks shown below. How many blocks will he stack in the next figure in his sequence?

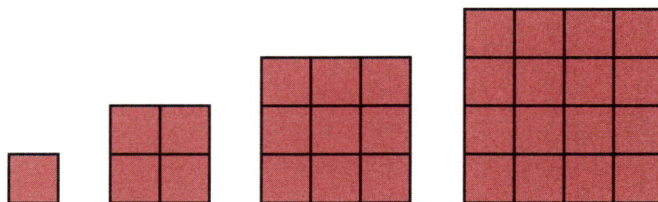

A 18
B 25
C 32
D 42
E 64

15 What is the next term in the sequence below?

−5, −10, −20, −40, −80, . . .

A −160
B −140
C −120
D −100
E −90

16 The function $f(x) = 50 - x^2$ was used to create the following function table. Which number is missing from the table?

| x | −2 | −1 | 0 | 1 | 2 |
|---|----|----|----|----|----|
| f(x) | 46 | 49 | 50 | | 46 |

A 54
B 51
C 50
D 49
E 46

17 Which value for *x* for the function $f(x) = \frac{1}{2}x$ results in a value of $f(x)$ that is equal to 1?

A −2
B −1
C 0
D 1
E 2

**18** What term in the sequence will have only one circle?

- **A** fourth
- **B** fifth
- **C** sixth
- **D** seventh
- **E** eighth

**19** What is the seventh term in the sequence?
−3, −6, −9, −12, −15, . . .
- **A** −18
- **B** −21
- **C** −24
- **D** −27
- **E** −30

**20** For the function $f(x) = \frac{8}{x}$, which of the following values for $x$ results in a value of $f(x)$ that is less than 1?
- **A** 5
- **B** 6
- **C** 7
- **D** 8
- **E** 9

**21** How many triangles will the next term of the sequence have?

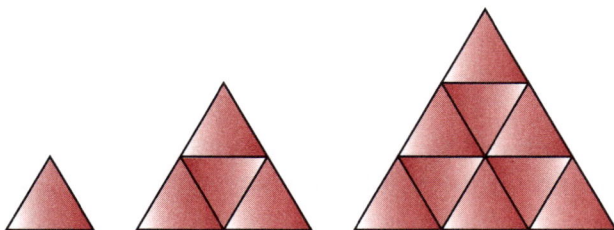

- **A** 13
- **B** 14
- **C** 16
- **D** 17
- **E** 18

**22** The formula $I = (1{,}000)(r)(5)$ shows the amount of interest earned on a $1,000 investment over 5 years with a certain interest rate. What must the interest rate be to earn $250 interest on the investment in 5 years?
- **A** 2%
- **B** 3%
- **C** 4%
- **D** 5%
- **E** 6%

**23** The function $d = 55t$ describes the distance $d$ a car will travel at a constant speed of 55 miles per hour in a certain amount of time $t$. For which of the following values of $t$ is the value of $d$ equal to 220?
- **A** 2.0
- **B** 2.5
- **C** 3.0
- **D** 3.5
- **E** 4.0

**24** The increase in a population that grows annually at 1% can be described by the function $I = 0.01N$, where $I$ is the increase and $N$ is the initial population size. By what number does the population increase in a year if the initial population is 5,000?
- **A** 5
- **B** 10
- **C** 25
- **D** 50
- **E** 55

**25** $F = \frac{9}{5}C + 32$ describes the relationship between degrees Fahrenheit (F) and degrees Celsius (C). For which of the following Celsius temperatures is the equivalent Fahrenheit temperature 80°?
- **A** 176.0°C
- **B** 62.2°C
- **C** 86.4°C
- **D** 44.5°C
- **E** 26.7°C

**Directions:** Questions 26 and 27 are based on the information below.

An archer shot an arrow at an angle of 45 degrees. The distance and height of the arrow along its path were recorded at several points. The data is shown in the table below.

| DISTANCE (d) in METERS | 1 | 2 | 3 | 4 | 5 |
|---|---|---|---|---|---|
| HEIGHT (h) in METERS | 0.8 | 1.2 | 1.2 | 0.8 | |

**26** Which of the following equations expresses the relationship between *h* and *d*?

A $h = d + 0.2d$

B $h = d^2$

C $h = d - 0.2$

D $h = d - 0.2d^2$

E $h = d^2 - 0.2d^2$

**27** What is the height of the arrow when it has traveled a distance of 5 meters?

A 1.8 meters

B 1.6 meters

C 1.2 meters

D 0.8 meters

E 0 meters

**28** Kara substituted 1 for *x* into the function $f(x) = 3x^2 + 1$. Henry substituted a different value for *x* but found the same output. Which value of *x* would give the same output?

A 4

B 2

C 0

D −1

E −2

**29** For the function $y = x^3$, which of the following values for *x* results in a value for *y* that is equal to the value for *x*?

A 1

B 2

C 4

D 8

E 16

**30** What is the eighth term in the sequence?

$$2, -4, 8, -16, 32, \ldots$$

A −512

B −256

C −128

D 128

E 256

**31** The function $y = x^2$ was used to create the following function table. Which number is missing from the table?

| x | −5 | −1 | 0 | $\frac{1}{2}$ | |
|---|---|---|---|---|---|
| y | 25 | 1 | 0 | $\frac{1}{4}$ | 4 |

A −4

B −1

C 0

D 1

E 2

**32** As you move below sea level, the pressure increases. Pressure at different levels below sea level is shown in the table.

| DISTANCE (d) in METERS BELOW SEA LEVEL | 0 | 10 | 20 | 30 | 40 |
|---|---|---|---|---|---|
| PRESSURE (p) in POUNDS PER SQUARE INCH | | 14.7 | 29.4 | 44.1 | 58.8 |

Which of the following equations expresses the relationship between *d* and *p*?

A $p = \frac{d}{10} + 14.7$

B $p = \frac{d}{10}(14.7) + 14.7$

C $p = 10d + 14.7$

D $p = \frac{d}{10}(14.7)$

E $p = 10d(14.7)$

# Factoring

## ① Learn the Skill

**Factors** are numbers or expressions that are multiplied together to form a **product**. In the term $4y$, 4 and $y$ are factors. A factor may have two terms, such as $(x + 5)$. You can multiply factors with two terms using the FOIL method, in which you multiply the *First*, *Outer*, *Inner*, and *Last* terms in that order. An expression or equation with more than one term also may be factored, or split into factors.

## ② Practice the Skill

A quadratic equation can be written in the form $ax^2 + bx + c = 0$, where $a$, $b$, and $c$ are integers and $a$ is not equal to zero. Knowing how to factor can help you solve for a missing value in a quadratic equation. Read the examples and strategies below. Use this information to answer question 1.

**A** To factor, work backward.
1. List all of the possible factors for the third term.
2. Next, find the two factors of the third term that have a sum equal to the coefficient of the middle term.
3. Use the variable as the first term in each factor and the integers from step 3 as the second terms.
4. Use the FOIL method to check your answer.

**B** To solve a quadratic equation, rewrite the equation to set the quadratic expression equal to 0. Then factor and set each factor equal to 0. Then solve. Check both values by substituting them in the original equation.

### The FOIL Method
Multiply $(x + 2)(x - 4)$

First $x(x) = x^2$    Outer $x(-4) = -4x$
Inner $2(x) = 2x$    Last $2(-4) = -8$

$$x^2 + (-4x) + 2x + (-8) = x^2 - 2x - 8$$

### Factor $4x + 12$

$$\frac{4x + 12}{4} = \frac{4x}{4} + \frac{12}{4} = x + 3, \text{ so } 4x + 12 = 4(x + 3)$$

### Factoring Quadratic Expressions
$$x^2 - 2x - 8$$
1. Factors of −8: (1, −8), (−1, 8), (2, −4), (−2, 4)
2. −4 + 2 = −2
3. $(x - 4)(x + 2)$
4. Check: $x^2 + 2x - 4x - 8 = x^2 - 2x - 8$

### Solving Quadratic Equations
$x^2 + 4x = 12 \longrightarrow x^2 + 4x - 12 = 0 \longrightarrow (x - 2)(x + 6) = 0$
If $x - 2 = 0$ and $x + 6 = 0$, then $x = 2$ and $x = -6$.

### ✓ TEST-TAKING TIPS

Each term in an expression or equation belongs with the sign that precedes it. In the expression $x^2 - 2x - 8$, the first term is positive, while the next two terms are negative.

**1** Which of the following are factors of $x^2 + 5x - 6$?
A $(x + 6)(x - 1)$
B $(x - 2)(x - 3)$
C $(x + 2)(x + 3)$
D $(x + 2)(x - 3)$
E $(x - 6)(x + 1)$

UNIT 4

**2** What is the product of $(x + 5)(x - 7)$?
   A $x^2 - 2x - 12$
   B $x^2 + 2x + 12$
   C $x^2 + 2x + 35$
   D $x^2 - 2x - 35$
   E $x^2 + 2x - 35$

**3** Which of the following is equal to $(x - 3)(x - 3)$?
   A $x^2 - 6x + 9$
   B $x^2 + 6x - 9$
   C $x^2 + 6x + 9$
   D $x^2 - 9x - 6$
   E $x^2 - 9x + 6$

**4** Which of the following is equal to $x^2 - 6x - 16$?
   A $(x + 4)(x - 4)$
   B $(x - 2)(x - 8)$
   C $(x - 2)(x + 8)$
   D $(x + 2)(x - 8)$
   E $(x - 4)(x - 4)$

**5** The dimensions of a rectangle are $2x - 5$ and $-4x + 1$. Which expression represents the area of the rectangle?

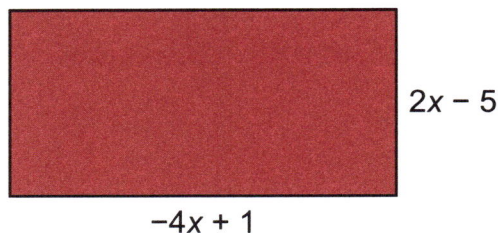

$2x - 5$

$-4x + 1$

   A $-8x^2 + 22x - 5$
   B $8x^2 + 22x - 5$
   C $-8x^2 - 18x - 5$
   D $8x^2 - 18x - 5$
   E $-8x^2 + 22x + 5$

**6** If $4x + 1$ is one factor of $4x^2 + 13x + 3$, which of the following is the other factor?
   A $x + 1$
   B $x + 3$
   C $x - 13$
   D $x + 12$
   E $x - 2$

**7** Which shows the solutions for the following equation?

$$2x^2 + 18x + 36 = 0$$

   A $-3$ and $6$
   B $3$ and $-6$
   C $6$ and $6$
   D $-6$ and $-6$
   E $-3$ and $-6$

**8** Miranda made a rectangular vegetable garden next to her house. She used the house as one side and fenced in the other three sides. She used 12 meters of fencing. The area of her garden is 32 square meters. To find a possible width of her garden, solve $w^2 - 12w = -32$.

Which of the following is a possible width of her garden?
   A 1 m
   B 4 m
   C 6 m
   D 7 m
   E 10 m

**9** Which expression has a product that has only two terms?
   A $(x + 7)(x - 1)$
   B $(x - 1)(x - 1)$
   C $(x - 7)(x + 7)$
   D $(x + 1)(x - 7)$
   E $(x - 7)(x - 7)$

UNIT 4

**10** What is the product of $(x + 5)(x - 4)$?

A  $x^2 + x + 20$
B  $x^2 + 9x - 20$
C  $x^2 + x - 20$
D  $x^2 - x - 20$
E  $x^2 - 9x + 20$

**11** The side of a square is represented by $x - 4$. What expression represents the area of the square?

A  $x^2 + 16$
B  $x^2 - 16$
C  $x^2 - 8x - 16$
D  $x^2 - 8x + 16$
E  $x^2 + 4x + 16$

**12** The number of students in a classroom is 5 less than the number of pencils each student has. Which expression represents the total number of pencils?

A  $x^2 - 5$
B  $x^2 - 5x$
C  $x - 5$
D  $25 - x$
E  $x^2 - 25$

**13** Which expression is the same as $x^2 - 4x - 21$?

A  $(x + 3)(x - 7)$
B  $(x - 3)(x + 7)$
C  $(x + 1)(x - 21)$
D  $(x - 1)(x + 21)$
E  $(x - 3)(x - 7)$

**14** In the equation $x^2 + 8x - 20 = 0$, which of the following is a possible value of $x$?

A  10
B  8
C  0
D  −8
E  −10

**15** If $x^2 - 5x - 6 = 0$, what is one possible value for $x$?

A  −1
B  −3
C  −5
D  −7
E  −9

**16** What are the solutions of the quadratic equation $x^2 - 7x - 30 = 0$?

A  −6 and 5
B  −3 and 10
C  −5 and 6
D  −10 and 3
E  −3 and 4

**17** The length of a rectangle is represented by $x + 2$ and the width of the rectangle is represented by $x - 5$. Which expression represents the area of the rectangle?

A  $x^2 - 3x - 10$
B  $x^2 - 10x - 3$
C  $x^2 + 3x - 10$
D  $x^2 + 3x + 10$
E  $x^2 + 10x - 3$

**18** Which pair of solutions makes the quadratic equation $x^2 - 16 = 0$ true?

A  −4 and 2
B  2 and −8
C  −4 and 1
D  −4 and 4
E  −2 and 8

**19** If the area of a square is represented by $x^2 + 6x + 9$, which expression represents the side of the square?

A  $x + 1$
B  $x + 3$
C  $x + 6$
D  $x + 9$
E  $x + 15$

**20** The product of two consecutive integers is 42. Which quadratic equation could be solved to find the value of the first integer?

   **A** $x^2 + 2x - 42 = 0$
   **B** $x^2 + x - 42 = 0$
   **C** $x^2 - 42x = 0$
   **D** $x^2 - x + 42 = 0$
   **E** $x^2 - 2x + 42 = 0$

**21** The area of the square shown below is 49 square feet. What is the value of $x$?

$x + 4$

   **A** −11
   **B** −3
   **C** 0
   **D** 3
   **E** 11

**22** The sum of the squares of two consecutive integers is 113. What are the two integers?

   **A** 4 and 5
   **B** 5 and 6
   **C** 6 and 7
   **D** 7 and 8
   **E** 8 and 9

**23** The product of two consecutive odd integers is 35. What is the second integer?

   **A** 1
   **B** 3
   **C** 5
   **D** 7
   **E** 9

**24** The length of the side of a square is $x - 3$. The area of the square is 81 square meters. What is the value of $x$?

   **A** 9
   **B** 11
   **C** 12
   **D** 13
   **E** 15

**Directions:** Questions 25 and 26 are based on the information below.

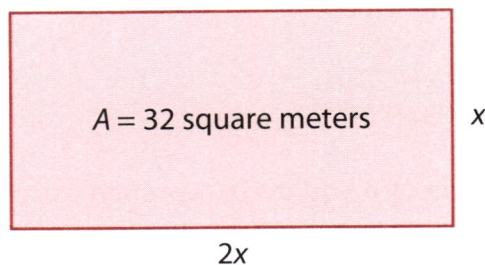

$2x$

**25** Which of the following equations could be used to find the value of $x$?

   **A** $2x^2 - 32 = 0$
   **B** $2x - 16 = 0$
   **C** $x^2 - 32 = 0$
   **D** $2(x^2 - 4) = 0$
   **E** $2x - 32 = 0$

**26** What is the length of the longer side of the rectangle?

   **A** 2 m
   **B** 4 m
   **C** 6 m
   **D** 8 m
   **E** 10 m

**27** Hank is standing on a hotel balcony. He throws a ball to his friend on the street. The equation for the ball's height $h$ at time $t$ seconds after being thrown is $h = t^2 - 2t - 8$. How many seconds does it take the ball to reach the ground, or a height of 0?

   **A** 1
   **B** 2
   **C** 3
   **D** 4
   **E** 5

**28** The product of two consecutive integers is 110. What are the two integers?

   **A** −12 and −11
   **B** −11 and −10
   **C** 9 and 10
   **D** 10 and 12
   **E** 11 and 12

Lesson 4 | Factoring

**29** What is the difference of $(x + 7)(x - 8) - 4(x - 1)$?

A $x^2 - 5x - 52$
B $x^2 - 4x - 52$
C $x^2 - 4x + 60$
D $x^2 - 3x - 52$
E $x^2 - 3x - 60$

**30** The product of two consecutive even integers is 10 more than 5 times their sum. What are the two integers?

A 4 and 6
B 6 and 8
C 7 and 8
D 10 and 12
E 14 and 16

**31** Joann drew a sketch of the garden she'd like to have. If she would like her garden to have an area of 160 square feet, how long should the shorter side of the garden be?

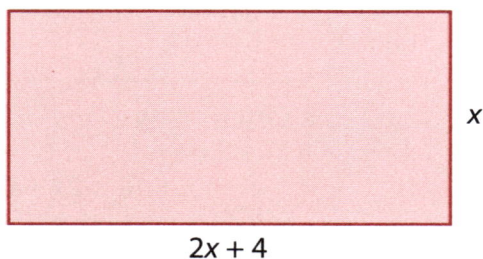

$2x + 4$

A 8
B 9
C 10
D 11
E 12

**32** The product of two consecutive even positive integers is 48. Which quadratic equation could be solved to find the value of the first integer?

A $x^2 + 4x - 48 = 0$
B $x^2 + x - 48 = 0$
C $x^2 - 48x = 0$
D $x^2 + 2x - 48 = 0$
E $x^2 - x + 48 = 0$

**33** Two consecutive integers have a product of 12. What are the two integers?

A −2 and 3
B −3 and −3
C −3 and 4
D −3 and −4
E −4 and −5

**34** In the quadratic equation $2x^2 - 8x - 10 = 0$, which of the following is a possible value of $x$?

A 1
B 2
C 5
D 7
E 10

**35** The area of a square is 64 square meters. If each side is $x - 2$, what is the value of $x$?

A 10
B 9
C 8
D 7
E 6

**Directions:** Questions 36 and 37 are based on the information below.

The area of the rectangle is 84 square feet.

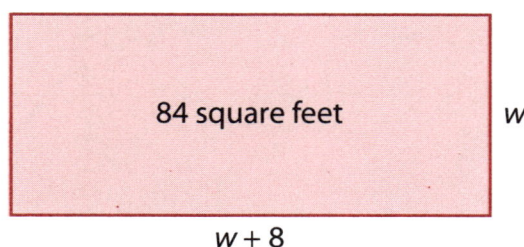

$w + 8$

**36** What is the width of the rectangle?

A 6
B 8
C 14
D 16
E 20

**37** What is the length of the rectangle?

A 8
B 12
C 14
D 16
E 18

# Solving and Graphing Inequalities

## ① Learn the Skill

An **inequality** states that two algebraic expressions are not equal. Inequalities are written with less than and greater than symbols ($<$, $>$) as well as two additional symbols. The $\geq$ symbol means "is greater than or equal to" and the $\leq$ symbol means "is less than or equal to."

## ② Practice the Skill

A solution to an inequality can include an infinite amount of numbers. For example, solutions to $b < 5$ include $b = 4.5, 4, 3.99, 3, 2, 1, 0, -3, -10$, and so on. When each individual solution is plotted as a point on a number line, a solid line is formed, which represents the solution set. Read the examples and strategies below. Use this information to answer question 1.

**A** Solve inequalities as you do equations. If you multiply or divide an inequality by a negative number, you must reverse the sign of the inequality. For example, if the inequality shown was $16 \leq -8x$, you would divide by $-8$ and reverse the sign, giving you $-2 \geq x$.

**B** For $x > 3$, every number to the right of 3 is in the solution set. Draw an open circle at 3 because 3 is *not* greater than 3 and therefore is not included in the solution set. Then draw a solid arrow to the right from 3.

For $x \leq 3$, each number to the left of 3 *as well as* 3 is included in the solution set. Draw a closed circle at 3 to show that 3 is included. Then draw a solid arrow pointing to the left from 3.

### Examples of Inequalities

$x \geq 4$ ⟶ A number is greater than or equal to 4.

$2x + 7 < 15$ ⟶ Two times a number plus seven is less than 15.

### Solving an Inequality

$$4 - 6(x - 3) \leq 2x + 6$$

| Simplify $4 - 6(x - 3)$ | $4 - 6x + 18 \leq 2x + 6$ |
| $22 - 6x \leq 2x + 6$ | Add 6x to both sides. |
| $22 \leq 8x + 6$ | Subtract 6 from both sides. |
| $16 \leq 8x$ | Divide by 8. |
| $2 \leq x$, or $x \geq 2$ | |

### Graphing an Inequality

$x > 3$

$x \leq 3$

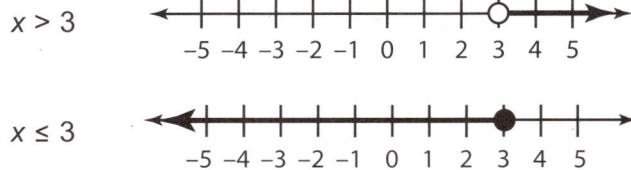

### ✓ TEST-TAKING TIPS

When graphing an inequality with a variable on the right side of the inequality symbol, read the inequality backward. For example, think of $7 < y$ as "$y$ is greater than 7."

**1** Five times a number is less than or equal to two times the number plus nine. What is the solution to the inequality?

   A   $x \geq 8$

   B   $x \geq 9$

   C   $x \leq 9$

   D   $x \geq 3$

   E   $x \leq 3$

**2** What is the solution to the inequality $x + 5 > 4$?

A $x > 1$

B $x < -1$

C $x < 1$

D $x > -1$

E $x > x - 1$

**3** What inequality is shown on the number line?

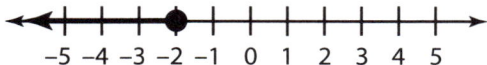

A $x \leq 2$

B $x \leq -2$

C $x > 2$

D $x < -2$

E $x > -2$

**4** The product of a number and 5, increased by 3, is less than or equal to 13. What is the inequality?

A $5x + 2 \leq 13$

B $5x \leq 13 + 3$

C $5x + 3 < 13$

D $5x + 3 \leq 13$

E $x + 3 \leq 5x + 13$

**5** The area of the following rectangle cannot be greater than 80 square centimeters. The length is 3 less than 3 times the width. Which inequality shows this relationship?

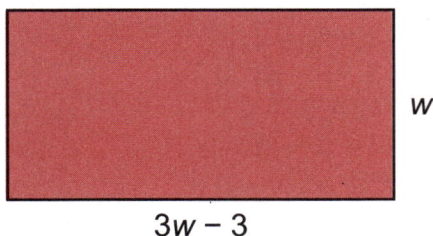

$w$

$3w - 3$

A $80 \leq 2(w + 3w - 3)$

B $80 \geq 2(w + 3w - 3)$

C $80 \geq w(3w) - 3$

D $80 \leq w(3w - 3)$

E $80 \geq w(3w - 3)$

**6** Kara has $15 and Brett has $22. Together, they have less than the amount needed to buy a pair of concert tickets. Which inequality describes their situation?

A $37 < x$

B $x + 15 < 22$

C $x \leq 37$

D $x + 22 \leq 15$

E $x + 37 \leq 22$

**7** A taxicab charges $2.00 as a base price and $0.50 for each mile. Josie needs to take a taxicab but only has $8. What is the greatest number of miles that Josie can ride in the cab?

A 8

B 9

C 11

D 12

E 13

**8** The sum of a number and 12 is less than or equal to 5 times the number plus 3. Which inequality represents this situation?

A $x + 12 \geq 5(x - 3)$

B $x + 12 \leq 5x + 3$

C $x + 12 > 5x + 3$

D $5x + 3 > x + 12$

E $x + 3 \leq 15$

**9** Which shows the solution for the inequality $8 - 3x > 2x - 2$?

A $x > 2$

B $x < 2$

C $x > 6$

D $x < 6$

E $x > 8$

**10** What is the solution to the inequality $-x - 4x > 30 - 3(x + 8)$?

A $-6\frac{3}{4} > x$

B $-3 < x$

C $-6\frac{3}{4} < x$

D $-3 > x$

E $3 > x$

**11** Which of the following inequalities is shown on the number line?

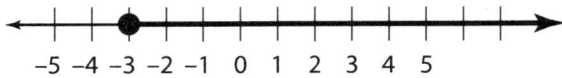

A  $x \geq 3$
B  $x \geq -3$
C  $x \leq 3$
D  $x \leq -3$
E  $x > -3$

**12** What is the solution to the inequality $x + 5 < 14$?

A  $x < -9$
B  $x \geq 19$
C  $x \leq 19$
D  $x > 9$
E  $x < 9$

**13** Which of the following inequalities is graphed on a number line using a closed circle?

A  $x < 5$
B  $x > -4$
C  $x \geq -3$
D  $x < -2$
E  $x > 0$

**14** Which of the following shows the solution to the inequality $4(x - 1) \geq 8$?

A
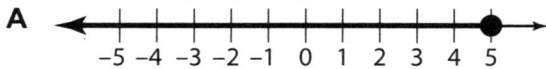

B

C
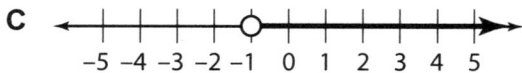

D
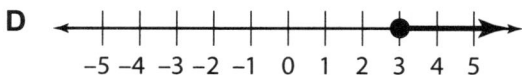

E

**15** What is the solution to the inequality $2x + 3 \geq 5x + 4$?

A  $x < -\dfrac{1}{3}$

B  $x \leq -\dfrac{1}{3}$

C  $x \geq \dfrac{1}{3}$

D  $x \geq -3$

E  $x < 3$

**16** Which inequality is shown on the number line?

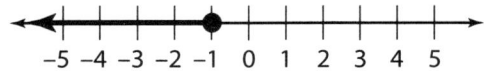

A  $x > 1$
B  $x \geq -1$
C  $x < -1$
D  $x \leq 1$
E  $x \leq -1$

**17** What is the solution to the inequality $2x - 7 \geq 15$?

A  $x \geq 11$
B  $x \geq 22$
C  $x \leq 11$
D  $x \leq 22$
E  $x > 11$

**18** When 4 times a number is added to 3, the result is greater than 2 less than 5 times that same number. Which of the following is the inequality?

A  $4x + 3 > 5x - 2$
B  $4x + 3 \geq 2 - 5x$
C  $4x + 12 > 5x - 2$
D  $4x + 3 \geq 5x - 10$
E  $4(x + 3) > 5(x - 2)$

UNIT 4

**19** Stacy pays $12 a month for basic cell phone service. Each minute she talks is an additional $.10. If she budgets $25 a month for her cell phone bill, what is the maximum number of minutes she can talk each month?

A  12
B  120
C  130
D  1,330
E  1,300

**20** Lydia purchased 3 gallons of milk. Her total was more than $9. What was the lowest possible price of 1 gallon of milk?

A  $2.00
B  $2.50
C  $3.01
D  $3.50
E  $4.00

**21** Brit scored 45, 38, and 47 on her first three math quizzes. What is the minimum score she must earn on the fourth quiz to have an average quiz score of at least 44?

A  43
B  44
C  45
D  46
E  47

**22** Colin earned $450 one week and $550 the next week. How much must he earn the third week to average $600 or more per week?

A  $600
B  $650
C  $700
D  $750
E  $800

**23** Admission to a park is $15. Each game in the park costs $.75. Cole takes $25 to the park. Which inequality represents the possible number of games that he can play?

A  $25 - 15 \le 0.75x$
B  $15 + 0.75x \le 25$
C  $0.75x - 15 \ge 25$
D  $15 - 0.75 \le 25$
E  $25 + 0.75x \ge 15$

**24** A store sold 156 football T-shirts. If the store has fewer than 34 of the T-shirts left, which expresses how many T-shirts the store had originally in terms of $x$?

A  $x \ge 190$
B  $x \le 190$
C  $x = 190$
D  $x > 190$
E  $x < 190$

**25** Gabe makes a base salary of $1,500 per month. He also earns a 3% commission on all of his sales. What must the amount of his monthly sales be for him to earn at least $3,000 per month?

A  $50
B  $500
C  $5,000
D  $50,000
E  $500,000

**26** Sara sold $14,000 worth of computer equipment in January. She sold $9,000 of computer equipment in February. What is the least amount of computer equipment Sara must sell in March to have average monthly sales greater than $10,000 for the first quarter?

A  $5,001
B  $7,001
C  $9,001
D  $12,001
E  $26,001

**27** A grocery store has spaghetti noodles on sale. The first box of noodles is $1.60. Each additional box is only $.95. What is the maximum number of boxes of noodles Jax can buy with $4.50?

A  1
B  2
C  3
D  4
E  5

28 Alia wants to buy a new winter jacket and boots. The jacket costs $2\frac{1}{2}$ times more than the boots. If Alia cannot spend more than $157.50, what is the most she can spend on the boots?
A  $37.50
B  $45.00
C  $75.50
D  $90.00
E  $112.50

29 The number of yards Michael swam on Tuesday was 400 less than 3 times the number of yards he swam on Monday. The number of yards he swam over the two days was less than 2,000. Which could be the number of yards Michael swam on Monday?
A  500
B  600
C  800
D  1,000
E  1,400

30 Jose's batting average last year was .266. Provided Jose has the same number of at-bats, what is the minimum batting average he can have this year to finish with a combined average of at least .300 over the two years?
A  .044
B  .144
C  .334
D  .366
E  .444

31 The number of students who are enrolled in first-semester biology at a university is 30 fewer than twice as many as last year. If the enrollment for the class cannot exceed 100 students, what was the maximum number of students enrolled in the class last year?
A  30
B  35
C  65
D  70
E  100

32 In the second round of a card game, Allen scored less than 2 times the number of points that he scored in the first round. If he scored 10 points in the first round, which inequality shows the number of points that he could have scored in the second round?
A  $y < 20$
B  $y > 10$
C  $y \leq 20$
D  $y \geq 20$
E  $y < -20$

33 A restaurant bill was less than $45. Three friends split the bill evenly. What is the greatest amount each friend could have paid?
A  $11.99
B  $12.99
C  $13.99
D  $14.99
E  $15.99

34 In an English class, students must earn an average score of 80% or above on their written papers to earn at least a B in the class. Leah's scores on her first four papers are shown below.

| LEAH'S ENGLISH PAPER SCORES | |
| --- | --- |
| PAPER | SCORE (%) |
| 1 | 78 |
| 2 | 85 |
| 3 | 82 |
| 4 | 74 |
| 5 | ? |

What is the minimum score Leah needs on her fifth and final paper to earn a B in the class?
A  78
B  79
C  80
D  81
E  82

# The Coordinate Grid and Vectors

## 1 Learn the Skill

A **coordinate grid** is made by the intersection of a horizontal line (*x*-axis) and a vertical line (*y*-axis). The point where the number lines meet (0, 0) is called the origin. In an **ordered pair**, the first value (*x*-value) tells how many spaces to move right (for positive) or left (for negative). The *y*-value tells how many spaces to move up (for positive) or down (for negative). A line segment is shown on a coordinate grid by plotting two points and connecting them. A **vector** is an object that has magnitude and direction. You can picture a vector as a line segment for which one point is its starting point and the magnitude and direction describe a path to the other point.

## 2 Practice the Skill

Examine the grid and strategies below. Use this information to answer question 1.

**A** To draw a line segment on the coordinate grid, plot the given points. Then draw a line to connect them.

**B** The pair of numbers that tell the magnitude of a vector differ from the ordered pair that identifies a point on the coordinate grid. Instead, they tell how the second point in the vector differs from the first. In this example the magnitude is <3, 4>. From the starting point E, the *x*-value increases by 3 while the *y*-value increases by 4.

**C** Changes to figures can be shown on a coordinate grid. They include translations, reflections, and dilations. In a *translation*, a figure slides to a new position. A translation has direction and magnitude, so it can be described by a vector. In a *reflection*, a figure is flipped over a line such as an axis. In a *dilation*, a figure is proportionally resized.

The coordinate grid below shows points *A, B, C, D,* and *E*. The coordinates for point *A* are (−1, 0). Point *B* is located at (−1, −5).

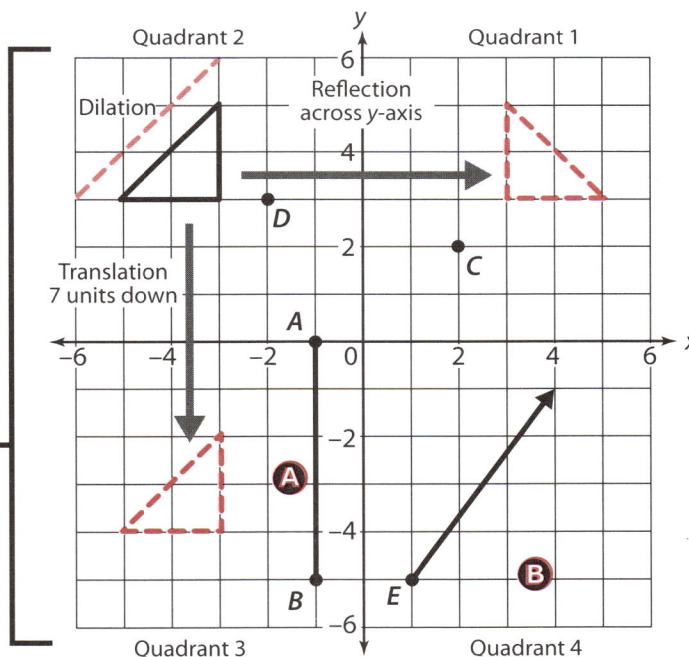

UNIT 4

### ✓ TEST-TAKING TIPS

You will respond to certain questions on mathematics tests using a coordinate grid.

**1  What are the coordinates of point C?**
  **A**  (2, 2)
  **B**  (−2, 2)
  **C**  (2, −2)
  **D**  (−2, 3)
  **E**  (3, −2)

**Directions:** Questions 2 through 4 are based on the information below.

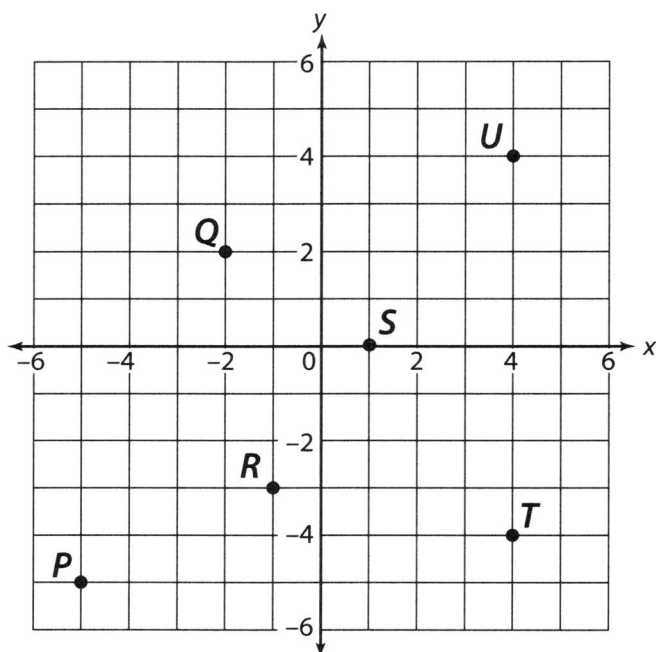

2   What are the coordinates of point *T*?
   A   (5, −4)
   B   (4, −4)
   C   (4, −5)
   D   (4, 4)
   E   (−4, 4)

3   Which of the following ordered pairs describes the location of point *S*?
   A   (1, −1)
   B   (1, 0)
   C   (−1, 0)
   D   (0, 1)
   E   (0, −1)

4   What are the coordinates of point *P*?
   A   (−5, −5)
   B   (−5, 5)
   C   (5, −5)
   D   (5, −6)
   E   (6, −5)

**Directions:** Questions 5 through 7 are based on the information below.

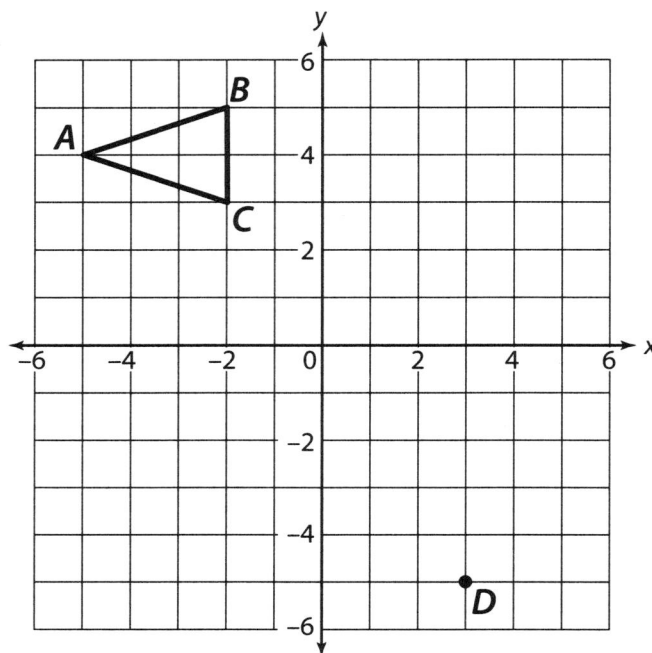

5   What are the coordinates of point *D*?
   A   (3, 5)
   B   (−3, −5)
   C   (−5, 3)
   D   (5, −3)
   E   (3, −5)

6   If triangle *ABC* were reflected across the *y*-axis, what would be the new location of point *C*?
   A   (1, 3)
   B   (2, 5)
   C   (2, 3)
   D   (3, 3)
   E   (2, 4)

7   If triangle *ABC* were translated 3 units down, what would be the new location of point *B*?
   A   (−2, 2)
   B   (−1, −5)
   C   (−2, 0)
   D   (−5, 1)
   E   (−5, 5)

**Directions:** Questions 8 through 12 are based on the information below.

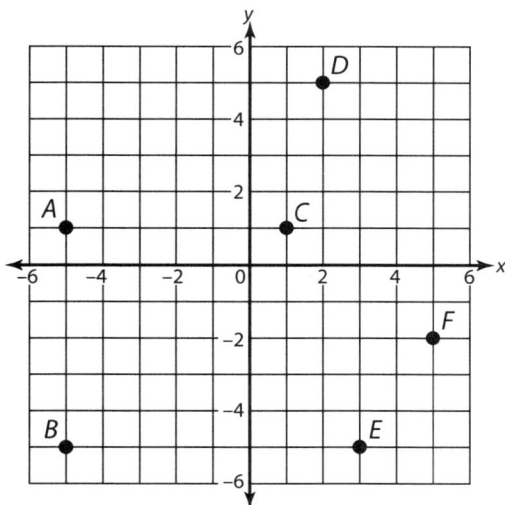

**8** Which of the following ordered pairs identifies the location of point *D*?
- **A** (−5, 2)
- **B** (−2, −5)
- **C** (−2, 5)
- **D** (5, −2)
- **E** (2, 5)

**9** Points *A*, *B*, and *C* mark three corners of a square. What is the location of the fourth corner needed to complete the square?
- **A** (−5, 1)
- **B** (1, −5)
- **C** (−5, −1)
- **D** (−1, −5)
- **E** (1, 5)

**10** Points *C*, *F*, and *E* mark the corners of a rectangle. What is the location of the fourth corner needed to complete the rectangle?
- **A** (−1, −3)
- **B** (−1, −2)
- **C** (−2, −1)
- **D** (0, −2)
- **E** (−2, −2)

**11** What is the new location of point *D* if it is translated 5 units down and 2 units to the right?
- **A** (4, 0)
- **B** (0, 4)
- **C** (0, 0)
- **D** (−3, 3)
- **E** (3, −3)

**12** If point *C* were the center of a circle and the circle were reflected in the *x*-axis, what would be the new location of point *C*?
- **A** (−1, 0)
- **B** (0, −1)
- **C** (0, 1)
- **D** (1, 1)
- **E** (1, −1)

**13** Which of the following points is found in quadrant 2 of the coordinate grid?
- **A** (2, 3)
- **B** (−4, −3)
- **C** (−2, 5)
- **D** (1, −6)
- **E** (−4, −5)

**14** Frank started at point (4, −3). He then moved down 1 and right 2. At which point did he land?
- **A** (6, −4)
- **B** (3, −1)
- **C** (−1, 3)
- **D** (−4, 6)
- **E** (−6, 4)

**15** If point (*x*, −6) is translated 3 units down, what would be the new coordinates?
- **A** (*x* − 3, −6)
- **B** (*x* − 3, −9)
- **C** (*x*, −9)
- **D** (*x*, −3)
- **E** Not enough information is given.

Directions: Questions 16 through 18 are based on the information below.

Directions: Questions 19 through 21 are based on the information below.

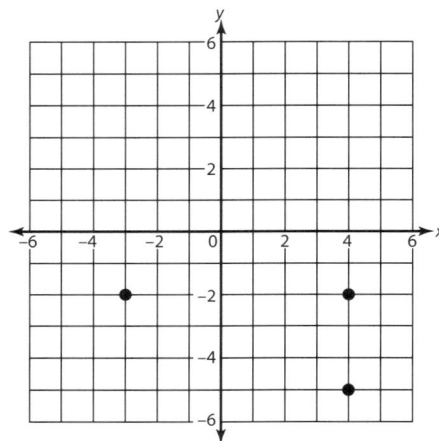

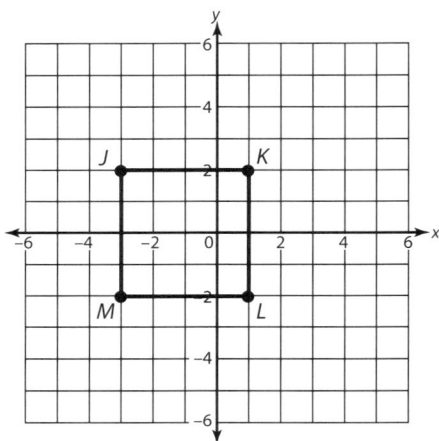

**16** When square *JKLM* is translated 2 units up and 3 units to the right, the new location of point *J* is (0, 4). Which shows the new location of point *M*?

**A** (−1, 1)

**B** (0, 0)

**C** (0, −1)

**D** (0, −5)

**E** (1, 5)

**17** Under a translation, the new point *M*, which is represented by *M'*, is in the same location as the original point *K*. Which of the following describes the translation?

**A** 2 units up and 3 units to the right

**B** 3 units up and 3 units to the right

**C** 4 units up and 4 units to the right

**D** 4 units up and 3 units to the right

**E** 5 units up and 4 units to the right

**18** When square *JKLM* is reflected across the *y*-axis, the new location of point *J* is (3, 2). What is the new location of point *K*?

**A** (0, −1)

**B** (−1, 0)

**C** (−1, 2)

**D** (1, 2)

**E** (2, −1)

**19** Which of the following ordered pairs describes the location of a point that lies in the third quadrant?

**A** (3, 2)

**B** (2, 3)

**C** (−2, 3)

**D** (−3, −2)

**E** (5, −2)

**20** The three points on the coordinate grid mark the corners of a rectangle. What is the location of the fourth corner needed to complete the figure?

**A** (−3, −5)

**B** (−3, −4)

**C** (5, −3)

**D** (4, −3)

**E** (3, 5)

**21** If you drew line segments to connect the three existing points on the coordinate grid, what figure would you draw?

**A** an equilateral triangle

**B** an obtuse triangle

**C** an equiangular triangle

**D** an isosceles triangle

**E** a right triangle

**Directions:** Questions 22 through 24 are based on the information below.

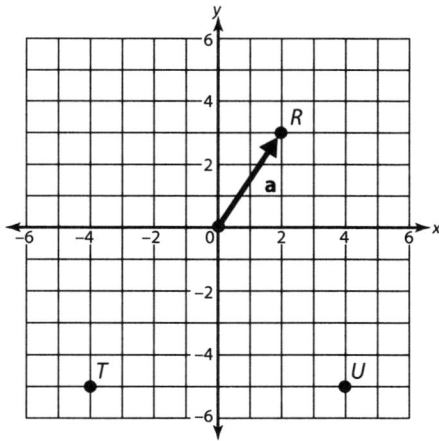

**Directions:** Questions 25 and 26 are based on the information below.

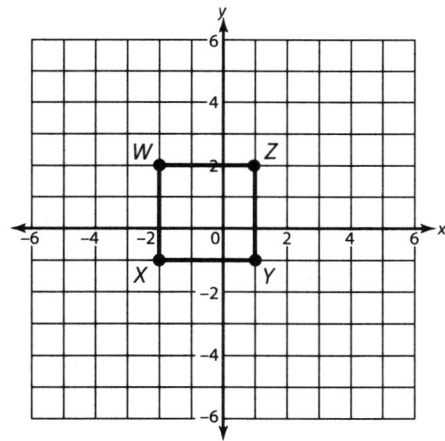

**22** Which pair tells the magnitude of the vector **a**?

   **A** <–2, 2>

   **B** <–1, 2>

   **C** <2, –3>

   **D** <0, 1>

   **E** <0, 0>

**23** If vector **a** were shown on the coordinate grid beginning at point *T*, which ordered pair identifies the point that the arrow tip would meet?

   **A** (4, –1)

   **B** (–2, –2)

   **C** (–4, –1)

   **D** (0, –4)

   **E** (–4, 0)

**24** If you drew a vector **b** <8, –6> on the grid beginning at point U, which direction would the arrow point?

   **A** straight up

   **B** up and to the left

   **C** up and to the right

   **D** down and to the left

   **E** down and to the right

**25** When square *WXYZ* is dilated by a factor of 2 using point *W* as the center of dilation, the location of point *W* remains the same. The new location of the other points is found by drawing a ray from *W* through the other points and marking a point 2 times the distance from *W* to the point. For example, the new location of point *X* would be (–2, –4).

Which point is *Z'*, the new location of point *Z*, under the same dilation?

   **A** (4, 2)

   **B** (4, 3)

   **C** (5, 1)

   **D** (4, –4)

   **E** (3, 2)

**26** If square *WXYZ* is dilated by a factor of $\frac{1}{3}$ using point *Z* as the center of dilation, what is the new location of point *Y*?

   **A** (–1, 1)

   **B** (1, 1)

   **C** (1, 2)

   **D** (1, 3)

   **E** (1, 4)

# Graphing Linear Equations

## ① Learn the Skill

Some equations have two variables. In this case, the value of one variable depends on the other. You can show the possible solutions for an equation with two variables on a graph. A **linear equation** is one that forms a straight line when graphed. All of the solutions of the equation lie on a line. To draw a line, you must find at least two points on the line and connect them.

## ② Practice the Skill

You will find questions relating to equations and graphing equations on high school equivalency mathematics tests. Read the examples and strategies below. Use this information to answer question 1.

**Ⓐ** An equation such as $4x - y = 3$ can be rewritten as $y = 4x - 3$ to make it a function of $x$. In this form, for any value of $x$ it is easy to determine the value of $y$. To find the $y$-intercept point, let $x$ be zero and it is easy to see that $y$ will be $-3$. To graph the equation, choose any other value for $x$ and solve for $y$, then draw a straight line through both points.

**Ⓑ** Use this formula to find the distance between two points:

$$\text{distance between points} = \sqrt{(x_2 - x_1)^2 + (y_2 - y_1)^2}$$

To find the distance between points $(0, -3)$ and $(2, 5)$, substitute the coordinates into the formula. Solve.

$$d = \sqrt{(2 - 0)^2 + (5 - (-3))^2}$$
$$= \sqrt{2^2 + 8^2}$$
$$= \sqrt{4 + 64}$$
$$= \sqrt{68}$$
$$\approx 8.2 \text{ (rounded the nearest tenth)}$$

**Graph** $y = 4x - 3$

Let $x = 0$.
$y = 4(0) - 3$
$y = 0 - 3$
$y = -3$
Plot $(0, -3)$.

Let $x = 2$.
$y = 4(2) - 3$
$y = 8 - 3$
$y = 5$
Plot $(2, 5)$.

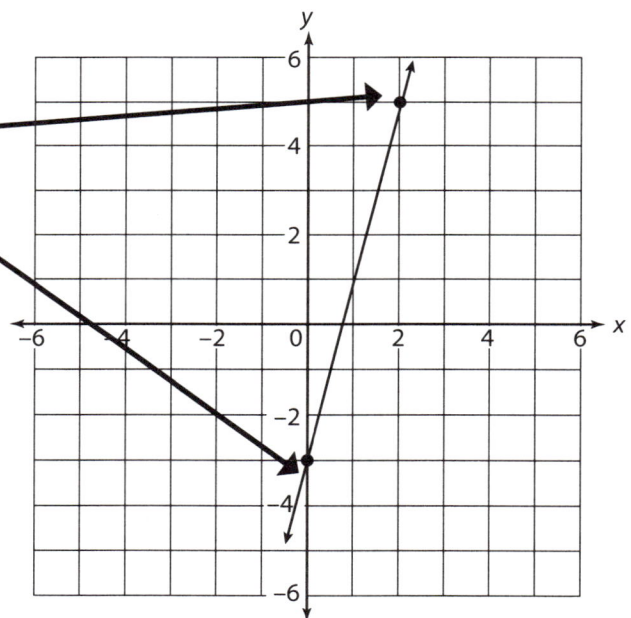

1   Which ordered pair is a solution to $2x + y = 5$?
   A   $(-1, 3)$
   B   $(3, -1)$
   C   $(0, -5)$
   D   $(-3, -4)$
   E   $(-2, 6)$

**2** Which of the following ordered pairs is a point on the line of the equation $x + 2y = 4$?
  A  $(-2, 0)$
  B  $(1, 3)$
  C  $(0, 2)$
  D  $(-3, -1)$
  E  $(2, -4)$

**3** Which ordered pair is a solution to $2x - y = 0$?
  A  $(0, 0)$
  B  $(1, -2)$
  C  $(0, 2)$
  D  $(-1, 2)$
  E  $(2, -2)$

**4** What is the missing $x$-value if $(x, 3)$ is a solution to the function $y = 2x + 2$?
  A  $-1$
  B  $-\dfrac{1}{2}$
  C  $0$
  D  $\dfrac{1}{2}$
  E  $1$

**5** A segment is drawn from the origin to $(-4, 3)$. What is the length of the segment?
  A  1.0
  B  2.6
  C  5.0
  D  7.0
  E  12.0

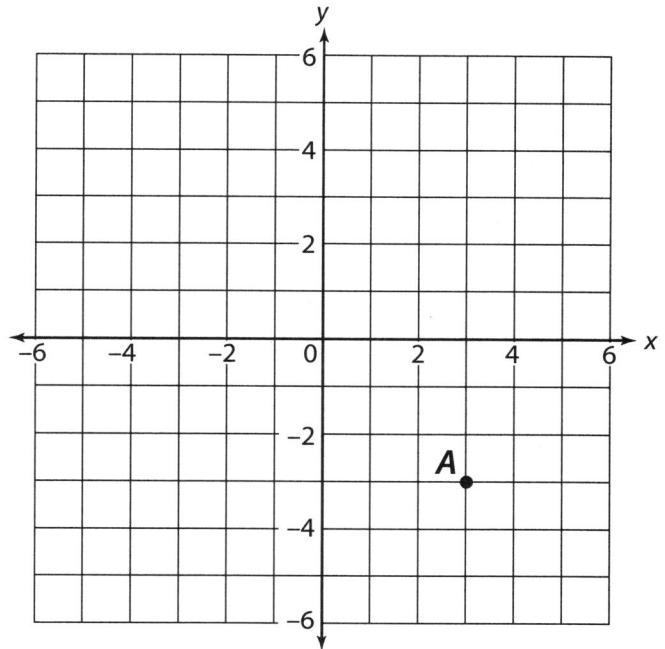

**6** Point $A$ lies on a line of the equation $x + 2y = -3$. Which of the following is another point on this line?
  A  $(0, -3)$
  B  $(-1, 2)$
  C  $(0, -2)$
  D  $(-5, 1)$
  E  $(4, -3)$

**7** Two points are located at $(2, 5)$ and $(4, 3)$. What is the distance between the points to the nearest hundredth?
  A  1.41
  B  2.00
  C  2.45
  D  2.65
  E  2.83

**8** Marvin walks a straight line from $(-5, 2)$ to $(-3, 1)$ and stops. Then he walks a straight line from $(-3, 1)$ to $(-1, -4)$. What distance did Marvin walk?
  A  14.94
  B  9.04
  C  7.62
  D  6.00
  E  5.83

**9** Which ordered pair is a solution of the function $y = \frac{1}{2}x$?

A  (4, 8)

B  (1, 3)

C  (4, 2)

D  (4, 1)

E  (1, 2)

**10** What is the missing $y$-value if (2, $y$) is a solution of $-x = y + 1$?

A  −3

B  −2

C  −1

D  0

E  1

**11** What is the missing $x$-value if ($x$, −3) is a solution of $2x + 2y = -8$?

A  −3

B  −1

C  0

D  1

E  3

**12** The graph of the equation $y = 4 - 3x$ would pass through which of the following points on the coordinate grid?

A  (1, −1)

B  (3, 3)

C  (4, 8)

D  (3, 1)

E  (2, −2)

**13** What is the missing $x$-value if ($x$, 1) is a solution of $3x - y = 5$?

A  2

B  3

C  6

D  7

E  9

**14** Which of the following shows the graph of the equation $x + 2y = 2$?

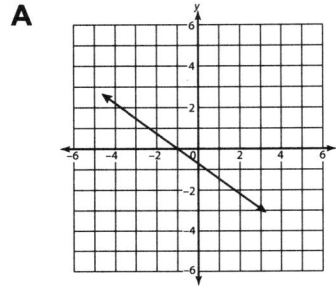

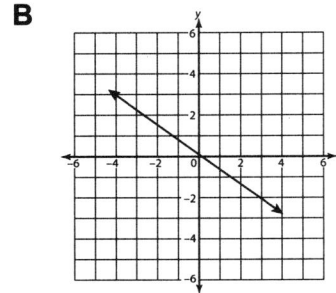

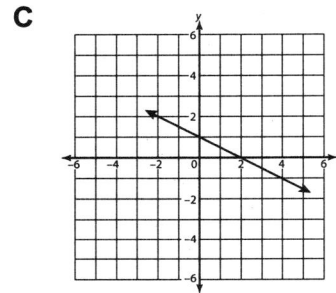

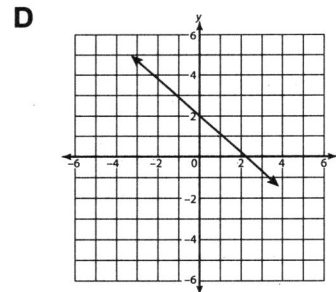

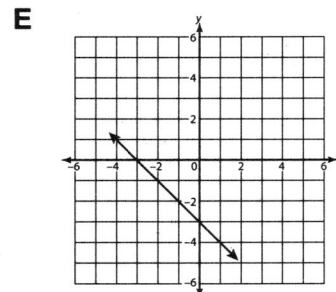

A

B

C

D

E

**Lesson 7 | Graphing Linear Equations**

**15** Point *B* is located at (4, 7) on a coordinate grid. If a line were drawn directly from the point to the origin, what would be the length of the line to the nearest tenth?

  A   3.0
  B   7.9
  C   8.0
  D   8.1
  E   8.2

**16** The points graphed on the grid below satisfy which of the following equations?

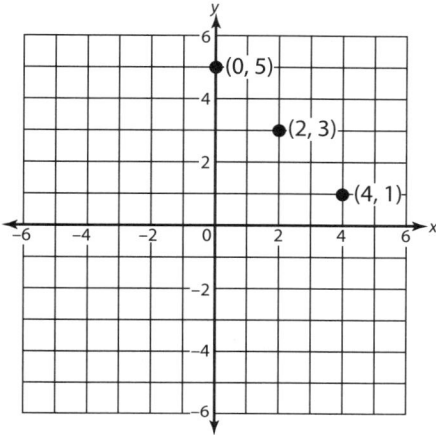

  A   $x + y = -3$
  B   $y - x = 5$
  C   $x + y = 2$
  D   $x - y = -1$
  E   $x + y = 5$

**17** Two points are located at (−2, −5) and (−3, −8). What is the distance between these two points?

  A   3.2
  B   4.6
  C   5.8
  D   8.1
  E   13.9

**18** The graph of the equation $y = -2x - 1$ passes through which of the following points?

  A   (1, −3)
  B   (1, −2)
  C   (0, 1)
  D   (0, 2)
  E   (−1, 2)

**Directions:** Questions 19 through 21 are based on the grid below.

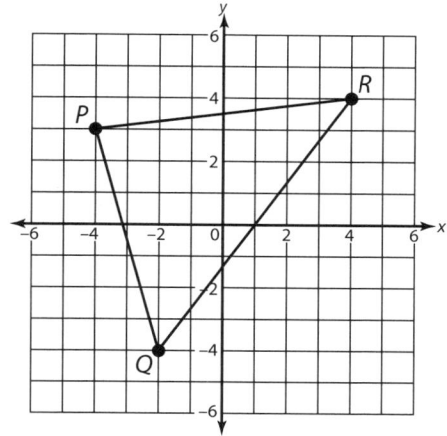

**19** What is the length of side *PQ* to the nearest tenth?

  A   6.4
  B   7.1
  C   7.3
  D   8.1
  E   9.2

**20** What is the length of side *PR* to the nearest tenth?

  A   6.4
  B   7.3
  C   8.1
  D   9.2
  E   10.0

**21** What is the perimeter of triangle *PRQ* to the nearest tenth?

  A   10.0
  B   12.5
  C   15.2
  D   25.4
  E   30.4

**22** Which ordered pair is a solution of $2x - 3 = y$?
  A  (0, 3)
  B  (1, 1)
  C  (−2, 1)
  D  (3, −3)
  E  (4, 5)

**23** Which equation of a line is shown on the graph?

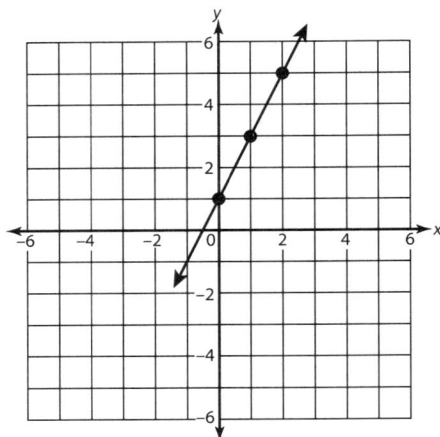

  A  $2x - y = 1$
  B  $\frac{1}{2}x = y$
  C  $2x + 1 = y$
  D  $\frac{1}{2}x + 1 = y$
  E  $2y - 1 = x$

**24** Which of the following points is found on the graph of the function $y = 4 - 2x$?
  A  (−2, 3)
  B  (4, 0)
  C  (−2, 0)
  D  (0, −4)
  E  (3, −2)

**Directions:** Questions 25 and 26 are based on the information below.

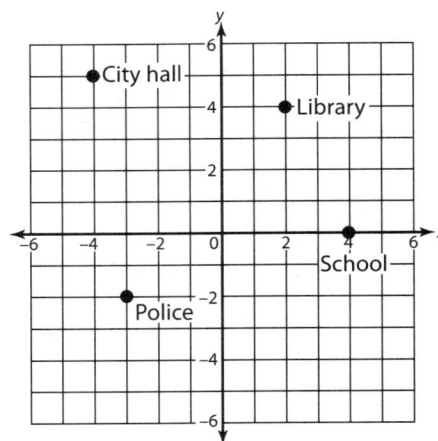

Jack sketched out a map of his city using a coordinate grid. Each unit on the grid represents 1 mile.

**25** A bike path was made that goes straight from the library to the school. How many miles long is the bike path to the nearest tenth of a mile?
  A  2.7
  B  3.2
  C  4.5
  D  5.7
  E  7.0

**26** A police officer drove from the police station to city hall by driving 1 mile straight west and then driving straight north. How many miles straight north did the officer drive?
  A  4
  B  5
  C  6
  D  7
  E  8

**27** What is the distance, to the nearest tenth, between two points on a graph that are located at (−5, 1) and (5, 5)?
  A  6.4
  B  10.8
  C  11.0
  D  11.7
  E  12.8

UNIT 4

# Slope and Intercepts

## ① Learn the Skill

The **slope** is a number that measures the steepness of a line. Slope can be positive, negative, or zero. You can find the slope of a line by counting spaces on a graph or by using an algebraic formula. By rewriting an equation as a function of $x$, you can find the slope and the **intercept point** where the line crosses the $y$-axis.

## ② Practice the Skill

The slope of a line is constant, meaning that the line always climbs or falls at the same rate. A line that rises from left to right has a positive slope. A line that falls from left to right has a negative slope. A horizontal line has a slope of zero, and a vertical line has an undefined slope, or no slope. If two lines are parallel, they have the same slope. Examine the grids and strategies below. Use this information to answer question 1.

**Ⓐ** Use two points to find a slope. Start at the lower point. How many units must you climb to reach the other point? This is the *rise*, or numerator. How many units must you move left or right to reach the point? This is the *run*, or denominator. If you move left, the value is negative. There is also an algebraic formula you can use to find slope.

**Ⓑ** To find the equation of a line, find the $y$-intercept (where the line crosses the $y$-axis). The line crosses the $y$-axis at $-2$. Next, find the slope. The slope of this line is $-1$. Substitute the values of $m$ and $b$ into the equation.

$$y = mx + b$$
$$y = -1x + (-2)$$
$$y = -x - 2$$

Run, Rise, (0, 4), (−3, 0)

$m = \dfrac{2}{-2} = -1$, $y$-intercept

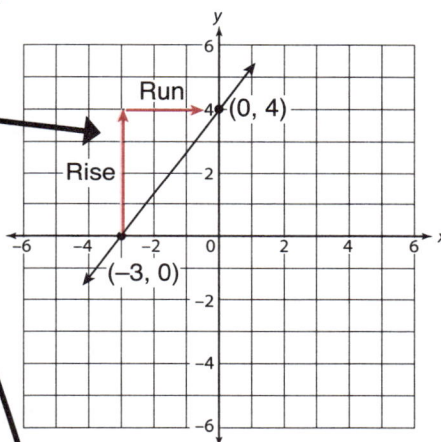

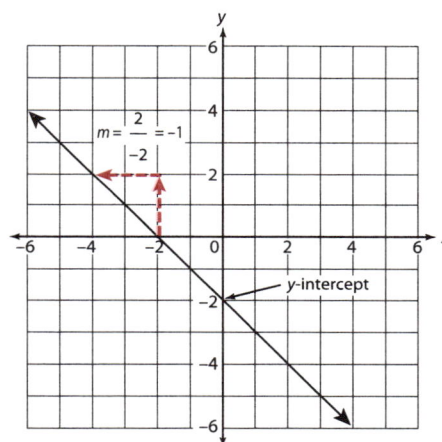

$$\text{Slope } (m) = \frac{y_2 - y_1}{x_2 - x_1} = \frac{4 - 0}{0 - (-3)} = \frac{4}{3}$$

**UNIT 4**

### ✓ TEST-TAKING TIPS

Notice that the slopes of $\frac{-1}{2}$ and $\frac{1}{-2}$ have the same value. Both show a negative slope.

However, $\frac{-1}{-2}$ actually shows a positive slope, because a negative divided by a negative equals a positive.

**1** What is the slope of a line that passes through (−1, 3) and (1, 4)?

A   $-\dfrac{1}{2}$

B   0

C   $\dfrac{1}{2}$

D   $\dfrac{2}{3}$

E   1

**Directions:** Questions 2 and 3 refer to the grid below.

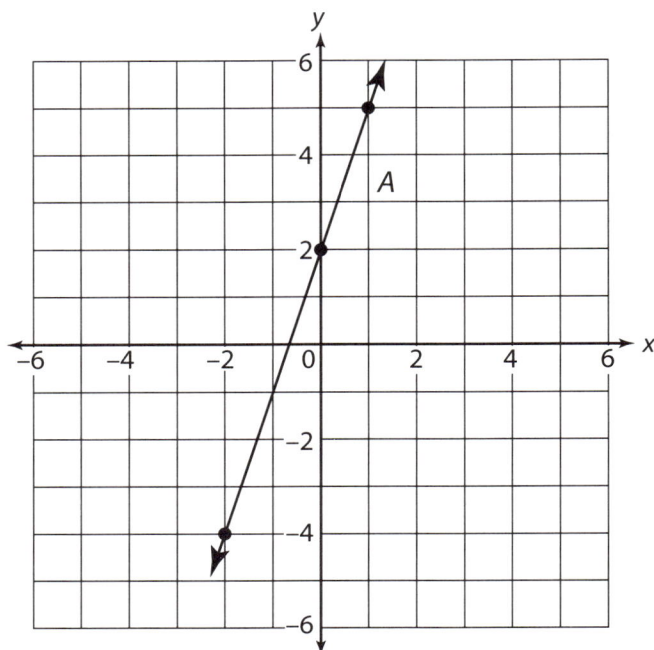

2   The following points lie on line *A*: (−2, −4), (0, 2), and (1, 5). What is the slope of line *A*?
   A   3
   B   2
   C   1
   D   −2
   E   −3

3   What is the equation of line *A*?
   A   $y = \frac{1}{2}x + 3$

   B   $y = \frac{1}{3}x + 2$

   C   $y = 2x + 3$

   D   $y = 3x + 2$
   E   $y = 3x + 3$

4   A ramp was built to allow wheelchair access to a front door. The ramp rises 2 feet, as shown in the diagram below.

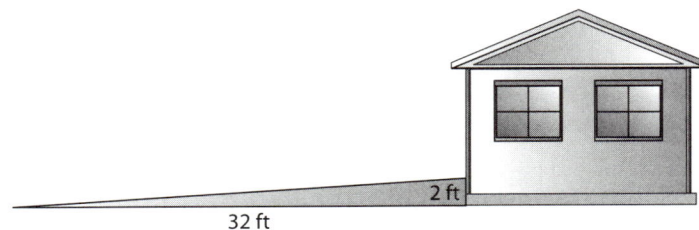

**What is the slope of the ramp?**
   A   $\frac{1}{32}$

   B   $\frac{1}{18}$

   C   $\frac{1}{16}$

   D   $\frac{1}{8}$

   E   $\frac{1}{4}$

5   A linear function is represented by $f(x) = 2$. What is the slope of the line?
   A   −2
   B   −1
   C   0
   D   1
   E   2

6   Which equation shows a line parallel to $4 - y = 2x$?
   A   $2 + y = 2x$

   B   $y - 2 = \frac{1}{2}x$

   C   $2 + \frac{1}{2}y = 2x$

   D   $-y = 2 - 2x$

   E   $y = -2x + 2$

**7** The points (−4, 4) and (2, 3) lie on line *H*. What is the slope of line *H*?

A $-\frac{1}{6}$

B $-\frac{1}{3}$

C $-\frac{1}{2}$

D 2

E 6

**8** What is the slope of a line that passes through points (−1, −2) and (−3, −4)?

A 1

B $\frac{1}{2}$

C 0

D $-\frac{1}{2}$

E −2

**9** A linear equation is represented by $y = \frac{1}{2}x + 3$. The graph of which of the following equations would be parallel to that of the equation above?

A $y = -\frac{1}{3}x + 2$

B $y = \frac{1}{2}x - 3$

C $y = 2x + 3$

D $y = x - 3$

E $y = x + 3$

**10** Line *B* has a slope of −1. It passes through point *K* at (4, −2), and it passes through point *L*, which has an *x*-coordinate of 2. What are the coordinates of point *L*?

A (2, −2)

B (0, 2)

C (2, 0)

D (−2, 0)

E Not enough information is given.

**11** What is the equation of the line shown in the grid?

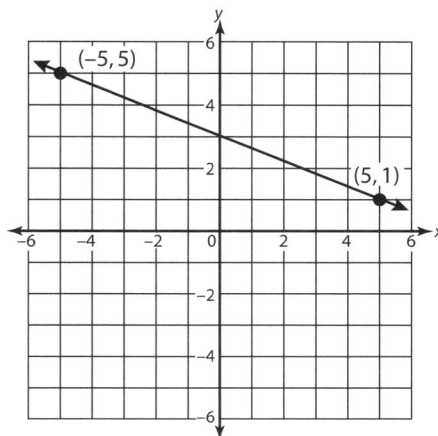

A $y = -\frac{2}{5}x + 3$

B $y = -\frac{1}{3}x + 3$

C $y = \frac{2}{5}x + 3$

D $y = \frac{1}{3}x + 3$

E $y = 3x + 3$

**12** Which of the following equations shows the slope-intercept form for a line with a slope of 2 that passes through point (4, 1)?

A $y = -2x + 5$

B $y = 2x - 4$

C $y = 2x + 3$

D $y = 2x + 1$

E $y = 2x - 7$

**13** Andrea paid an initial fee of $20 to set up her cell phone. Now she pays $30 per month for service. The amount she pays for cell phone service for a certain number of months can be graphed in the first quadrant of a coordinate grid.

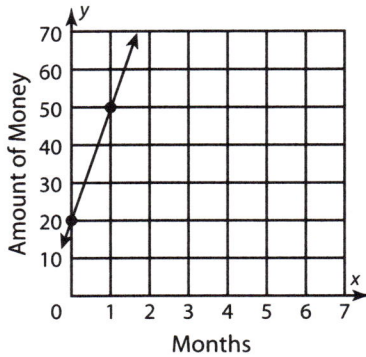

Which of the following is the equation of the line on the graph?

**A** $y = 30x - 20$

**B** $y = \frac{1}{3}x + 20$

**C** $y = 20x + 30$

**D** $y = 30x + 20$

**E** $y = 20x - 30$

**14.** What is the slope of line $T$?

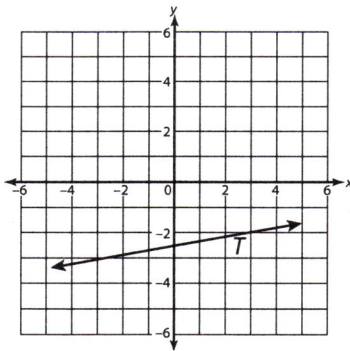

**A** $-\frac{1}{6}$

**B** $\frac{1}{6}$

**C** $\frac{1}{3}$

**D** $-\frac{1}{3}$

**E** $\frac{2}{3}$

**15** The slope of the roof shown on the house is $\frac{1}{3}$.

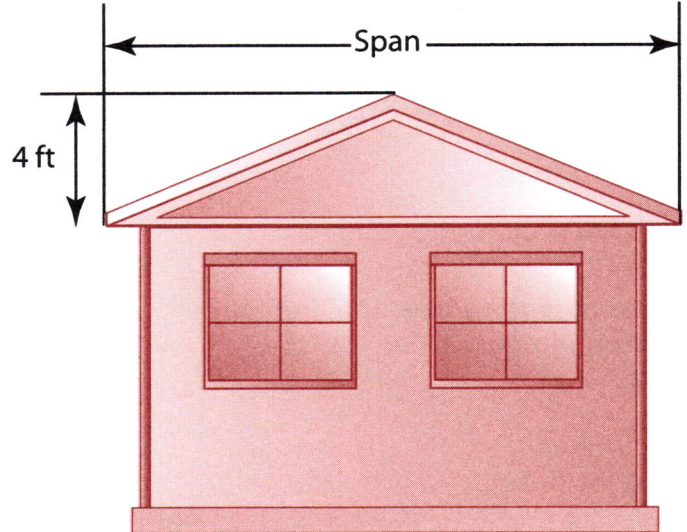

What is the span of the roof?

**A** 4 ft

**B** 12 ft

**C** 18 ft

**D** 24 ft

**E** 30 ft

**16** Line $J$ has a slope of $-\frac{1}{2}$. It passes through point $Q$ at $(-4, -3)$ and through point $R$, which has an $x$-coordinate of 0. What is the equation of line $J$?

**A** $y = -\frac{1}{2}x - 5$

**B** $y = -\frac{1}{2}x - 4$

**C** $y = -\frac{1}{2}x - 3$

**D** $y = -\frac{1}{2}x + 4$

**E** $y = -\frac{1}{2}x + 5$

**Directions:** Questions 17 through 20 are based on the information below.

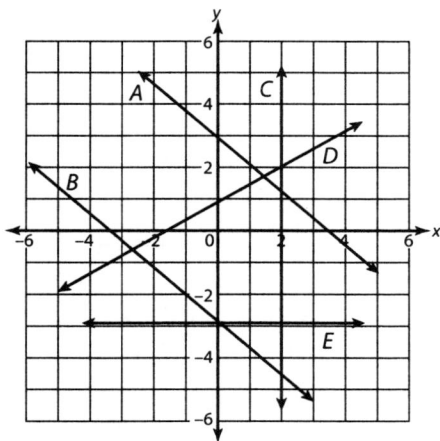

**17** Which of the lines has a positive slope?

   **A** line *A*
   **B** line *B*
   **C** line *C*
   **D** line *D*
   **E** line *E*

**18** Which of the lines has a slope of zero?

   **A** line *A*
   **B** line *B*
   **C** line *C*
   **D** line *D*
   **E** line *E*

**19** Which of the lines has no slope?

   **A** line *A*
   **B** line *B*
   **C** line *C*
   **D** line *D*
   **E** line *E*

**20** Which two lines are parallel to each other?

   **A** line *A* and line *B*
   **B** line *B* and line *C*
   **C** line *C* and line *D*
   **D** line *D* and line *E*
   **E** line *E* and line *A*

**Directions:** Questions 21 through 23 are based on the information below.

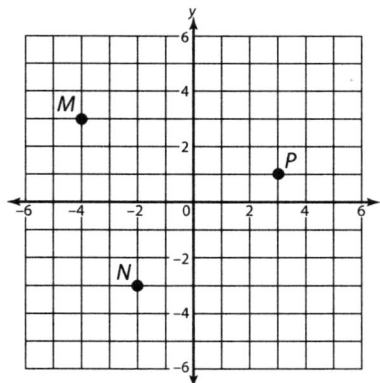

**21** What would be the slope of a line drawn through points *M* and *N*?

   **A** $-3$
   **B** $-\frac{1}{3}$
   **C** $1$
   **D** $\frac{1}{3}$
   **E** $3$

**22** Which of the following could be the equation of a line that is parallel to a line drawn through points *N* and *P*?

   **A** $y = \frac{5}{4}x + 4$
   **B** $y = -\frac{1}{5}x + 4$
   **C** $y = -4x + 3$
   **D** $y = 4x + 2$
   **E** $y = \frac{4}{5}x - 2$

**23** What would be the slope of a line that is parallel to a line drawn through points *M* and *P*?

   **A** $2$
   **B** $-\frac{1}{2}$
   **C** $-\frac{1}{7}$
   **D** $-\frac{2}{7}$
   **E** $-2$

# Graphing Quadratic Equations

## ① Learn the Skill

An equation for which the graph does not form a straight line is called a nonlinear equation. **Quadratic equations** are equations set in the form $ax^2 + bx + c = 0$, where $a$ does not equal 0. Quadratic equations exhibit various characteristics when shown graphically. These include zero, one or two points where the plot of such an equation crosses the $x$-axis, one point where it crosses the $y$-axis, either a maximum when $a < 0$ or a minimum when $a > 0$, and symmetry with respect to that maximum or minimum. The coefficients in the equation—$a$, $b$, and $c$—can quantify these characteristics. For example, larger values of $a$ will contract a curve, while smaller values of $a$ will expand a curve. Negative values of $a$ will turn it upside down. You also can use plotted data and knowledge of characteristics such as $x$- and $y$-intercepts to help determine the coefficients.

## ② Practice the Skill

By practicing graphing quadratic equations, you will improve your test-taking abilities. Study the information and graph below. Use this information to answer question 1.

**Ⓐ** Curves cross the $x$-axis when $y = 0$; here $\frac{1}{3}x^2 + x - 4 = 0$. Values can be found by factoring, use of the quadratic formula, and so on.

**Ⓑ** Curves cross the $y$-axis when $x = 0$; the constant term, in this case $-4$, is the $y$-intercept.

**Ⓒ** Curves go through a minimum when $a > 0$, and a maximum when $a < 0$. Here, $a = \frac{1}{3} > 0$, so the curve has a minimum; $y$ increases as one moves away from the minimum in either direction. Curves are symmetric about the maximum or minimum; for example, the two points where the curve crosses the $x$-axis are the same distance from the minimum.

The function $y = ax^2 + bx + c = \frac{1}{3}x^2 + x - 4$ is plotted in the following graph.

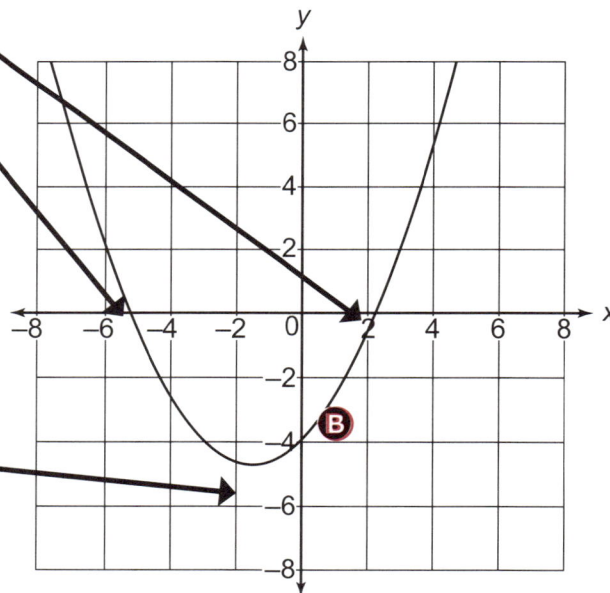

### USING LOGIC

The formula for the $x$-value of the maximum or minimum is $x = \frac{-b}{2a}$.

**1** What are the coordinates of the minimum in the curve?
A  (−1.45, −4.50)
B  (−1.45, −4.75)
C  (−1.50, −4.50)
D  (−1.50, −4.75)
E  (−1.75, −4.50)

**Directions:** Questions 2 through 4 are based on the quadratic equation below.

$$y = x^2 + 2x - 8$$

**2** Which of the following pairs of *x*-values represent where the curve crosses the *x*-axis?
A $x = -8, x = 8$
B $x = -4, x = 2$
C $x = -4, x = 4$
D $x = -2, x = 4$
E $x = -2, x = 2$

**3** Which of the following values of *y* represents where the curve crosses the *y*-axis?
A $y = -8$
B $y = -4$
C $y = 2$
D $y = 4$
E $y = 8$

**4** Which of the following values of *x* represents where the curve goes through a minimum?
A $x = -2$
B $x = -1$
C $x = 0$
D $x = 1$
E $x = 2$

**5** Which value of *y* corresponds to the point where the curve defined by $y = 2x^2 - 5x + 3$ crosses the *y*-axis?
A $y = -5$
B $y = -3$
C $y = 3$
D $y = 5$
E $y = 8$

**6** Indicate whether the curve defined by $y = -3x^2 + 12x - 5$ goes through a maximum or a minimum and the value of *x* at which it occurs.
A Minimum, $x = -2$
B Minimum, $x = +2$
C Maximum, $x = -2$
D Maximum, $x = +2$
E Maximum, $x = 4$

**Directions:** Questions 7 through 10 are based on the information below.

The following graph features plots of five different quadratic equations of the form $y = ax^2 + bx + c$, identified by letters *A* through *E*.

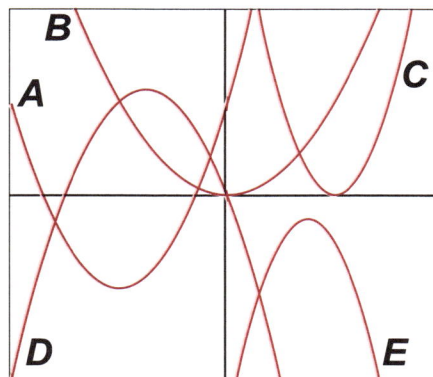

**7** Which curves correspond to equations with $a < 0$?
A Curve *D* only
B Curves *A* and *D*
C Curves *B* and *C*
D Curves *C* and *E*
E Curves *D* and *E*

**8** Which curves correspond to equations with $b = 0$?
A only Curve *B*
B only Curve *C*
C Curves *B* and *C*
D Curves *B* and *D*
E Curves *B* and *E*

**9** Which curves correspond to equations with $\dfrac{b}{2a} < 0$?
A Curves *A* and *D*
B Curves *A* and *E*
C Curves *B* and *C*
D Curves *C* and *E*
E Curves *D* and *E*

**10** Which curves correspond to equations with $c = 0$?
A Curve *B* only
B Curve *C* only
C Curves *B* and *C*
D Curves *B* and *D*
E Curve *E*

**11** If the point at (−2, 7) is the maximum, what negative value of $x$ corresponds to a $y$-value of −2?

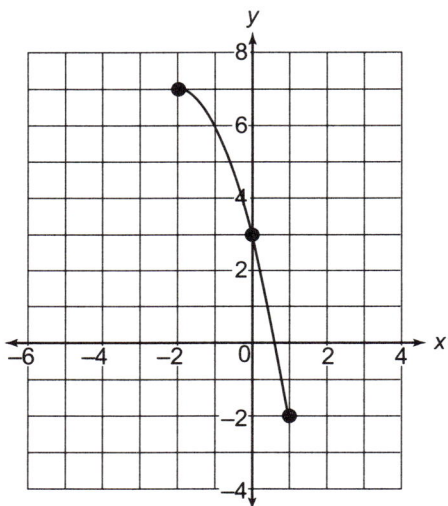

    **A** −3
    **B** −4
    **C** −5
    **D** −6
    **E** −8

**Directions:** Questions 12 through 14 are based on the information below.

    Two quadratic equations are plotted on the graph below. The one to the left is $y = \frac{1}{2}x^2 + 4x + 4$, while the one to the right is $y = -2x^2 + 16x - 24$.

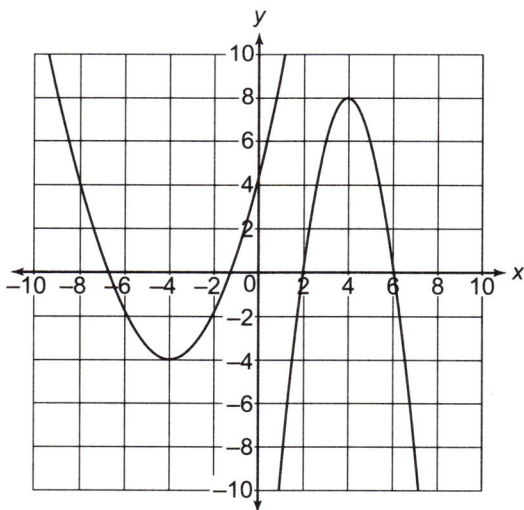

**12** At what $y$-value does the right-hand curve cross the $y$-axis?

    **A** −16
    **B** −24
    **C** −32
    **D** −40
    **E** −48

**13** At what $x$-values does the left-hand curve cross the $x$-axis, expressed to the nearest tenth?

    **A** $x = -6.8$, $x = -1.2$
    **B** $x = -6.7$, $x = -1.2$
    **C** $x = -6.8$, $x = -1.1$
    **D** $x = -6.7$, $x = -1.1$
    **E** $(-7.6, -2.1)$

**14** The following graph shows the path of a ball thrown horizontally from a height of 36 feet. The ball travels a horizontal distance of 6 feet by the time it lands.

**What is the quadratic equation corresponding to the path of the ball?**

    **A** $y = -x^2 - 36$
    **B** $y = -x^2 + 36$
    **C** $y = -x^2 + 6x - 36$
    **D** $y = -x^2 + 6x + 36$
    **E** $y = -x + 6x - 36$

UNIT 4

**Directions:** Questions 15 through 18 are based on the information below.

The following graph features plots of five different quadratic equations of the form $y = ax^2 + bx + c$, identified by letters *A* through *E*.

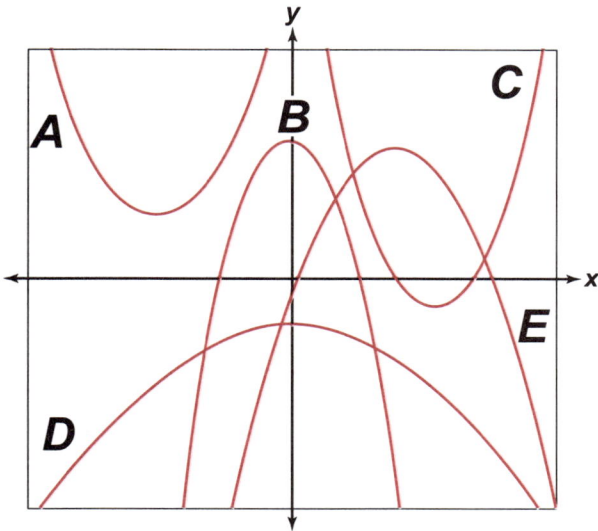

**15** Which curves correspond to equations with $a < 0$?
   A   Curve *A* only
   B   Curve *D* only
   C   Curves *A* and *C*
   D   Curves *B*, *D*, and *E*
   E   Curves *C*, *D*, and *E*

**16** Which curves correspond to equations with $b = 0$?
   A   Curve *D* only
   B   Curves *B* and *D*
   C   Curve *C* only
   D   Curves *A* and *C*
   E   Curves *A* only

**17** Which curves correspond to equations with $\frac{b}{2a} > 0$?
   A   Curve *A* only
   B   Curve *C* only
   C   Curve *D* only
   D   Curves *C* and *D*
   E   Curves *C* and *E*

**18** Which curve corresponds to the equation with the most negative value of *a*?
   A   Curve *A*
   B   Curve *B*
   C   Curve *C*
   D   Curve *D*
   E   Curve *E*

**Directions:** Questions 19 through 21 are based on the quadratic equation below.

$$y = -2x^2 - 4x + 6$$

**19** Which pair of *x*-values represent where the curve crosses the *x*-axis?
   A   $x = 3, x = 1$
   B   $x = 1, x = 3$
   C   $x = -1, x = 3$
   D   $x = -3, x = 1$
   E   $x = -3, x = -1$

**20** Which value of *y* represents where the curve crosses the *y*-axis?
   A   $y = -6$
   B   $y = -2$
   C   $y = 2$
   D   $y = 6$
   E   $y = 12$

**21** Which value of *x* represents where the curve goes through a minimum?
   A   $x = 2$
   B   $x = 1$
   C   $x = 0$
   D   $x = -1$
   E   $x = -2$

**22** If the point at (2, −5) is the minimum, what positive value of x corresponds to a y-value of +3?

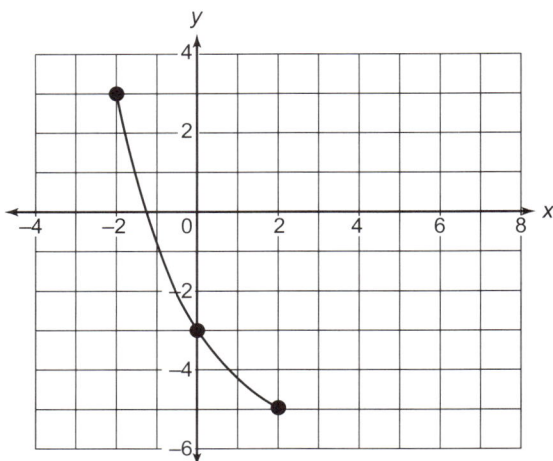

**A** 7
**B** 6
**C** 5
**D** 4
**E** 3

**Directions:** Questions 23 through 26 are based on the information below.

The diagram shows a ball player, standing at the origin, throwing a ball to another player. The ball's path, expressed in feet, obeys the equation $y = -\frac{1}{144}x^2 + x + 6$, which assumes the ball is 6 ft off the ground when it is thrown and caught.

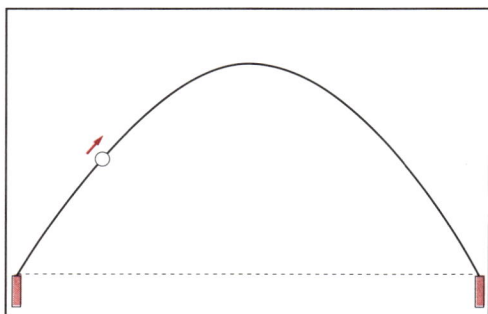

**23 How far apart are the two ball players?**
**A** 36 feet
**B** 72 feet
**C** 108 feet
**D** 144 feet
**E** 148 feet

**24 When the ball reaches its highest point, how far has it traveled horizontally?**
**A** 36 feet
**B** 72 feet
**C** 108 feet
**D** 144 feet
**E** 148 feet

**25 How high off the ground is the ball at its highest point?**
**A** 30 feet
**B** 36 feet
**C** 42 feet
**D** 72 feet
**E** 108 feet

**26 Suppose the ball were thrown harder, so that it reached its maximum height at a horizontal distance of 92 feet. How much greater would the *total* horizontal distance traveled be, assuming its path is still represented by a quadratic function?**
**A** 0 feet
**B** 20 feet
**C** 40 feet
**D** 60 feet
**E** 80 feet

Lesson 9 | Graphing Quadratic Equations

# Comparison of Functions

## ① Learn the Skill

A **function** is a relation in which each input has exactly one output. Functions can be represented by sets of ordered pairs in tables, in graphs, algebraically, or by verbal descriptions. Two or more functions can be compared based on their slopes or rates of change, intercepts, the locations and values of minimums and maximums, and other features. You can compare two linear functions, two quadratic functions, or a linear function and a quadratic function.

When functions are compared, they may be presented in the same way or in different ways. For example, you may want to compare the rates of change of one function represented by a table of values and another function represented by an algebraic expression.

## ② Practice the Skill

By practicing the skill of comparing functions, you will improve your study and test-taking abilities, especially as they relate to the mathematics high school equivalency tests. Study the graph, table, and information below. Use this information to answer question 1.

**Ⓐ** The rate of change of a linear function is also known as its slope. In a graph, the rate of change is the ratio of the vertical change, or *rise*, to the horizontal change, or *run*. The function represented by this graph has a rate of change of $\frac{2}{3}$, meaning it rises 2 spots and runs 3 spots. Its intercepts are $y = -2$ and $x = 3$. In a table, the rate of change is the ratio of the change in $y$-value to the change in $x$-value. The function represented by this table has a rate of change of 2 and a $y$-intercept of 1.

**Ⓑ** The intercepts of a function are the values of one coordinate when the other coordinate is zero. In a graph, look for points that cross the axes. In a table, look at the rows in which one value is 0.

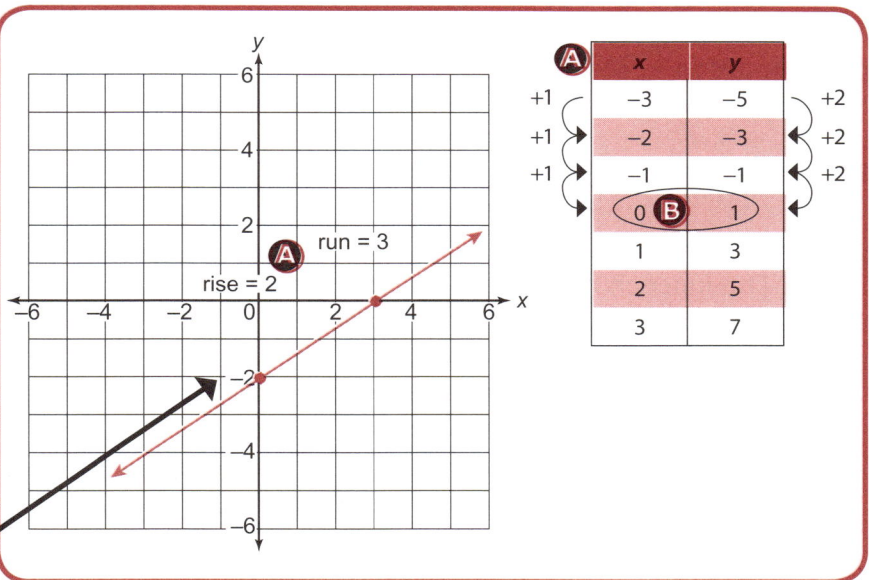

| x | y |
|---|---|
| −3 | −5 |
| −2 | −3 |
| −1 | −1 |
| 0 | 1 |
| 1 | 3 |
| 2 | 5 |
| 3 | 7 |

### ✓ TEST-TAKING TIPS

When functions are represented differently (one in a graph, the other in a table), you may want to change the representation of one or both functions.

**1** A function has a rate of change that is greater than the rate of change shown in the graph above and less than the rate of change shown in the table above. Which equation could represent the function?

**A** $f(x) = 3x + 2$

**B** $f(x) = \frac{1}{2}x - 1$

**C** $f(x) = x + 3$

**D** $f(x) = \frac{5}{2}x + 2$

**E** $f(x) = 3x - 2$

**Directions:** Questions 2 through 4 are based on the information below.

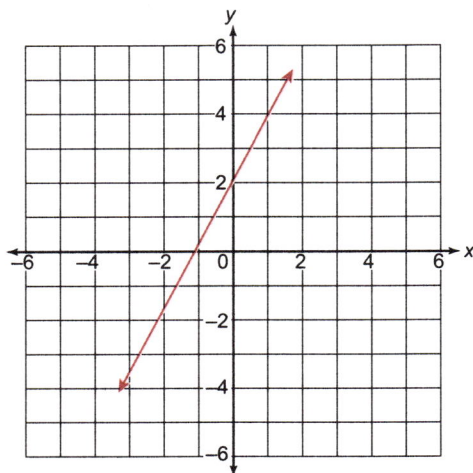

**2** Which function has the same *y*-intercept as the function represented in the graph?
A   $f(x) = x - 2$
B   $f(x) = 2x + 3$
C   $f(x) = -3x + 2$
D   $f(x) = -6x - 1$
E   $f(x) = 3x - 2$

**3** A function is represented by the set of ordered pairs below.

{(−2, 2), (0, 6), (2, 10), (4, 14)}

Which statement is true?
A   The function has the same rate of change and *y*-intercept as the function represented in the graph.
B   The function has the same rate of change and a different *y*-intercept from the function represented in the graph.
C   The function has the same *y*-intercept and a different rate of change from the function represented in the graph.
D   The function has a different rate of change and *y*-intercept from the function represented in the graph.
E   None of the statements are true.

**4** For *x* = −2, which function has the same value as the function represented in the graph?
A   $f(x) = -x$
B   $f(x) = \frac{x}{2}x + 1$
C   $f(x) = x + 4$
D   $f(x) = 6x + 10$
E   $f(x) = x - 4$

**5** Which function has the same *x*-intercepts as the function represented in the table?

| x | y |
|---|---|
| −3 | −5 |
| −2 | 0 |
| −1 | 3 |
| 0 | 4 |
| 1 | 3 |
| 2 | 0 |
| 3 | −5 |

A   $f(x) = \frac{1}{2}x^2 - 2$
B   $f(x) = \frac{1}{2}x^2 + 2$
C   $f(x) = 2x^2 - 2$
D   $f(x) = 2x^2 + 2$
E   $f(x) = 3x - 2$

**6** For the function represented in the graph below, which is true?

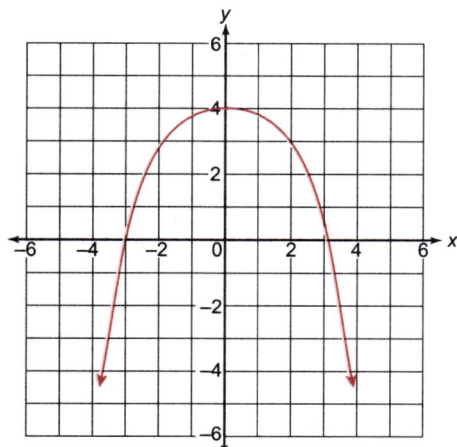

| x | y |
|---|---|
| −3 | −5 |
| −2 | 0 |
| −1 | 3 |
| 0 | 4 |
| 1 | 3 |
| 2 | 0 |
| 3 | −5 |

**A** The function represented in the graph has the same maximum value as the function represented in the table.

**B** The function represented in the graph has the same minimum value as the function represented in the table.

**C** The function represented in the graph has both the same minimum value and the same maximum value as the function represented in the table.

**D** The function represented in the graph has neither the same minimum value nor the same maximum value as the function represented in the table.

**E** None of the statements are true.

**Directions:** Questions 7 and 8 are based on the information below.

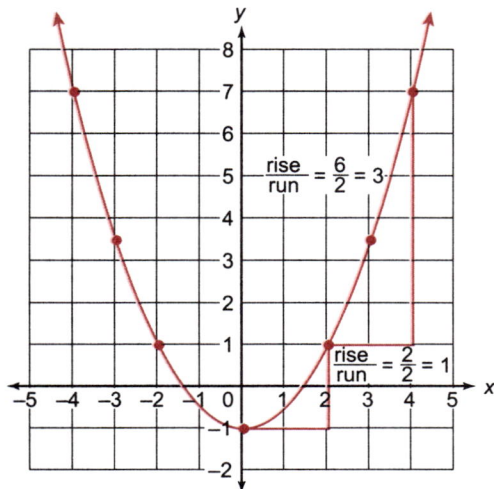

| x | y |
|---|---|
| −4 | −4 |
| −3 | −2 |
| −2 | 0 |
| −1 | 2 |
| 0 | 4 |
| 1 | 6 |
| 2 | 8 |
| 3 | 10 |
| 4 | 12 |

**7** Which function has a lesser rate of change than the average rate of change of the quadratic function over the interval $x = 0$ to $x = 3$?

**A** $f(x) = 2x - 1$
**B** $f(x) = 0.5x + 3$
**C** $f(x) = 7x + 2$
**D** $f(x) = 1.5x - 4$
**E** $f(x) = 3x - 2$

**8** Over what interval is the rate of change of the quadratic function above the same as the rate of change of the function $f(x) = -2x - 3$?

**A** $x = -4$ to $x = 0$
**B** $x = -4$ to $x = -2$
**C** $x = -3$ to $x = 2$
**D** $x = 0$ to $x = 4$
**E** $x = 1$ to $x = 4$

**Directions:** Questions 9 and 10 are based on the information below.

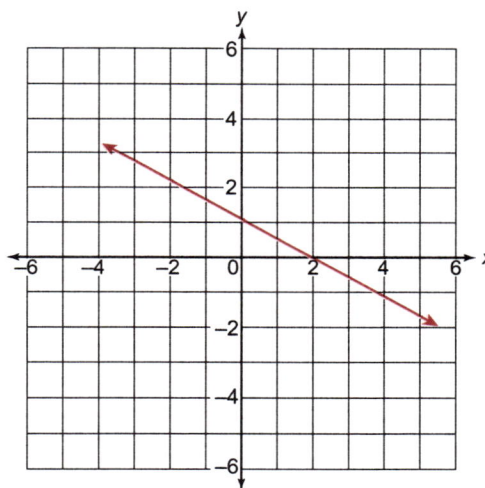

**9** Which function has the same y-intercept as the function represented in the graph?

**A** $f(x) = x - 2$
**B** $f(x) = 2x^2 + 2$
**C** $f(x) = 3x - 1$
**D** $f(x) = 4x^2 + 1$
**E** $f(x) = 3x - 2$

**10** Which function has the same x-intercept as the function represented in the graph?

**A** $f(x) = x + 1$
**B** $f(x) = \frac{3}{2}x - 1$
**C** $f(x) = -x + 2$
**D** $f(x) = -\frac{2}{2}x - 2$
**E** $f(x) = 3x - 2$

**11** Which function has the same maximum value as the function represented in the graph?

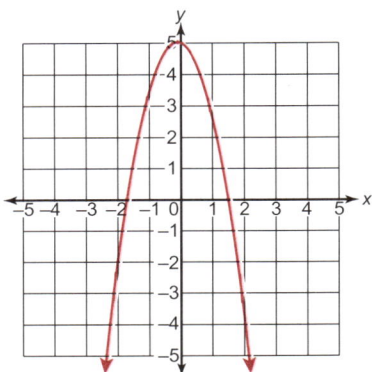

A   $f(x) = x^2 + 5$
B   $f(x) = 2x^2 - 3$
C   $f(x) = -2x^2 + 3$
D   $f(x) = -3x^2 + 5$
E   $f(x) = 3x - 2$

**12** The graph shows the distance traveled by Car 1, traveling at constant speed. The equation $d = 45t$ represents the distance, $d$, that Car 2 travels in $t$ hours. Which statement explains which car is traveling at a greater speed?

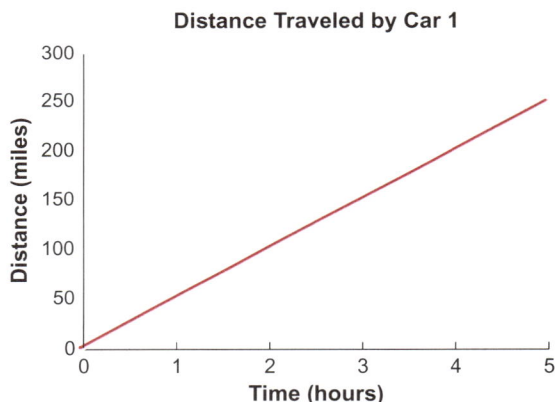

**Distance Traveled by Car 1**

A   Car 1 is traveling at a greater speed, because the slope of the line in the graph is less than 45.
B   Car 1 is traveling at a greater speed, because the slope of the line in the graph is greater than 45.
C   Car 2 is traveling at a greater speed, because the slope of the line in the graph is less than 45.
D   Car 2 is traveling at a greater speed, because the slope of the line in the graph is greater than 45.
E   The cars are traveling at the same speed.

**Directions:** Questions 13 and 14 are based on the information below.

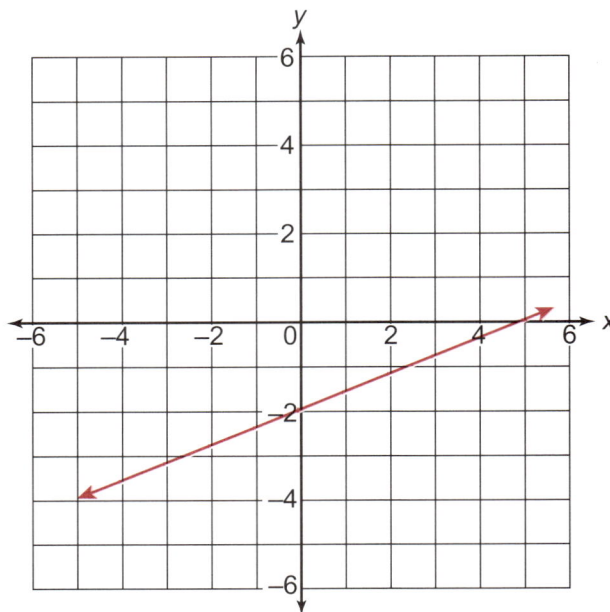

**13** Which statement is true?

The function represented in the graph has
A   a greater rate of change and a greater $y$-intercept than the function $f(x) = 0.75x - 1$.
B   a greater rate of change and a lesser $y$-intercept than the function $f(x) = 0.75x - 1$.
C   a lesser rate of change and a greater $y$-intercept than the function $f(x) = 0.75x - 1$.
D   a lesser rate of change and a lesser $y$-intercept than the function $f(x) = 0.75x - 1$.
E   the same rate of change and $y$-intercept as the function $f(x) = 0.75x - 1$

**14** A graph of a linear function $g(x)$ has the same $y$-intercept as the function represented in the graph and a slope that is twice as steep as that of the function represented in the graph. Which point is on the graph of the function $g(x)$?
A   $(-10, -6)$
B   $(-5, -6)$
C   $(5, -2)$
D   $(10, -10)$
E   $(10, 5)$

UNIT 4

**Directions:** Questions 15 and 16 are based on the information below.

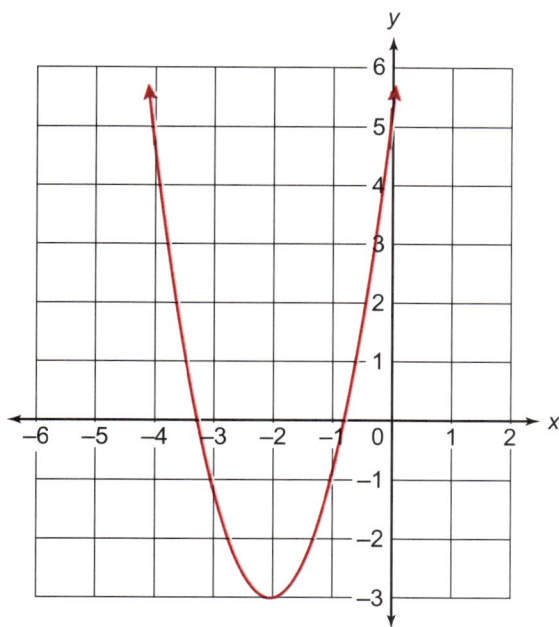

**15** Alaina says the function $f(x) = x^2 - 2$ has the same minimum value as the function represented in the graph. Which explains Alaina's conclusion?

A Alaina is correct, because both functions have a minimum value of −2.

B Alaina is correct, because both functions have a minimum value of −2.

C Alaina is correct, because both functions have a minimum value of −3.

D Alaina is incorrect, because the function represented in the graph has a minimum value of −3 and the function $f(x) = x^2 - 2$ has a minimum value of 2.

E Alaina is incorrect, because the function represented in the graph has a minimum value of −3 and the function $f(x) = x^2 - 2$ has a minimum value of −2.

**16** Which linear function has a rate of change that is equal to the average rate of change of the function represented in the graph over the interval from $x = -3$ to $x = 0$?

A $f(x) = \frac{1}{4}x - 4$

B $f(x) = \frac{1}{2}x + 1$

C $f(x) = 2x - 6$

D $f(x) = 4x + 7$

E $f(x) = 2x - 6$

**Directions:** Questions 17 through 19 are based on the information below.

| x | y |
|---|---|
| −4 | 6 |
| −2 | 5 |
| 0 | 4 |
| 2 | 3 |
| 4 | 2 |

**17** A linear function has a rate of change of −0.5 and a $y$-intercept of 5. Which statement is true?

Compared to the function in the table, this function has

A the same rate of change and $y$-intercept.

B a greater rate of change and $y$-intercept.

C a greater rate of change and the same $y$-intercept.

D the same rate of change and a greater $y$-intercept

E a lesser rate of change and $y$-intercept.

**18** James says the function $f(x) = x + 8$ has the same $x$-intercept as the function represented in the table. Sam says that the function $f(x) = 0.25x - 2$ has the same $x$-intercept as the function represented in the table. Who is correct?

A James

B Sam

C both James and Sam

D neither James nor Sam

E Not enough information is provided.

**19** The ordered pairs below represent a linear function.

$\{(-6, -1), (-2, 1), (4, 4), (8, 6)\}$

Which is true of the function represented above?

A The function is increasing at the same rate that the function represented in the table is increasing.

B The function is increasing at the same rate that the function represented in the table is decreasing.

C The function is decreasing at the same rate that the function represented in the table is increasing.

D The function is decreasing at the same rate that the function represented in the table is decreasing.

E None of these statements are true.

**UNIT 4**

# Unit 4 Review

The Unit Review is structured to resemble mathematics high school equivalency tests. Be sure to read each question and all possible answers very carefully before choosing your answer.

To record your answers, fill in the letter of the circle that corresponds to the answer you select for each question in the Unit Review.

Do not rest your pencil on the answer area while considering your answer. Make no stray or unnecessary marks. If you change an answer, erase your first mark completely.

Mark only one answer space for each question; multiple answers will be scored as incorrect.

**Sample Question**

**The number of students at a large university can be written as $8^4$. How many students are at the university?**

A  512

B  4,096

C  10,024

D  32,028

E  32,768

Ⓐ⬤ⒸⒹⒺ

---

1  **A painter charges $20 per hour for herself and $15 per hour for her assistant. In painting a living room, the assistant worked 5 hours more than the painter. The total charge for labor was $355.**

**Let $h$ be the number of hours that the painter worked. Which of the following equations can be used to find $h$?**

A  $20h + 15(h + 5) = 355$

B  $20(h + 5) + 15h = 355$

C  $20h + 15(h - 5) = 355$

D  $20h - 15(h + 5) = 355$

E  $20h - 15(h - 5) = 355$

Ⓐ Ⓑ Ⓒ Ⓓ Ⓔ

2  **If $x^2 = 36$, then $2(x + 5)$ could equal which of the following numbers?**

A  6

B  11

C  12

D  22

E  28

Ⓐ Ⓑ Ⓒ Ⓓ Ⓔ

3  **Sara has $1,244 in her checking account. She deposits a check for $287 and withdraws $50 cash. What is her new balance?**

A  $1,294

B  $1,394

C  $1,481

D  $1,531

E  $1,581

Ⓐ Ⓑ Ⓒ Ⓓ Ⓔ

**4** What is the next term in the sequence?

3, 2.5, 2, 1.5, 1, 0.5, 0, . . .

A  −1
B  −0.5
C  0
D  1
E  1.5

ⒶⒷⒸⒹⒺ

**5** The number of men acting in a theater production is five more than half the number of women. Which of the following expressions describes the number of men in the production?

A  $2w + 5$

B  $\frac{1}{2}w + 5$

C  $2w - 5$

D  $\frac{1}{2}w - 5$

E  $-\frac{1}{2}w + 5$

ⒶⒷⒸⒹⒺ

**6** If $3x + 0.15 = 1.29$, what is the value of $x$?

A  0.24
B  0.28
C  0.32
D  0.34
E  0.38

ⒶⒷⒸⒹⒺ

**Directions:** Questions 7 and 8 are based on the information below.

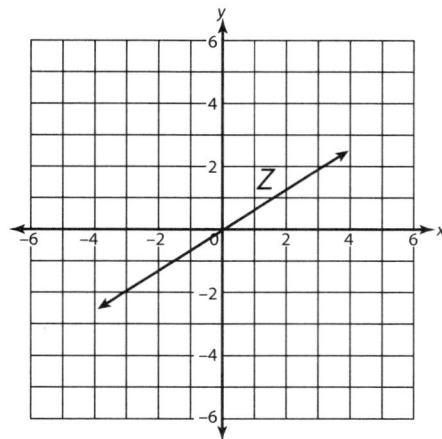

**7** What is the slope of line Z?

A  $-\frac{2}{3}$

B  $-\frac{1}{2}$

C  1

D  $\frac{1}{2}$

E  $\frac{2}{3}$

ⒶⒷⒸⒹⒺ

**8** What is the equation of line Z in slope-intercept form?

A  $y = -\frac{2}{3}x + 1$

B  $y = \frac{1}{2}x$

C  $y = \frac{2}{3}x$

D  $y = \frac{3}{2}x$

E  $y = x + \frac{2}{3}$

ⒶⒷⒸⒹⒺ

**9** The number of people voting in an election who were over 25 years old was 56 less than twice the number of those voting who were under 25 years old. Which expression represents the number of people who were over 25 years old who voted in the election?

A  $56x - 25$
B  $2x - 56$
C  $x + 56$
D  $56x + 25$
E  $2x + 56$

ⒶⒷⒸⒹⒺ

**10** Ellie has a pass that allows her to drive through tolls without stopping to pay. The amount of the toll is automatically charged to her credit card. She pays a fee of $5 per month for this service. Each toll she pays is $1.25. She budgets $65 a month for her total toll bill. What is the maximum number of tolls she can pass through each month and still stay within her budget?

A  12
B  24
C  36
D  48
E  60

ⒶⒷⒸⒹⒺ

**11** The Earth is 149,600,000 kilometers from the sun. What is this distance written in scientific notation?

A  $1.496 \times 10^{-7}$ km
B  $1.496 \times 10^{-8}$ km
C  $1.496 \times 10^{8}$ km
D  $1.496 \times 10^{9}$ km
E  $1.496 \times 10^{10}$ km

ⒶⒷⒸⒹⒺ

**Directions:** Questions 12 and 13 are based on the information below.

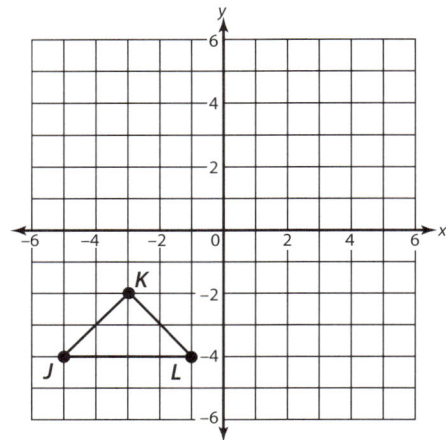

**12** If triangle *JKL* is reflected across the *y*-axis, then what is the new location of point *K*?

A  $(3, 2)$
B  $(3, -2)$
C  $(-3, 2)$
D  $(-3, -2)$
E  $(-2, 3)$

ⒶⒷⒸⒹⒺ

**13** What is the slope of side *JL* in triangle *JKL*?

A  $-2$
B  $-1$
C  $0$
D  $1$
E  $2$

ⒶⒷⒸⒹⒺ

**14** The product of two consecutive negative integers is 19 more than their sum. What is the greater integer?

A  0
B  $-1$
C  $-2$
D  $-3$
E  $-4$

ⒶⒷⒸⒹⒺ

**15** Line *H* passes through points (−5, 4) and (0, 1). What is the distance between these two points to the nearest hundredth unit?

**A** 5.66
**B** 5.83
**C** 6.00
**D** 6.16
**E** 6.32

Ⓐ Ⓑ Ⓒ Ⓓ Ⓔ

**16** Each day for three days, Emmit withdrew $64 from his account. Which number shows the change in his account after the three days?

**A** −$192
**B** −$128
**C** −$64
**D** $128
**E** $192

Ⓐ Ⓑ Ⓒ Ⓓ Ⓔ

**17** The function $y = \frac{3}{4}x$ was used to create the following table. Which number is missing from the table?

| *x* | −2 | −1 | 0 | 1 | 2 |
|-----|-----|-----|-----|-----|-----|
| *y* | $-\frac{3}{2}$ | $-\frac{3}{4}$ | 0 | $\frac{3}{4}$ | |

**A** $-\frac{1}{4}$

**B** $-\frac{1}{2}$

**C** 1

**D** $\frac{3}{2}$

**E** $\frac{5}{2}$

Ⓐ Ⓑ Ⓒ Ⓓ Ⓔ

**18** A skier takes a chairlift 786 feet up the side of a mountain. He then skis down 137 feet and catches a different chairlift 542 feet up the mountain. What is his position when he gets off the chairlift relative to where he began on the first chairlift?

**A** −1,191 feet
**B** −649 feet
**C** +679 feet
**D** +1,191 feet
**E** +1,465 feet

Ⓐ Ⓑ Ⓒ Ⓓ Ⓔ

**19** Which of the following inequalities is shown on the number line?

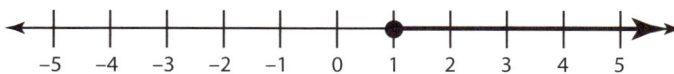

**A** $x \geq 1$
**B** $x \leq 1$
**C** $x < -1$
**D** $x > -1$
**E** $x > 1$

Ⓐ Ⓑ Ⓒ Ⓓ Ⓔ

**20** A summer family camp costs $230 for adults. The cost for a child is $30 less than one-half the cost for adults. What is the cost for 3 children?

**A** $200
**B** $220
**C** $248
**D** $255
**E** $262

Ⓐ Ⓑ Ⓒ Ⓓ Ⓔ

**Directions:** Questions 21 and 22 are based on the information below.

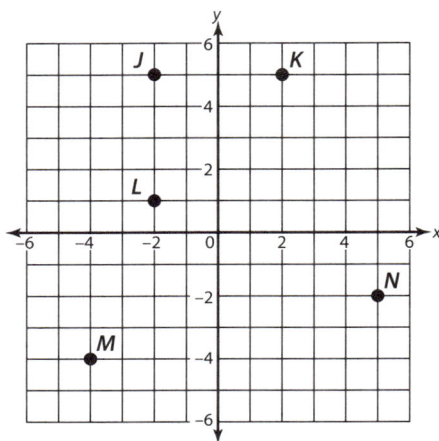

**21** Points *J*, *K*, and *L* mark the corners of a rectangle. What is the location of the fourth corner needed to complete the rectangle?

  **A** (2, 2)
  **B** (2, 1)
  **C** (1, 2)
  **D** (−1, −2)
  **E** (−1, 1)

Ⓐ Ⓑ Ⓒ Ⓓ Ⓔ

**22** Which of the following is the equation of a line that passes through points *L* and *K*?

  **A** $y = \frac{1}{3}x$
  **B** $y = 3x$
  **C** $y = x + 3$
  **D** $y = x - 3$
  **E** $y = 3x + 3$

Ⓐ Ⓑ Ⓒ Ⓓ Ⓔ

**23** The equation $h = -16t^2 - 48t + 160$ represents the height *h* of a ball above ground at time *t* seconds after being dropped. How many seconds does it take the ball to reach the ground?

  **A** 1
  **B** 2
  **C** 3
  **D** 4
  **E** 5

Ⓐ Ⓑ Ⓒ Ⓓ Ⓔ

**24** The weight of a mother elephant is 200 kg more than 4 times the weight of her newborn calf. Which of the following expressions represents the mother elephant's weight?

  **A** $4c + 200$
  **B** $200 - 4c$
  **C** $4(c - 200)$
  **D** $4c - 200$
  **E** Not enough information is given.

Ⓐ Ⓑ Ⓒ Ⓓ Ⓔ

**25** Keenan purchased solar lights for his front walkway. The total amount he paid for 8 lights was $73.36, including $4.16 tax. What was the cost per light before tax?

  **A** $7.92
  **B** $8.18
  **C** $8.65
  **D** $9.17
  **E** $9.69

Ⓐ Ⓑ Ⓒ Ⓓ Ⓔ

# Glossary

## A

**acute triangle:** a triangle in which all three angles are less than 90°

**adjacent angles:** two angles that share a common side and have the same vertex

**algebraic expression:** a mathematical phrase made up of numbers, variables, and operations; variables in the expression may change in value, causing the overall expression to represent different values

**alternate exterior angles:** a pair of congruent angles formed by two parallel lines cut by a transversal; located outside the parallel lines and on opposite sides of the transversal

**alternate interior angles:** a pair of congruent angles formed by two parallel lines cut by a transversal; located inside the parallel lines and on opposite sides of the transversal

**analysis:** a careful examination of data to identify relationships and meaning

**angle:** two rays that share the same endpoint; represented as $\angle$

**area:** the amount of space within a two-dimensional figure; measured in square units

## B

**bar graph:** a visual display that uses vertical or horizontal bars, aligned side-by-side for easy comparison, to represent data quantities

**base (geometry):** the bottom line segment of a figure that forms a 90° angle when paired with a height segment

**base (number):** the whole amount in a percent problem; the number that is multiplied by itself when associated with an exponent

**box plot:** a visual display of a data set that shows and allows comparison of the median, the upper and lower quartile values, and the maximum and minimum values

## C

**census:** a collection of data from an entire population, as opposed to collection from a sample

**circle:** a completely round shape in which all points on the perimeter (outside) are the same distance from the center

**circle graph:** a visual display that shows how parts compare to a whole

**circumference:** the distance around a circle; $C = \pi d$

**coefficient:** a number multiplied by a variable in an expression, such as the number 3 in the expression $3x + 4$

**combination:** a selection of objects or options in which the order of options does not matter

**common denominator:** the same bottom number shared by two or more fractions; for example, $\frac{1}{3}$ and $\frac{2}{3}$

**complementary angles:** two angles for which their sum equals 90° (a right angle)

## cone

**cone:** a three-dimensional figure with a circular base and one vertex, which are connected by a curved surface

**congruent:** exactly the same, such as line segments that have the same length or angles that have the same measurement

**congruent figures:** two or more figures that have the same shape and size

**convenience sample:** a non-random sample selected from a population because of accessibility

**coordinate grid:** a two-dimensional visual representation of points, or ordered pairs

**coordinate system:** a system used to describe the position of a point in space

**correlation:** the relationship between data sets

**corresponding angles:** congruent angles (have the exact same measurement) that are in the same relative location (i.e., both are to the left of the intersecting line and on top of a parallel line); also, lines or angles of two similar figures that are in the same position

**cross product:** method by which, in a proportion, the numerator of one fraction is multiplied by the denominator of the other fraction, producing equal products

**cube:** a three-dimensional figure with six congruent (exactly the same) square faces

**cube root:** that number which, when cubed, equals the given number

**cubing:** multiplying a number by itself three times, such as $(2 \times 2 \times 2)$; represented as $2^3$

## D

**data:** a collection of factual values or measurement

**data table:** a tool for organizing and analyzing information in which data are arranged in rows and columns in order to make the information easier to identify and compare; includes row labels and column headers

**decimal:** another way to write a fraction based on the base-ten place value system; for example, $\frac{1}{4}$ would become .25

**denominator:** the bottom number in a fraction that tells the number of equal parts in a whole

**diameter:** the distance across the center of a circle, from one side to the other; always equal to two times the radius

**difference:** the answer to a subtraction problem

**dilation:** the proportional resizing of a figure on a coordinate grid

**dimension:** the number of rows by the number of columns in a matrix; a matrix with 4 rows and 5 columns is a $4 \times 5$ matrix

**distributive property:** the principle that multiplying a number by a group of numbers added together is the same as doing each multiplication separately, such as $2(x + y) = 2x + 2y$

**dividend:** the initial quantity in a division problem (the number being divided)

**divisor:** the number by which the initial quantity, or dividend, is divided in a division problem

**dot plot:** an expanded number line that displays a data set by showing each occurrence of a value as a dot

# E

**endpoint:** the point at which a line segment or a ray ends

**equation:** an algebraic expression that states that two things are equal

**equilateral triangle:** a triangle with three congruent sides

**event:** a result or outcome of an experiment or trial

**experimental probability:** a ratio that expresses an actual number of outcomes as a part of a total number of trials

**experimental study:** a method of collecting data in which a researcher controls a factor that influences outcomes for one group within the population or sample

**exponent:** a superscript indicator that tells how many times a base number is used in a multiplication; for example, $4^3 = 4 \times 4 \times 4$

**exterior angles:** angles that fall outside two parallel lines cut by a transversal

# F

**factors:** numbers or expressions that are multiplied together to form a product

**FOIL method:** the multiplication of factors with two terms by multiplying the first, outer, inner, and last terms in that order

**fraction:** part of a whole or part of a group; indicated by separating two numbers with a fraction bar

**frequency:** the number of values for a given interval in a data set

**frequency table:** a table that displays data as numbers of occurrences

**function:** an equation in which each input has exactly one output; can be written as the solution for one variable as in $y = 3x + 4$, or can be written in terms of the inputs as in $f(x) = 4x - 2$

**fundamental counting principle:** the basic rule used to count possible outcomes; if one choice has $m$ options and another choice has $n$ options, then the number of possible outcomes for the two choices is $m \cdot n$

# H

**height:** a perpendicular line segment from a vertex to its opposite side; forms a 90° angle when paired with a base segment; in a prism, the perpendicular distance between its bases

**hemisphere:** half of a sphere

**histogram:** a visual display of bars that correspond to an associated scale to show frequency in ranges; may be used with any size data set

**hypotenuse:** the longest side of a right triangle

# I

**improper fraction:** a fraction in which the numerator is larger than denominator; should be written as a mixed number, for example $\frac{5}{4}$ converted to $1\frac{1}{4}$

**indirect measurement:** using proportions and corresponding parts of similar figures to find a measurement that you cannot find directly

**inequality:** an algebraic expression that states that one side is greater or less than the other side; similar to an algebraic equation except that $<$, $>$, $\leq$ or $\geq$ are used in place of an $=$, as in $y > 3x + 5$

**integers:** another name for whole numbers, including positive whole numbers (1, 2, 3, . . .), their opposites (–1, –2, –3, . . .), and zero (0)

**interior angles:** angles that fall between two parallel lines cut by a transversal

**inverse operations:** operations that undo each other, such as addition and subtraction or multiplication and division; used in algebra to solve equations by grouping variable terms on one side of an equation and constant terms on the other side

**irregular figures:** figures made up of several plane or solid shapes

**isolate a variable:** to perform operations to an equation in order get one variable alone on one side of the equation

**isosceles triangle:** a triangle with two congruent sides

# L

**legs:** the two shorter sides of a right triangle (can be different lengths)

**like terms:** algebraic terms that have the same variables raised to the same power

**line:** in geometry, a line is straight and continues forever in both directions; represented as ——

**line graph:** a visual display used to show how a data set changes over time

**line segment** the line formed by connecting two points on a coordinate grid

**linear equation:** an equation that makes a straight line when it is graphed; all of the solutions to the equation lie on the line

# M

**mathematical pattern:** an arrangement of numbers and terms created by following a specific rule

**matrix:** a tool for organizing and analyzing information in which data are arranged in rows and columns in order to make the information easier to identify and compare; no units of measure are included; often do not include labels and headers

**maximum of a curve:** the point on a curve for which $y$ has the highest value

**mean:** the average value of a data set

**measures of variation:** measures that describes "how scattered" a set of data is; range, variance, and standard deviation

**median:** the middle number in a set of data when the values are ordered from least to greatest

**metric system:** a base-ten system of measurement that includes length (such as centimeter and meter), mass (such as gram and kilogram), and capacity (such as milliliter and liter)

**minimum of a curve:** the point on a curve for which $y$ has the lowest value

**mixed number:** a fraction that includes a whole number, such as $1\frac{2}{5}$

**mode:** the value that occurs most frequently in a set of data

# N

**negative slope:** a line that falls from left to right on a coordinate grid

**normal distribution:** an arrangement of data values that produce a mean, median, and mode that coincide

**number pattern:** a list of numbers that follow a certain rule, such as a series where each number increases in value by 5

**numerator:** the top number in a fraction that tells the number of equal parts being considered

# O

**observational study:** a method of collecting data in which a researcher observes behavior or outcomes in a systematic manner without influencing or interfering with the behavior or outcome

**obtuse triangle:** a triangle in which the largest angle is greater than 90°

**order of operations:** the rule for which action is performed first in a series of steps to solve a mathematical expression; parentheses, exponents and roots, multiplication and division, addition and subtraction

**ordered pair:** a pair of $x$- and $y$-values that can be represented on a coordinate grid; the $x$-value is always shown first

**origin:** the point (0, 0) where the $x$-axis and $y$-axis cross on a coordinate grid

**outcome:** the single result of a trial or probability experiment

# P

**parallel:** two lines on a plane that never meet; similar line segments that are the same distance apart

**parallelogram:** a four-sided figure with opposite pairs of congruent and parallel sides

**part:** a piece of the whole or base in a percent problem

**percent:** a way to show part of a whole, comparing the partial amount to 100; for example, 50% is 50 parts out of a whole 100

**perimeter:** the distance around a polygon (a multi-sided figure with straight sides)

**permutation:** an arrangement of a group of objects, or the same options for one choice, in a particular order

**point-slope form:** the equation of a straight line in the form $y - y_1 = m(x - x_1)$ where $m$ is the slope of the line and $(x_1, y_1)$ are the coordinates of a given point on the line

**population:** the entire group of individuals in a study

**positive slope:** a line that rises from left to right on a coordinate grid

**power:** tells the number of times a base number is used in a multiplication, expressed by an exponent

**prism:** a solid figure with two parallel bases

**probability:** the likelihood that an event will happen

**product:** the answer to a multiplication problem

**proper fraction:** a fraction, such as $\frac{3}{4}$, in which denominator is larger than the numerator

**proportion:** an equation with equal ratios on each side

**pyramid:** a three-dimensional figure that has a polygon as its single base and triangular faces

**Pythagorean theorem:** a description of the relationship between the sides of a right triangle, named for the ancient Greek mathematician Pythagoras; the sum of the squares of the two smaller legs equals the square of the hypotenuse; $a^2 + b^2 = c^2$

# Q

**quadrant:** one of the four sections on a coordinate grid created by the intersection of the $x$- and $y$-axis; all $x$-values within a quadrant will be either positive or negative, and similarly all $y$-values will be either positive or negative

**quadratic equation:** an equation in which one or more of the terms is squared but raised to no higher power; they are set in the form $ax^2 + bx + c = 0$

**quadrilateral:** a closed four-sided figure with four angles; the sum of the four interior angles is always 360°

**qualitative:** descriptive details that cannot be measured

**quantitative:** characterized by measurable numbers

**quotient:** the answer to a division problem

# R

**radius:** the distance from the center of a circle to any point on the perimeter; half the diameter

**random sample:** a sample, selected from a population, in which every individual in the population has an equal chance of being chosen

**range:** the spread of a data set, expressed by the difference between the greatest value and the least value

**rate:** tells how the base and the whole are related in a percent problem

**rate of change of a linear function:** the slope of a line shown through the relationship between the rise and run

**ratio:** a comparison of two numbers; written as a fraction, by using the word *to*, or with a colon (:)

**ray:** a part of a line that has one endpoint and extends forever in the opposite direction; represented as $\rightarrow$

**rectangle:** a closed, four-sided figure with two pairs of parallel sides in which the corners all are 90° angles

**rectangular prism:** a box-shaped, three-dimensional figure with six faces that are rectangles

**reflection:** the flipping of a figure over a line, such as an axis, on a coordinate grid

**relation:** a connection between the commonalities of two patterns

**rhombus:** a parallelogram with four equal sides

**right triangle:** a triangle in which the largest angle is 90°

**rise:** the vertical increase between two points on a line; represented as the numerator in an equation for slope

**run:** the horizontal increase between two points on a line; represented as the denominator in an equation for slope

# S

**sample population:** a portion of a population selected for study when data cannot be collected for the entire group

**scale drawing:** a proportional drawing that represents an actual, larger object

**scalene triangle:** a triangle with no congruent sides

**scatter plot:** a line graph that shows how one set of data affects another

**scientific notation:** method of writing very large or very small numbers in a compact form through the use of exponents and powers of 10

**similar figures:** two or more figures that have the same shape and congruent angles, but have proportional sides, such as two right triangles, with one being 2 inches high and one being 1 inch high

**skewed distribution:** a "lopsided" arrangement of data values, for which the mean is not equal to the mode

**slope:** a number that measures the steepness of a line

**slope-intercept form:** the equation of a straight line in the form $y = mx + b$ where $m$ is the slope of the line and $b$ is its $y$-intercept

**solid figure:** a three-dimensional figure, such as a cube, pyramid, or cone

**solution:** the set of all values for the variables in an equation that make the equation true

**sphere:** a round, three-dimensional figure with no bases or faces, in which every point on the surface is equidistant from the center

**square pyramid:** a pyramid that has a square base and four congruent triangular faces

**square root:** a number that, when multiplied by itself, equals a given number

**squaring:** multiplying a number by itself one time; for example, $(2 \times 2)$ which is represented as $2^2$

**standard deviation:** a measure of how spread out numbers are; the square root of variance

**statistics:** the study of how to collect, summarize, and present data

**sum:** the answer to an addition problem

**supplementary angles:** two angles that produce a sum equal to 180° (a straight line)

**surface area:** the sum of a figure's two bases and its lateral surfaces

**survey:** a method of gathering data by collecting specific information from individuals

# T

**theoretical probability:** a ratio that expresses the likelihood of an outcome as a part of the number of ways an event can occur

**translation:** the sliding of a figure to a new position on a coordinate grid

**transversal:** a line crossing two or more parallel lines

**trapezoid:** a four-sided figure with only one pair of parallel sides

**trial:** a performance of a random experiment, such as a coin toss, that results in a single outcome

**triangle:** a closed, three-sided figure with three interior angles, the sum of which is always 180°

**triangular prism:** a prism with triangular bases

# U

**unit rate:** a ratio with a denominator of 1

**U.S. customary system:** system of measurement that includes length (such as inch and foot), weight (such as ounce and pound), and capacity (such as pint and quart)

# V

**validity:** based on reliable, factual data and free of bias

**variable:** a letter used to represent a value in algebraic expressions, such as $x$ in the expression $2x + 1$

**variance:** the average of squared differences from the mean

**vector:** a line segment that has magnitude and direction; one point is its starting point and the magnitude and direction describe a path to the other point

**vertex:** the endpoint shared by two rays; the points at which the faces of a three-dimensional figure connect

**vertical angles:** congruent, nonadjacent angles (opposite one another) that share a vertex formed by intersecting lines

**volume:** the amount of space inside a three-dimensional figure, measured in cubic units

# W

**whole numbers:** positive numbers written with the digits 0 through 9 and without fraction or decimal parts

# X

**$x$-intercept:** the point at which a line crosses the $x$-axis on a coordinate grid and the $y$-coordinate is 0

# Y

**$y$-intercept:** the point at which a line crosses the $y$-axis on a coordinate grid and the $x$-coordinate is 0

# Answer Key

## UNIT 1 NUMBERS AND OPERATIONS ON NUMBERS

### LESSON 1, pp. 2–6

**1. D,** To round to the nearest dollar, look at the tenths place. The tenths digit in 30,237.59 is 5, so the 5 rounds 7 ones to 8.

**2. D,** To round to the nearest tenth, look at the hundredths digit. Seven hundredths rounds the 2 to a 3.

**3. A,** All numbers have 1 hundred thousand. January and February both have 5 ten thousands, but January has 5 thousands and February has none.

**4. B,** January and February have the two highest values, and these months are in the winter.

**5. A,** 1.59 and 1.76 have the least amount of ones. 1.59 has 5 tenths and 1.76 has 7 tenths, so 1.59 is less.

**6. D,** There are four numbers less than 2.25: 1.59, 2.07, 1.76, and 2.15.

**7. C,** The two greatest numbers are 25.98 and 25.57 because they both have 5 ones. 25.98 has 9 tenths, while 25.57 only has 5 tenths, so 25.98 is greatest. Therefore, 24.30 is the least.

**8. D,** 1,107 is greater than 1,001, but less than 1,250.

**9. C,** Rows D, E, and F all have at least one seat in the thousands.

**10. C,** Morgan's time could be 218.45 because it is greater than 218.15, but less than 218.65.

**11. B,** 1,100 is greater than 1,097, but less than 1,105.

**12. D,** The information shows that S can be found on pages 1178 to 1360. The numbers 1234 through 1287 can be found within this range.

**13. A,** Twelve thousand is 12,000. Eight hundred two is 802. Twenty hundredths is .20.

**14. E,** 75 has 7 tens, so it is the least, and 84 has 8 tens, so it comes next. Then focus on ones to order 92, 95, and 98.

**15. E,** The first three digits are read as "one hundred fifty thousand," and the next three digits are read as "two hundred eighteen."

**16. B,** The 4 in the tens place is not large enough to round up to the hundreds place.

**17. B,** Compare the whole numbers and then the decimal places to find the order from least to greatest is 37.5, 37.7, 37.8, 38.1, and 38.3.

**18. C,** The scores 15.965 and 15.97 are less than Natalia's score of 15.975, so she came in third place.

**19. C,** The 8 is in the hundreds place, and the 7 is in the ones place, so 807 is read as "eight hundred seven."

**20. E,** All numbers have 1 hundred thousand. The greatest digit in the ten thousands place is 9 in 191,000.

**21. B,** 82 is written as eighty-two.

**22. D,** The digit 6 in the hundredths place of 42.468 is greater than 5, so 42.468 rounds up to 42.5 pounds.

**23. C,** Sample C weighs 1.121 kg, which is greater than the maximum weight of 1.103 kg, so Isaiah would reject it.

**24. B,** Compare the whole numbers and then the decimal places to find that Player B has the highest batting average.

**25. E,** Compare the whole numbers and then the decimal places to find that Players D, C, and B have the highest batting averages in order from least to greatest.

### LESSON 2, pp. 7–11

**1. B,** 6 ones minus 0 ones is 6 ones, 5 tens minus 4 tens is 1 ten, and then regroup 1 thousand as 10 hundreds to see that 12 hundreds minus 3 hundreds is 9 hundreds.

**2. E,** 7 ones plus 3 ones is 10 ones, or 1 ten, and 6 tens plus 8 tens plus the extra ten is 15 tens, or 1 hundred and 5 tens. 4 hundreds plus 5 hundreds plus the extra hundred is 10 hundreds.

**3. B,** Regroup 307 as 2 hundreds, 9 tens, and 17 ones, and then subtract 1 ten and 9 ones.

**4. D,** 40 hours × $9 per hour = $360 earned in one week.

**5. B,** $45 per month × 12 months per year = $540 for one year.

**6. D,** $64 divided evenly among 4 people is $16 per person.

**7. B,** Claire must find how many groups of 12 are in 504, so 504 ÷ 12 = 42 with no remainder.

**8. C,** $630 × 18 months = $11,340 in all.

**9. B,** 4,000 – 3,518 = 482 yards left, and then 482 yards divided by 2 games = 241 yards.

**10. E,** Multiply $325 by 6 to find a product of $1,950.

**11. C,** Subtract $2,750 – $1,560. Regroup 7 hundreds and 5 tens as 6 hundreds and 15 tens.

**12. D,** Multiply $72 by 12 to find a product of $864.

**13. A,** A total of $1,080 divided by 4 people is $270 per person.

**14. E,** Add $567,800 + $258,900 to find a sum of $826,700.

**15. D,** Add 22 + 14 + 12 to find a sum of 48.

**16. C,** To find the difference of 1,500 – 892, regroup 1 thousand and 5 hundreds as 0 thousands, 14 hundreds, 9 tens, and 10 ones.

**17. C,** Add $50 + $335 + $80 + $75 to find a total cost of $540. Then divide $540 by 5 people to find a quotient of $108.

**18. E,** Add $825 + $220 + $285 to find a sum of $1,330.

**19. C,** To subtract $285 – $179, regroup $285 as 2 hundreds, 7 tens, and 15 ones.

**20. E,** Multiply $179 by 12 to find a product of $2,148.

**21. A,** Annette works 5 × 6, or 30, hours per week. 30 hours times $13 per hour = $390 per week, and $390 per week times 4 weeks = $1,560.

**22. D,** Add 54 + 39 to find a total of 93 hours worked. Multiply 93 by $11 to find a product of $1,023.

**23. E,** Add $458 + $397 + $492 to find a sum of $1,347.

**24. D,** Divide $23,870 in sales by a price of $385 to find a quotient of 62.

**25. C,** A seamstress needs 2 × 5 = 10 yards for the shirts and 5 × 5 = 25 yards for the dresses. Subtract 25 – 10 to find a difference of 15.

**26. D,** Since 37 is the round-trip mileage, this figure includes driving to and from work. Multiply 5 days per week by 4 weeks to get 20 days. Then multiply 20 days by 37 miles to get a product of 740 miles.

**27. B,** Round 12,500 to 13,000. Round 4,020 to 4,000 and 2,902 to 3,000. The charity has already collected about $4,000 + $3,000, or $7,000. Subtract 7,000 from 13,000 to get an estimate of about $6,000.

## UNIT 1 (continued)

**28. A,** Divide the cost of $1,445 by 289 square feet to find a quotient of $5 per square foot.

**29. B,** Since there are 12 months in a year, there are $3 \times 12 = 36$ months in 3 years. Divide $13,392 by ($3 \times 12$) months.

**30. D,** Divide $270 by the cost of each share, which is $15.

**31. A,** Multiply $7 profit per share by 25 shares to find a product of $175.

### LESSON 3, pp. 12–16

**1. D,** 64 hot dogs × $3 = $192, and 38 yogurts × $2 = $76, so $192 + $76 = $268 in all.

**2. B,** A sale price of $42 – a $5 coupon = $37; the original price of $54 is a distraction.

**3. D,** Estimate because the problem asks for "about" how many. Round the amounts to the tens place, so round $645 to $650, $78 to $80, and $25 to $20. Add $650 + $80 + $20 to get an estimate of $750.

**4. A,** 143 students in all –67 soccer students = 76 non-soccer students; the number of students who signed up to play softball were only part of those playing a sport other than soccer.

**5. C,** Multiply 144 by $6 because the store is paying for the shirts; the product is $864.

**6. A,** The profit for one men's sport shirt is $16 – $9, which is $7, and 508 shirts × $7 per shirt = $3,556.

**7. B,** A budget of $5,500 – $5,000 spent = $500 remaining. Then $500 ÷ 4 = $125.

**8. C,** Juan has eaten 1,250 + 780, or 2,030, calories so far. Subtract this from 3,500. Regroup 3,500 as 3 thousands, 4 hundreds, 10 tens, and 0 ones.

**9. B,** 2,012 production employees + 157 shipping employees = 2,169 employees. Subtract this from 2,391 to find a difference of 222.

**10. D,** Multiply $126 by 12 months to find a product of $1,512. The $57 for monthly auto insurance is not needed to solve the problem.

**11. D,** Round $17.85 to $20 and 52 weeks to 50 weeks. Multiply $50 \times $20 to get an estimate of $1000.

**12. E,** Add 6 + 5 + 7 + 8 + 8 to find a total of 34 hours for the week. Multiply 34 hours by $8 to find a product of $272.

**13. D,** 34 hours last week + 6 more hours = 40 hours in all, and 40 hours × $9 = $360.

**14. E,** The amount of the tip is unknown, so the total cost cannot be determined.

**15. A,** After the deposit, Naomi has $913 + $130, or $1,043, in her account. The amount of checks and withdrawals is $75 + $75 + $50, or $200. Subtract 200 from 1,043 by regrouping 1,043 as 0 thousands, 10 hundreds, 4 tens, and 3 ones.

**16. A,** Subtract 97,634 – 41,868 to find a difference of 55,766, which is the number of miles Marla drove using the new tires. Subtract 40,000 from this number to find the number of miles driven above 40,000.

**17. B,** 16 gallons × $3 per gallon = $48 to fill the tank. To drive 800 miles, she needs to fill her tank 800 ÷ 400, or 2, times. Multiply $48 by 2 to find a product of $96.

**18. E,** Multiply $950 by 3 to find a product of $2,850.

**19. A,** Subtract 4,598 – 1,354 to find a difference of 3,244. Then divide this number by 4 (Trisha + 3 siblings) to find a quotient of 811.

**20. C,** Round 4,208 to 4,000 and 983 to 1,000. Subtract 4,000 – 1,000 to find a difference of 3,000.

**21. B,** Add 3,156 + 2,634 + 4,208 to find a total of 9,998 red, blue, or white skeins. Add 1,920 + 983 + 732 + 531 + 1,828 + 935 to find a total of 6,929 skeins of other colors. The difference of 9,998 – 6,929 is 3,069, which rounds to 3,100. Or, first round 9,998 to 10,000 and 6,929 to 6900. Then subtract the rounded numbers 10,000 – 6,900 to get 3,100.

**22. C,** The payment plan will cost $150 × 12, or $1,800. Subtract $1,620 from this amount to get a difference of $180.

**23. B,** Cody earned $50 × 4, or $200, on televisions and $30 × 3, or $90, on speakers. He earned $200 + $90 = $290 in all.

**24. E,** The cost of mulch is unknown, so the cost of mulch and blocks cannot be determined.

**25. B,** $1,476 divided by 12 months = $123 per month.

**26. C,** Gretchen ordered 3 + 8 + 14 = 25 boxes of men's shirts and 5 + 6 + 5 = 16 boxes of women's shirts. Multiply 25 boxes by 25 men's shirts per box to find a product of 625 men's shirts. Multiply 16 boxes by 35 women's shirts per box to find a product of 560 women's shirts. The difference of 625 – 560 is 65.

**27. D,** Women's shirts: 6 + 5 = 11 boxes; 11 boxes × 35 shirts per box = 385 shirts. Men's shirts: 3 + 8 = 11 boxes; 11 boxes × 25 shirts per box = 275 shirts. 385 women's shirts + 275 men's shirts = 660 shirts in all.

**28. B,** The difference of 245 girls who play soccer – 218 girls who play basketball = 27 girls.

### LESSON 4, pp. 17–21

**1. E,** The change in temperature was 12 – (–3) = 12 + 3 = 15°F.

**2. B,** Her net gain is 3 – 4 + 8 = –1 + 8 = 7 spaces forward.

**3. D,** Uyen's new balance is $154 – $40 = $114.

**4. A,** Since Sasha descended, her change in position is –212 + (–80) = –292 feet.

**5. A,** Multiply each score by 3 holes: 3(+3) = +9; 3(–4) = –12; 3(–1) = –3. Add the scores 9 + (–12) + (–1) to find Scott's final score of –6, or 6-under par.

**6. D,** There were 3,342 – 587 – 32 + 645 = 3,368 students enrolled in the fall.

**7. C,** The change in the number of students between May and the fall is –587 – 32 + 645 = –619 + 645 = 26.

**8. B,** Add Melanie's sixth round score to the scores in the table. Melanie's overall score was 8 – 6 – 4 + 3 + 4 – 8 = –3.

**9. C,** The football team has gone 8 – 10 + 43 = 41 yards.

**10. E,** The change is 11,416 – 11,498 = –82.

**11. D,** She is at 8,453 – 2,508 + 584 = 6,529 feet.

**12. D,** The diver's position is 3 + 2 – 8 = –3 meters.

**13. D,** Deshon has a score of –145 + 80 + 22 = –43 points.

**14. E,** Solve x + (–7) = 12 for x to find that x = 19.

**15. B,** Donna's score was 5 + (–10) + (–10) = –15 points.

**16. E,** Dorothy scored 15 + 5 + 0 = 20 points. Nikki scored 0 + 10 + (–15) = –5 points. Dorothy scored 20 – (–5) = 25 more points than Nikki.

**17. D,** The submarine's new position is –3,290 + 589 – 4,508 = –7,209.

**18. D,** The mountain is 10,549 + 872 = 11,421 feet above sea level.

# UNIT 1 (continued)

**19. A,** The balance of Jordan's account is $890 - ($45 \cdot 3) =$ $890 - $135 = $755.

**20. D,** Solve $-2x + 2 = 0$ for $x$ to find that the original number is $x = 1$.

**21. A,** The team members have $(-120) \cdot 4 = -480$ points.

**22. B,** The rock climbers descended $363 \div 3 = 121$ feet in each phase. Since the rock climbers were descending, their change in height in each phase was $-121$ feet.

**23. A,** Erik was $12 - 8 = 4$ miles from home.

**24. A,** Solve $3 + x = -10$, where $x =$ points scored in the second round. Don scored $-10 - 3 = -13$ points.

**25. E,** The final product is $(-7)(-1)(-1)(-1) = (7)(-1)(-1) = (-7)(-1) = 7$.

**26. E,** Since $156 is taken out of her account each month, the change is the product $(-$156) \cdot 12 = -$1,872$.

**27. C,** Karin will pay her sister $1,554 \div 6 = $259.

**28. E,** The difference $A - B$ is $-4 - 7 = -11$. The absolute value of $-11$ is $11$, which is the distance on the number line between the two points.

**29. C,** Janet is on floor $54 - 22 + 5 = 37$.

**30. D,** Cheryl receives $527 \cdot 6 = $3,162.

**31. A,** Connor's monthly payment is $3,228 \div 12 = $269, which is taken out of his account, so the change is $-$269 each month.

**32. B,** Solve $x - (-10) = 6$ for $x$ to find that $x = -4$.

---

## LESSON 5, *pp. 22–26*

**1. C,** To add the fractions, rename them as $\frac{12}{20}$ and $\frac{15}{20}$. Rename the sum of $\frac{27}{20}$ as $1\frac{7}{20}$.

**2. E,** To add the fractions, rename them as $3\frac{2}{6}$ and $1\frac{3}{6}$. Add the numerators and whole numbers to find a sum of $4\frac{5}{6}$.

**3. C,** To find the difference, solve $5\frac{3}{12} - 2\frac{2}{12}$ to get $3\frac{1}{12}$.

**4. E,** Divide 23 by 4 to get 5 with a remainder of 3. Use the remainder as the numerator of the mixed number to get $5\frac{3}{4}$.

**5. A,** To find the number of curtains, divide the total length by the amount needed for each curtain. $11\frac{1}{4} \div 2\frac{1}{4} = \frac{45}{4} \div \frac{9}{4} = \frac{45}{4} \times \frac{4}{9}$. Simplify by dividing the numerator and denominator by 4 and 9. $\frac{45}{4} \times \frac{4}{9} = \frac{5}{1} = 5$

**6. A,** 8 stacks of $\frac{2}{3}$ means $8 \times \frac{2}{3} = \frac{8}{1} \times \frac{2}{3} = \frac{16}{3} = 5\frac{1}{3}$.

**7. B,** The sum is $3\frac{3}{8} + 1\frac{2}{8} + 2\frac{4}{8} = 6\frac{9}{8} = 6 + 1\frac{1}{8} = 7\frac{1}{8}$.

**8. A,** To find half of something, multiply by one-half. $1\frac{3}{4} = \frac{7}{4}$, $\frac{7}{4} \times \frac{1}{2} = \frac{7}{8}$

**9. C,** Rename both numbers as improper fractions: $\frac{8}{1}$ and $\frac{7}{2}$. Then find the common denominator and subtract: $\frac{16}{2} - \frac{7}{2} = \frac{9}{2}$. So, Chandra has $4\frac{1}{2}$ sick days left.

**10. E,** Find the common denominator of the fractions and then add the mixed numbers: $3\frac{3}{4} + 5\frac{2}{4} + 4\frac{1}{4} = 12\frac{6}{4}$. Then simplify to see that Todd used $12\frac{6}{4} = 13\frac{1}{2}$ gallons of water.

**11. D,** Fractions less than $\frac{1}{2}$ round down. So, round $\frac{1}{10}$ to 0, round $\frac{7}{8}$ to 1, and round $2\frac{3}{16}$ to 2, to get $0 + 1 + 2$.

**12. C,** Divide $29,400 by 24 to get $1,225. Elias makes $1,225 on his semi-monthly paycheck.

**13. D,** Divide 24 by $\frac{1}{4}$ to find that the chef can make $24 \div \frac{1}{4} = 24 \times \frac{4}{1} = 96$ specials.

**14. D,** It takes Mike 15 minutes, or $\frac{15}{60} = \frac{1}{4}$ of an hour, to mow one lawn, so he should be able to mow $30 \div \frac{1}{4} = 30 \times \frac{4}{1} = 120$ lawns in 30 hours.

**15. B,** It takes Amy 35 minutes, or $\frac{35}{60} = \frac{7}{12}$ of an hour, to edit one page of a textbook. So, it would take her $300 \times \frac{7}{12} = 175$ hours to edit a 300-page book.

**16. B,** To find $\frac{2}{3}$ of $\frac{1}{2}$, multiply $\frac{2}{3}$ by $\frac{1}{2}$ to find a product of $\frac{1}{3}$.

**17. B,** Find a common denominator and rename each fraction, $25\frac{7}{8} = 25\frac{21}{24}$ and $17\frac{5}{6} = 17\frac{20}{24}$, and then subtract. Ed has $25\frac{21}{24} - 17\frac{20}{24} = 8\frac{1}{24}$ yards of fencing left.

**18. B,** Find the common denominator and then add each fraction: $\frac{1}{5} + \frac{1}{2} = \frac{2}{10} + \frac{5}{10} = \frac{7}{10}$. Subtract $\frac{7}{10}$ from 1, or $\frac{10}{10}$, to find that $\frac{3}{10}$ of the group sat in red seats.

**19. D,** Using the common denominator of 30, rename the fractions as $\frac{15}{30}$, $\frac{30}{30}$, $\frac{18}{30}$, $\frac{10}{30}$ and $\frac{24}{30}$. One-half $= \frac{15}{30}$, so the only fraction less than one-half is $\frac{10}{30}$, which corresponds to Team 4.

**20. E,** Since the amounts of water in the bowls are combined, use addition. Use common denominators to find $\frac{1}{2} + \frac{1}{3} = \frac{3}{6} + \frac{2}{6} = \frac{5}{6}$.

**21. D,** Write the fraction of members who ordered cake $\frac{3}{15}$ and the fraction who ordered a milkshake $\frac{3}{15}$. Then add the fractions to get the $\frac{3}{15} + \frac{3}{15} = \frac{6}{15} = \frac{2}{5}$.

**22. C,** The board should be $3\frac{1}{4} + \frac{3}{8} = 3\frac{2}{8} + \frac{3}{8} = 3\frac{5}{8}$ feet long.

**23. D,** Multiply the whole number by the denominator, and then add the product to the numerator: $50 \times 2 + 1 = 101$. Use this as the numerator with the same denominator to make the improper fraction $\frac{101}{2}$.

**24. A,** Rename each mixed number as an improper fraction, and then divide to find $4\frac{3}{4} \div 2\frac{1}{4} = \frac{19}{4} \div \frac{9}{4} = \frac{19}{4} \times \frac{4}{9} = \frac{19}{9}$. So, it took them $\frac{19}{9} = 2\frac{1}{9}$ hours to hike the trail.

**25. C,** Since it takes $\frac{3}{4}$ hour to write a 200-word blog, it takes $\frac{3}{4} \times \frac{1}{2} = \frac{3}{8}$ hour to write a 100-word blog. Multiply $\frac{3}{8}$ by 5 to see that it takes $\frac{15}{8}$, or $1\frac{7}{8}$, hours to write a 500-word blog.

**26. C,** Rename each mixed number as an improper fraction, and then find a common denominator. Subtract to find that Mario needs to work $32\frac{5}{6} - 19\frac{7}{8} = \frac{197}{6} - \frac{159}{8} = \frac{788}{24} - \frac{477}{24} = \frac{311}{24} = 12\frac{23}{24}$ more hours this week.

**27. D,** The drive-through staff can process $18 \div \frac{2}{3} = 18 \times \frac{3}{2} = 27$ orders in 18 minutes.

---

## LESSON 6, *pp. 27–31*

**1. D,** Solve $\frac{3}{$12} = \frac{5}{x}$ to find that $x = $20.

**2. D,** To simplify $\frac{92}{64}$, divide the numerator and denominator by 4 to get $\frac{23}{16}$.

## UNIT 1 (continued)

**3. C,** Solve $\frac{558}{9 \text{ hr}} = \frac{x}{1 \text{ hr}}$ to find that $x = 62$ miles.

**4. A,** To simplify $\frac{2}{10}$, divide the numerator and denominator by 2 to get $\frac{1}{5}$.

**5. B,** Solve $\frac{4}{20} = \frac{x}{120}$ to find that $x = 24$ miles.

**6. E,** Solve $\frac{2}{7} = \frac{14}{x}$ to find that $x = 49$ children.

**7. B,** Solve $\frac{65}{1 \text{ hr}} = \frac{260}{x}$ to find that $x = 4$ hours.

**8. E,** Solve $\frac{3}{2} = \frac{144}{x}$ to find that $x = 96$ trucks.

**9. B,** To simplify $\frac{25}{5}$, divide the numerator and denominator by 5 to get $\frac{5}{1}$.

**10. B,** Solve $\frac{1}{12} = \frac{x}{36}$ to find that $x = 3$ teachers.

**11. A,** Olive performed $12 + 4 = 16$ total skills, and her ratio of incorrect to total skills is $\frac{4}{16} = \frac{1}{4}$.

**12. E,** The ratio of games lost to games won is 4:38, which can be simplified to 2:19.

**13. C,** Divide 16 by 8 to get $2 per can of soup.

**14. E,** Solve $\frac{1 \text{ in.}}{3 \text{ ft}} = \frac{4 \text{ in.}}{x \text{ ft}}$ for $x$ to find $x = 12$ feet.

**15. D,** The ratio of miles driven on Monday to miles driven on Tuesday is $\frac{96}{60} = \frac{8}{5}$.

**16. A,** Solve $\frac{12}{30} = \frac{x}{10}$ to find that $x = 4$ eggs, so the ratio is 4 to 10, or 2 to 5.

**17. C,** Solve $\frac{8}{3} = \frac{24}{x}$ for $x$ to find that $x = 9$ games were lost.

**18. B,** The ratio of full-time to part-time employees is 30:12, or 5:2.

**19. C,** Solve $\frac{2}{150} = \frac{6}{x}$ for $x$ to find that $x = 450$ miles.

**20. B,** Solve $\frac{5}{15} = \frac{275}{x}$ for $x$ to find that $x = \$825$.

**21. D,** Solve $\frac{1}{5} = \frac{x}{35}$ for $x$ to find that $x = 7$ dogs were adopted.

**22. A,** Trevor drove $\frac{48 \text{ miles}}{3 \text{ hours}} = 16$ miles per hour.

**23. C,** The recipe calls for $2 + 3 = 5$ teaspoons of sauce. Solve $\frac{3}{5} = \frac{x}{20}$ for $x$ to find that $x = 12$ teaspoons of caramel sauce.

**24. C,** Subtract $\frac{4}{5}$ from 1 to find that $\frac{1}{5}$ of the students do not ride the bus. Solve $\frac{1}{5} = \frac{x}{460}$ for $x$ to find that $x = 92$.

**25. C,** Solve $\frac{2}{5} = \frac{x}{30}$ to find that $x = 12$ people.

**26. B,** Solve $\frac{14}{1} = \frac{406}{x}$ for $x$ to find that $x = 29$ teachers.

**27. E,** Solve $\frac{110}{1} = \frac{x}{4.5}$ for $x$ to find that $x = 495$ calories.

**28. A,** Subtract 16 from 30 to find that 14 people do not drive to work. The ratio of drivers to non-drivers is 16:14 = 8:7.

**29. C,** Solve $\frac{2}{3} = \frac{26}{x}$ for $x$ to find that $x = 39$ parking spots.

**30. C,** Solve $\frac{414}{18} = \frac{x}{1}$ for $x$ to find that $x = 23$ miles per gallon.

**31. E,** Add Leila's mileage to find that she drove a total of 2,040 miles over 5 weeks. The ratio of Week 1 to total weeks is 420:2,040 = 7:34.

**32. D,** Solve $\frac{1}{22} = \frac{x}{176}$ for $x$ to find that $x = 8$ lifeguards.

**33. A,** Solve $\frac{3}{8} = \frac{387}{x}$ for $x$ to find that $x = 1,032$ people.

**34. E,** Solve $\frac{8}{24} = \frac{12}{x}$ to find that $x = \$36$.

**35. D,** The ratio of applicants to openings for a staff writer is 48:4 = 12:1.

**36. C,** Solve $\frac{27}{3} = \frac{x}{5}$ for $x$ to find that $x = 45$ applicants for the art researcher position. Subtract 27 from 45 to find that 18 additional people will apply for the position.

---

## LESSON 7, pp. 32–36

**1. A,** The cost of the coffee and muffin is $2.95 + $1.29, which is $4.24. Subtract $4.24 from $5.00 to get a difference of $0.76.

**2. D,** Multiply $3.50 by 6 to get a product of $21.00.

**3. C,** Ben paid $9.32 for bread because $2.33 × 4 = $9.32. He paid $2.58 for butter because $1.29 × 2 = $2.58. The difference of $9.32 − $2.58 = $6.74.

**4. B,** The cost of 6 gallons of milk is 6 × $3.50, or $21.00. The cost of 5 boxes of cereal is 5 × $3.85, or $19.25. The sum of $21.00 + $19.25 is $40.25.

**5. C,** The answer is an estimate. Round $12.25 to $12 and $3.85 to $4. Divide 12 ÷ 4 to get about 3 boxes he can buy.

**6. B,** The difference in cost between reams is $5.25 − $3.99 = $1.26, so $1.26 saved per ream × 15 reams = $18.90.

**7. D,** Divide the length of the whole rope by the number of equal pieces. 14.4 ÷ 4 = 3.6 meters

**8. E,** Coach Steve will spend $17 × 12, or $204, on uniforms. He will spend $12.95 × 6, or $77.70, on soccer balls. The sum of the two products is $281.70.

**9. C,** Coach Steve will spend $10.95 × 12, or $131.40, on shin guards. He will spend $8.95 × 12, or $107.40, on knee pads. The difference of $131.40 − $107.40 is $24.00.

**10. E,** The answer is an estimate. Round $2.99 to $3 and 5.17 to 5. Multiply $3 × 5 to get about $15 for the cost of the strawberries.

**11. B,** Divide $5.89 by 8 to find that a can of beans costs $0.74.

**12. D,** Evan spent $589.45 + 82.32 + 14.99 = $686.76 in all.

**13. B,** Subtract $75.45 from $89.79 to find that chairs at In the Woods cost $14.34 less. William saved $14.34 × 4 = $57.36.

**14. B,** The cost of a single package is $7.62 ÷ 6 = $1.27.

**15. D,** Ariana spent $17.95 + $3.27 = $21.22 total. She has $15.78 remaining.

**16. C,** Carmen has $163.60 ÷ 8 = $20.45 deducted from her paycheck each week.

**17. A,** Russell spent $5.69 + $3.98 + $1.99 = $11.66 on lunch.

**18. A,** Timothy pays $143 ÷ 52 = $2.75 per week.

**19. D,** A deposit means addition so add $217.98 to the amount in the amount, $597.16. Taking out money from the account means subtraction, so subtract $45 from the sum: $597.16 + $217.98 − $45.

**20. A,** Without sales tax, 2.3 pounds of salami cost $3.95 × 2.3 = 9.085 ≈ $9.09.

**21. A,** Divide $675 by 12 to find the monthly payment of $56.25.

**22. A,** Terese spent a total of $14.89 + $2.38 + $0.79 = $18.06. She should receive $20.00 − $18.06 = $1.94 in change.

**23. D,** Lisa's total cost is $22.95 + $ 66.25 = $89.20.

**24. C,** Subtract $14.85 from $22.95 to find that Mark would save $8.10.

**25. C,** Tim has a combined total of 95.75 + 92.5 + 98.25 = 286.5 on his three exams.

## UNIT 1 (continued)

**26. B,** Subtract $85.04 from $124.53 to find the difference of $39.49.

**27. E,** Jonah spent $124.53 + $118.92 + $95.41 + $88.73 + $85.04 + $86.29 = $598.92 from July to December.

**28. A,** The difference in prices of one box is $2.29 − $2.05. Multiply that difference by 5 boxes: 5 × ($2.29 − $2.05). Use parentheses to indicate that the subtraction is done first.

**29. C,** The total score of all three judges is 8 + 8.5 + 7.5 = 24, so his total score is 24 × 3.2 = 76.8.

**30. D,** The answer is an estimate. Round $7.90 to $8 and $1.79 to $2. Divide $8 ÷ $2 to get about 4 drinks Dylan could buy.

**31. D,** Alexis spent $2.65 × 6 = $15.90 in all and should receive $20.00 − $15.90 = $4.10 in change.

**32. B,** Walt pays $1,556.28 ÷ 12 = $129.69 per month.

**33. E,** Multiply 60.2 by 3.5 to get a product of 210.7 miles.

---

### LESSON 8, pp. 37–41

**1. D,** $\frac{27}{45} = 27 \div 45 = 0.6$, and 0.6 = 60%.

**2. D,** $25\% = \frac{25}{100} = \frac{25 \div 25}{100 \div 25} = \frac{1}{4}$.

**3. B,** $7\frac{1}{2} = 7\frac{5}{10} = 7.5$.

**4. A,** $\frac{3}{8} = 3 \div 8 = 0.375$, and 0.375 = 37.5%.

**5. C,** $\frac{1}{8} = 1 \div 8 = 0.125$.

**6. D,** Since $0.22 = \frac{22}{100}$, 22 people said yes. Since 22 people said yes, 100 − 22, or 78, people said no. $\frac{78}{100} = \frac{78 \div 2}{100 \div 2} = \frac{39}{50}$.

**7. C,** $\frac{9}{13} = 9 \div 13 \approx 0.6923$, and 0.6923 = 69.23%, which rounds to 69.2%.

**8. E,** Either find the product of 0.88 × 25 or solve $\frac{88}{100} = \frac{x}{25}$ to get 22 questions.

**9. D,** 75% of 300 is the same as $\frac{3}{4} \times 300$, which equals 225.

**10. D,** There is a percentage of $\frac{1}{8} = 1 \div 8 = 0.125 = 12.5\%$ first-graders dropped off at school.

**11. C,** Since $\frac{2}{50} = \frac{4}{100}$, 4% of the students work full time while attending college.

**12. E,** Since $85\% = \frac{85}{100}$, this can be simplified to $\frac{17}{20}$.

**13. C,** Divide 3 by 25 to get 0.12 = 12%.

**14. E,** Customers will pay $90\% = \frac{90}{100} = \frac{9}{10}$ of the original price.

**15. B,** Employees pay 70% = 0.70 of the original price, so Sam will pay $580 × 0.7 = $406.

**16. A,** It will cost 500 × $2 = $1,000 for the flyers. Marie's down payment will be $1,000 × 0.3 = $300.

**17. B,** The fraction of science-related questions is $45\% = \frac{45}{100} = \frac{9}{20}$.

**18. C,** The unit price is $22\frac{1}{2}$ cents = 22.5 cents = $0.225.

**19. B,** Subtract 22.8 from $54\frac{1}{2} = 54.5$ to find that Nina biked 31.7 miles on the second day.

**20. B,** Self-employment taxes are $\frac{5.5}{100} = \frac{55}{1,000} = \frac{11}{200}$ of her income.

**21. B,** Divide 82 by 100 to find that 82% = 0.82.

**22. B,** The Panthers won $\frac{22}{34} = 0.647 = 64.7\%$ of their games.

**23. C,** Ted answered $\frac{41}{50} = 0.82 = 82\%$ of the questions correctly.

**24. D,** $1,230 × 0.2 = $246 down payment. Subtract $246 from $1,230 = $984 owed on the computer after the down payment.

**25. A,** Jim makes a profit of $10.50 − $7 = $3.50 per knapsack. His percentage of profit margin is $3.50 ÷ $7 = 0.5 or 50%.

**26. D,** Maya paid $70\% = \frac{70}{100} = \frac{7}{10}$ of the original price.

**27. D,** Carlos will save $40\% = \frac{40}{100} = \frac{4}{10} = \frac{2}{5}$ of the original price.

**28. C,** Add $\frac{1}{6} = \frac{10}{60}$ for food and $0.35 = \frac{35}{100} = \frac{21}{60}$ for housing to get a total of $\frac{31}{60}$. Subtract this from 1, or $\frac{60}{60}$, to find the fraction of her remaining earnings is $\frac{29}{60}$.

**29. C,** Elliott's friends pay $90\% = \frac{9}{10}$ of the original price.

**30. D,** The Kickers won 24 × 0.75 = 18 matches.

**31. A,** The interest rate $5\frac{1}{2}\%$ can be written as 5.5 ÷ 100 = 0.055.

---

### LESSON 9, pp. 42–46

**1. B,** Using the formula $I = prt$, $I = 1,000(0.05)(\frac{1}{2}) = 25$, which means Kirsten owes $25 in interest. She owes $25 plus the $1,000 she borrowed, which equals $1,025.

**2. D,** Solve $\frac{12}{100} = \frac{x}{552}$ to find that x = $66.24.

**3. D,** Divide 309 by 824 to get 0.375. Then multiply by 100 to get 37.5%.

**4. C,** Andrew received a raise of $25,317.40 − $24,580.00, or $737.40. Then solve $\frac{737.40}{24,580} = \frac{x}{100}$ to find that x = 3%.

**5. D,** Write the percent 6% as 0.06. Add the sales tax (0.06 × $425) to the price of the bicycle ($425), so use the expression $425 + (0.06 × $425) to find the total paid.

**6. B,** Multiply 0.45 by 420 to get 189.

**7. B,** To find the discount, multiply 0.20 by $659 to get $131.80. Then subtract $659.00 − $131.80 to get a sale price of $527.20.

**8. E,** Using the formula $I = prt$, $I = 5,000(0.05)(\frac{9}{12}) = 187.50$.

**9. B,** Multiply 0.20 by $2,250 to get a product of $450.

**10. A,** Multiply 0.40 by 25 to find that 10 new employees will be hired. 25 current employees + 10 new employees = 35 employees in all.

**11. B,** Theo now pays $615 − $585 = $30 more for his rent. Solve $\frac{30}{585} = \frac{x}{100}$ for x to find the percent increase is x ≈ 5.128 ≈ 5%.

**12. E,** Solve $\frac{2.86}{x} = \frac{8}{100}$ for x to find that x = $35.75.

**13. B,** Use the formula: $I = prt$. Ezra will pay 3,000 × 0.03 × 1.5 = $135 in interest.

**14. C,** Dan's down payment was $16,584 × .2 = $3,316.80. He has to pay $16,584 − $3,316.80 = $13,267.20 over the next 24 months. Dan's monthly payment is $13,267.20 ÷ 24 = $552.80.

**15. D,** The discount for the jacket is $152.60 × 0.4 = $61.04. Noelle paid $152.60 − $61.04 = $91.56 for the jacket.

**16. C,** Solve $\frac{2,025}{x} = \frac{4.5}{100}$ for x to find that x = $45,000.

**17. C,** The interest on Remy's loan is $10,000 × 0.056 × 3 = $1,680. She paid a total of $10,000 + $1,680 = $11,680.

**18. B,** $22,000 × 0.2 = $4,400. Subtract $4,400 from $22,000 to find an owed balance of $17,600. Divide $17,600 by 36 months to find a monthly car payment for Jae of $488.89.

**19. D,** Solve $\frac{72}{x} = \frac{6}{100}$ to find that x = $1,200.

## UNIT 1 (continued)

**20. E,** $\dfrac{308{,}205 \text{ parts}}{1{,}158{,}675 \text{ in all}} \approx 0.2659$, and $0.2659 \approx 27\%$.

**21. D,** $\dfrac{545{,}380}{1{,}158{,}675} \approx 0.47069$, which is about 47%.

**22. B,** The answer is an estimate. Round $6.10 to $6 and $40.66 to $40. Divide $6 \div 40$ to get 0.15, or about a 15% tip.

**23. D,** The increase is $11 \times 0.04 = \$0.44$ per hour. $11.00 + an increase of $0.44 = $11.44.

**24. C,** The decrease in passengers was $5{,}478 - 4{,}380 = 1{,}098$. Solve $\dfrac{1{,}098}{5{,}478} = \dfrac{x}{100}$ to find that there was about a 20% decrease.

**25. B,** Michelle's new salary = 115% × her old salary, so $\$86{,}250 = 1.15x$; $x = \$75{,}000$.

**26. C,** Solve $\dfrac{x}{35} = \dfrac{20}{100}$ to find that the class size increased by 7 students. There are now $35 + 7 = 42$ students.

**27. C,** Using the formula $I = prt$, $787.50 = 10{,}500(3)r$; $r = 0.025$, which is 2.5%.

**28. A,** Solve $324 = 0.72x$ to find that there were 450 teachers in the union. Then subtract $450 - 324$ to find that 126 teachers were against it.

**29. E,** He pays 95% of the first shirt, 85% of the second, and 75% of the third. $0.95(12) + 0.85(12) + 0.75(12) = \$30.60$.

**30. D,** In the formula base • rate = part, the base is the number of original employees (850) and the rate is 8%. So, the equation $0.08(x) = 850$ is not set up correctly and cannot be used to find the number of new employees (the part).

**31. B,** The population decreased by $756 - 711 = 45$ students. Solve $\dfrac{45}{756} = \dfrac{x}{100}$ to find a decrease of about 6%.

**32. E,** The interest rate is unknown.

**33. B,** In Maryland, the computer game would cost $31.80, which is less than Delaware's price of $32.

---

### LESSON 10, pp. 47–51

**1. D,** The length of the square is $\sqrt{81} = 9.0$ m.

**2. B,** The length of one side of Meredith's garden is $\sqrt{121} = 11$ ft.

**3. D,** Use the answer from question 2 (11 ft) and double it to get 22 ft. Then square the side length to find the new area: $22 \times 22 = 484$ sq ft.

**4. D,** Carlos completed $5^3 = 5 \times 5 \times 5 = 125$ squats.

**5. A,** The length of one side is the cube root of 64, which is 4, because $4 \times 4 \times 4 = 64$.

**6. A,** The solution is $\dfrac{\sqrt{64}}{4} = \dfrac{8}{4} = 2$.

**7. C,** Solve $x^2 = 30$ by taking the square root of both sides, so $\sqrt{30} \approx 5.477 \approx 5.5$.

**8. C,** The length of one side is $\sqrt{50} \approx 7$ ft, so the perimeter is $7 \times 4 = 28$ ft.

**9. D,** Solve $(8 - x)^2 = 64$ to get $8 - x = \sqrt{64}$. The value $\sqrt{64}$ can be either +8 or −8. So, $8 - x = -8$ gives $x = 16$, OR $8 - x = 8$ gives $x = 0$. Because the student was told the car is moving, the distance $x$ cannot be 0, so the car traveled 16 miles.

**10. C,** There are $\sqrt{144} = 12$ questions.

**11. B,** The number is $\sqrt{7{,}788} \approx 88.24965$, which rounds to 88.25.

**12. C,** The volume of the cube is $(29)(29)(29) = 24{,}389$ cm³.

**13. B,** The length of the square is $\sqrt{6.7} \approx 2.5884$, which rounds to 2.6.

**14. B,** The square root of 25 is 5, and the square root of 36 is 6. Since 33 falls between 25 and 36, the square root of 33 falls between 5 and 6.

**15. A,** The width is $x$, and the length is 50% more, or $1.5x$. The area of the room is $x \cdot 1.5x$, so solve $1.5x^2 = 216$ to get $x^2 = 144$. Take the square root of both sides to get $x = 12$, the width of the room. The length is $1.5x$, or 18 feet.

**16. B,** The integers 4 and −4 both square to equal 16.

**17. B,** Simplify $19^2 - \sqrt{169} = 361 - 13 = 348$ to get a $348 car payment.

**18. D,** The width is $x$, and the length is 60% more, or $1.6x$. The area of the field is $x \cdot 1.6x$, so solve $1.6x^2 = 4{,}000$ to get $x^2 = 2{,}500$. Take the square root of both sides to get $x = 50$, the width of the field. The length is $1.6x$, or 80 yards.

**19. B,** The volume of the table is $4^3 = 64$. The volume of the hollow cube is $1.5^3 = 3.375$. Subtract $64 - 3.375$ to get 60.625 cubic feet.

**20. D,** Represent the side lengths as $a$, $b$, and $c$, so the original volume is $abc$. Each new side is three times as long, so the new volume is $(3a)(3b)(3c) = 3^3abc = 27abc$. The volume has increased by a factor of 27.

**21. B,** Represent the side lengths as $a$ and $b$, so the original area is $ab$. Each new side is twice as long, so the new area is $(2a)(2b) = 2^2ab = 4ab$. The area has increased by a factor of 4.

**22. A,** Solve the equation to find $x = 5$ and $x = -5$. (Both, when squared, give 25 as the result.) The product of 5 and −5 is −25.

**23. E,** Take the square root of both sides to find that $x - 1$ can equal 8 or −8. Solve $x - 1 = 8$ to get $x = 9$. Solve $x - 1 = -8$ to get $x = -7$. The product $9 \times (-7) = -63$.

**24. D,** The square root of 64 is 8, so the cube of 8 is $8^3 = 512$. The cube root of 64 is 4, so the square of 4 is $4^2 = 16$. Divide $512 \div 16$ to get 32 times greater.

**25. D,** Take the square root of both sides to find that $x - 6$ can equal 2 or −2. Solve $x - 6 = 2$ to get $x = 8$. Solve $x - 6 = -2$ to get $x = 4$. The product $8 \times 4 = 32$.

**26. B,** Take the square root of 30.25 to get the side length of 5.5 inches. The volume is $(5.5)(5.5)(5.5) = 166.375$, which rounds to 166 cubic inches.

**27. C,** The square root of $2{,}000 \approx 44.721$, which rounds to 45 inches.

**28. D,** The width 18 inches is 1.5 feet. The volume is $(1.5)(1.5)(1.5) = 3.375 \approx 3.4$ cubic feet.

**29. D,** Since each block is 2 inches wide, it takes $(6)(6)$ blocks to make a square that is a foot long and a foot high. It takes 6 of these squares to make a cube, so there are $6^3$, or 216 blocks in the cubic stack.

**30. A,** Substitute the value $F = 41$ into the equation. Find $(41 - 32)$ is 9, and $9^2$ is 81, to get the equation $(25)(81) = (81)C^2$. Divide both sides by 81 to get $25 = C^2$. Take the square root to find $C$ can be 5 or −5. Only the positive value appears among the choices.

**31. E,** Take the square root of both sides of the equation to get $x = \sqrt{-16}$. Since there is no real number, when squared, that equals a negative number, $x$ is undefined.

**32. C,** Take the cube root of both sides of the equation to get $x$ = cube root of $-64 = -4$, because $(-4)(-4)(-4) = -64$.

# UNIT 1 (continued)

**33. C,** The length is $x$, and the width is $\frac{3}{4}x$. The area of the room is $x \cdot \frac{3}{4}x$, so solve $\frac{3}{4}x^2 = 192$ to get $x^2 = 256$. Take the square root of both sides to get $x = 16$ feet, the length of the room. The width is $\frac{3}{4}x$, or 12 feet.

**34. A,** The width is $x$, the depth is $x$, and the length is $3x$. The volume of the box is $x \cdot x \cdot 3x$, or $3x^3$, so solve $3x^3 = 192$ to get $x^3 = 64$. Take the cube root of both sides to get $x = 4$ inches, the width and the depth of the room. The length is $3x$, or 12 feet.

**35. D,** The expression is undefined when the quantity under the square root sign is a negative number. That will happen for any number $x$, when $x^2$ is less than 1.5. The set of values –1, 0, 1 gives $(-1)^2 = 1$, $(0)^2 = 0$, and $(1)^2 = 1$, all less than 1.5.

**36. C,** The square of $14 = 14^2 = 196$. The square of 21 minus the square root of $49 = 21^2 - \sqrt{49} = 441 - 7 = 434$. Add $196 + $434 to get $630.

**37. C,** The square root of a negative number is not defined, so the value of $x$ for $x^2 = -49$ is undefined.

---

## LESSON 11, pp. 52–56

**1. B,** Move the decimal point seven places to the left to find $5.8 \times 10^7$.

**2. B,** Move the decimal point seven places to the left to find $2.54 \times 10^7$.

**3. B,** The area of the rectangle is $2^6 \cdot 2^5 = 2^{11}$.

**4. D,** The width is $(1.5 \times 10^{-3})(2.0 \times 10^5) = (1.5 \times 2.0)(10^{-3} \times 10^5) = 3.0 \times 10^2$ cm.

**5. C,** The expression $5^1 + 4^0$ is equal to $5 + 1 = 6$.

**6. A,** The numbers with exponents all have the same base, so the first term is $5(7^4)$, the second term is $5(7^0) = 5$, and the third term is $-(7^4)$. Add the three terms, combining the first and third terms because they have a base to the same power: $5(7^4) + 5 - (7^4) = (5 - 1)7^4 + 5 = 4(7^4) + 5$.

**7. C,** Write each term with the negative exponents as fractions: $\frac{6}{2^3} + \frac{5}{2^4} + \frac{4}{2^5} = \frac{6}{8} + \frac{5}{16} + \frac{4}{32}$. Write the fractions with a common denominator of 32, and simplify: $\frac{24}{32} + \frac{10}{32} + \frac{4}{32} = \frac{38}{32} = \frac{19}{16}$.

**8. E,** Multiply the factor of 2 in the second term to get $2x^2 - 10x - 4$. Add $3x^2 + 3x + 2 + 2x^2 - 10x - 4$ and combine like terms to get $5x^2 - 7x - 2$.

**9. A,** Multiply the factor of 2 in the second term to get $2x^2 - 10x - 4$. Subtract $3x^2 + 3x + 2 - (2x^2 - 10x - 4) = 3x^2 + 3x + 2 - 2x^2 + 10x + 4$ and combine like terms to get $x^2 + 13x + 6$.

**10. B,** Simplify the numerator first by multiplying in the second term: $6x^2 + 4 - 4 + 6x^2$. Combine like terms to get $12x^2$. Divide that numerator by the denominator, $4x$, to get $3x$.

**11. B,** Write the expressions for $x^2$ and $x^{-2}$, then simplify. $x^2 = \left(\frac{1}{3}\right)^2 = \frac{1}{9}$ and $x^{-2} = \left(\frac{1}{3}\right)^{-2} = \frac{1}{\left(\frac{1}{3}\right)^2} = 1 \div \frac{1}{9} = 9$, so $x^{-2}$ can be greater than $x^2$.

**12. A,** Move the decimal point four places to the right to find $5.0 \times 10^{-4}$.

**13. D,** There are $3 \times 3 \times 3 \times 3 = 81$ items, which is approximately 80.

**14. C,** The total number of students is $3^4 + 2^6$.

**15. A,** Since any number raised to the zero power is equal to 1, $4^0 = 5^0$.

**16. E,** Multiply each term in the parentheses by $4x$ to find that $4x(x^2 + 2y) = 4x^3 + 8xy$.

**17. C,** The number with the greatest power of ten is Jupiter.

**18. A,** The numbers with the least power of ten are Mercury, $3.3 \times 10^{23}$, and Mars, $6.42 \times 10^{23}$. Since $3.3 < 6.42$, Mercury has the least mass.

**19. B,** Divide the masses: $\frac{1.899 \times 10^{27}}{6.42 \times 10^{23}} = 0.29 \times 10^4$. Round 0.29 to 0.3 and write in scientific notation, $3.0 \times 10^3$.

**20. A,** Write each mass in terms $10^{27}$ kg by moving the decimal point the appropriate number of places to the left. The mass of Mercury, for example, is $0.00033 \times 10^{27}$. The sum of the masses is $2.667612 \times 10^{27}$, which is about $2.7 \times 10^{27}$.

**21. B,** The greatest common factor of the terms is $5x^3$. Divide each term by $5x^3$ to get $5x^5 \div 5x^3 = x^2$, $-15x^4 \div 5x^3 = -3x$, and $10x^3 \div 5x^3 = 2$. So, the equivalent expression is $5x^3(x^2 - 3x + 2)$.

**22. E,** Simplify the numerator first by multiplying in the second term: $x^4 + 5x^3 + x^4 - 2x^2$. Combine like terms to get $2x^4 + 3x^3$. Divide that numerator by the denominator, $x^3$, to get $2x + 3$.

**23. A,** Expand the first term by multiplying to get $6x^2 - 3$. Subtract $6x^2 - 3 - 5x^2 - x - 3$ and combine like terms to get $x^2 - x - 6$.

**24. C,** Substitute –1 for $x$. The numerator is $(-1)^3 = (-1)(-1)(-1) = -1$. The denominator is $(-1)^3 - 1 = -1 - 1 = -2$. Since the numerator and denominator are both negative, the quotient is positive and, therefore, greater than zero.

**25. D,** Substitute –1 for $x$. The numerator is $(-1)^3 = -1$. The denominator is $(-1)^3 + 1 = -1 + 1 = 0$. Since the denominator is zero, the expression is undefined.

**26. D,** Substitute –1 for $x$. The numerator is $(-1 - 1)^4 = (-2)^4 = 16$. The denominator is $(-1)^2 - 1 = 1 - 1 = 0$. Since the denominator is zero, the expression is undefined.

**27. D,** Move the decimal point nine places to the right to find Saturn is 1,433,500,000 km from the sun.

**28. B,** Jupiter is $(7.786 \times 10^8) - (1.082 \times 10^8) = (7.786 - 1.082) \times 10^8 = 6.704 \times 10^8$ km farther from the sun.

**29. C,** Since 10 is to the fifth power, place 5 zeros after 1 to find that $10^5 = 100,000$.

**30. B,** The expression $b^{-4}$ is equal to $\frac{1}{b^4}$.

**31. E,** Move the decimal point the left 10 spaces to find that the marbled lungfish weighs 0.00000000013283 grams.

**32. C,** Move the decimal point nine places to the left to find $4.435 \times 10^9$.

**33. C,** When the decimal point is moved one place to the left, the exponent becomes one digit greater, so $4.404 \times 10^9 = 0.4404 \times 10^{10}$.

**34. D,** Since $81 = 9 \times 9 = 3 \times 3 \times 3 \times 3$, then $x = 4$.

**35. C,** The number of students at Sunnyside High School is $3 \times 2^9 = 3 \times 512 = 1,536$.

**36. B,** A negative $\times$ a negative = a positive. A negative $\times$ a negative $\times$ a negative = a negative. So, if $x$ equals an even number, the answer will be positive.

**37. C,** Simplify the left side to get $(3^x)(3^x) = 3^{x+x} = 3^{2x}$. The right side is $3 = 3^1$. Since $3^{2x} = 3^1$, then $2x$ must equal 1. Solve $2x = 1$ to get $x = \frac{1}{2}$.

# UNIT 1 (continued)

**38. B,** There are 60 seconds in 1 minute and 60 minutes in 1 hour, so there are $60 \times 60 = 3,600$ seconds in 1 hour. The number of operations is $3,600 \times 5 \times 10^{13} = 18,000 \times 10^{13} = 1.8 \times 10^{17}$.

## UNIT 1 REVIEW, pp. 57–61

**1. E,** If 35% were in favor, then $100\% - 35\% = 65\%$ objected. Multiply $0.65 \times 1,200 = 780$.

**2. D,** The chairs cost $\$65.30 \times 4$, or $\$261.20$. The total cost is $\$261.20 + \$764.50 = \$1,025.70$.

**3. B,** Subtract $210.5 - 135.8$ to get a difference of 74.7.

**4. C,** Divide $4\frac{1}{2}$ by $1\frac{1}{2}$ to get $4\frac{1}{2} \div 1\frac{1}{2} = \frac{9}{2} \div \frac{3}{2} = \frac{9}{2} \times \frac{2}{3} = 3$.

**5. D,** To find the total number of students, add $118 + 54 + 468 + 224$ to get 864. Simplify $\frac{54}{864} = \frac{54 \div 54}{864 \div 54} = \frac{1}{16}$.

**6. D,** Add $468 + 224$ to get 692. Then simplify $\frac{692}{864}$ by dividing the numerator and denominator by 4 to get $\frac{173}{216}$.

**7. E,** Using the formula $I = prt$, $I = (1,250)(0.06)(3) = 225$. Kara received $\$225$ in interest plus the original $\$1,250$ she invested, which equals $\$1,475$.

**8. C,** The population changed by $45,687 - 43,209$, or 2,478 people. Solve $\frac{2,478}{43,209} = \frac{x}{100}$ to find that $x \approx 5.73$, which rounds to 6%.

**9. B,** There are $8 \cdot 8 \cdot 8 \cdot 8 = 4,096$ students at the university.

**10. A,** The pretzels cost $\$1.95 \times 2$, or $\$3.90$. The soft drinks cost $\$0.99 \times 2$, or $\$1.98$. The pretzels and soft drinks cost $\$3.90 + \$1.98$, or $\$5.88$ in all. Subtract $\$10.00 - \$5.88$ to find a difference of $\$4.12$.

**11. C,** Rename the fractions as $\frac{10}{60}, \frac{3}{60}, \frac{20}{60}, \frac{12}{60}$, and $\frac{15}{60}$. Since $\frac{20}{60}$ is the greatest, soccer has the greatest participation.

**12. C,** Add $\frac{1}{4} + \frac{1}{6} = \frac{3}{12} + \frac{2}{12} = \frac{5}{12}$.

**13. D,** Since $x^2 = 36$, $\sqrt{x} = 6$ or $-6$. So, $2(6 + 5) = 2(11) = 22$.

**14. D,** His new position is $786 - 137 + 542 = +1,191$ feet.

**15. E,** Solve $\frac{301.5}{4.5 \text{ hr}} = \frac{x}{1 \text{ hr}}$ to find that $x = 67$ and Benjamin drove 67 miles per hour.

**16. B,** The profit for one share is $\$52 - \$43$, or $\$9$. Multiply $\$9$ by 20 shares to get $\$180$ that Scarlett made.

**17. D,** 426 people divided by 65 per bus is 6 buses, with 36 people that cannot fit in a bus. One more bus is needed, so $6 + 1 = 7$ buses are needed in all.

**18. C,** Fifty-six thousand is written as 56,000, and two hundred twenty-eight is written as 228. So, the income was 56,228.

**19. C,** Move the decimal point eight places to the left to find $1.496 \times 10^8$.

**20. A,** The change in Emmit's account is $(-\$64)(3) = -\$192$.

**21. B,** 9.65 and 9.19 both have 9 ones; 9.65 has 6 tenths, but 9.19 only has 1 tenth, so $\$9.65$ is the greatest price, which is for an elk sandwich.

**22. B,** The kid's platters cost $3 \times \$3.50$, or $\$10.50$. Add $\$10.50 + \$9.19 + \$5.89$ to get $\$25.58$. Subtract $\$50.00 - \$25.58$ to find $\$24.42$ that Kurt has left.

**23. D,** The elk sandwiches cost $\$9.65 \times 2$, or $\$19.30$. The kid's platters cost $\$3.50 \times 3$, or $\$10.50$. The difference of $\$19.30 - \$10.50 = \$8.80$.

**24. C,** Solve $\frac{2}{3} = \frac{x}{180}$ to find that $x = 120$.

**25. A,** Find the number of groups of $1\frac{2}{3}$ in 4 hours. $\frac{4}{1} \div 1\frac{2}{3} = \frac{4}{1} \div \frac{5}{3} = \frac{4}{1} \times \frac{3}{5} = \frac{12}{5}$, and $\frac{12}{5} = 2\frac{2}{5}$.

**26. D,** Multiply 0.84 by 175 to get a product of 147.

**27. E,** Multiply: $\frac{2}{3} \times \frac{24}{1} = \frac{2}{1} \times \frac{8}{1} = 16$.

**28. A,** Simplify $\frac{10}{2}$ by dividing the numerator and denominator by 2 to get $\frac{5}{1}$.

**29. B,** Solve $\frac{78}{200} = \frac{x}{400}$ to find that $x = 156$.

# UNIT 2  MEASUREMENT / GEOMETRY

## LESSON 1, pp. 64–68

**1. A,** Convert 30 mL to 3 cL. There are $3 + 2 = 5$ cL altogether.

**2. D,** The next train leaves at 2:35 P.M., which is 2 hours and 43 minutes later.

**3. C,** Add 1 hour and 45 minutes to 3:20 to find that he will arrive at 5:05 P.M.

**4. E,** Convert 1 cup to 8 fl oz and 1 pint to 16 fl oz. Mr. Trask needs $(2)(6) + 8 + 16 = 12 + 8 + 16 = 36$ fl oz of liquid food.

**5. A,** Divide the amounts in milligrams by 10. The students use $2 + 0.5 + (2)(1.5) + 0.5 = 6$ cg of Chemical A.

**6. E,** Shantell had 60 cg = 600 mg of Chemical C. She had $600 - 5 = 595$ mg more of Chemical C.

**7. C,** Convert 50 cg to 500 mg and 1 g to 1,000 mg. Diego used $15 + 500 + 1,000 = 1,515$ mg of chemicals. He used $1,515 - 535 = 980$ mg more than Dana.

**8. C,** Since 2 cups = 1 pint, Cedric will need $3 \times 1$ pint = 3 pints of orange juice.

**9. D,** Since 4 pints = 2 quarts, Cedric will need $3 \times 2$ quarts = 6 quarts of carbonated water.

**10. E,** Since 5 pints = 10 cups, Cedric will need $3 \times 10$ cups = 30 cups of ginger ale.

**11. D,** Divide 700 by 1,000 to find that Sabrina marked 0.7 km on Day 1.

**12. B,** Sabrina marked $700 + 600 + 800 + 1,000 = 3,100$ m by Day 4. Divide 3,100 by 1,000 to find that she marked 3.1 km.

**13. C,** The runners must race $700 + 600 + 800 + 1,000 + 900 = 4,000$ m = 4 km.

**14. A,** Jason threw the javelin $63 - 54 = 9$ feet more than Hector. Divide 9 by 3 to find 3 yards.

**15. C,** Hannah walked $(875)(3) + (2,625)(2) = 7,875$ yards. Divide 7,875 by 1,760 to find $4.47 \approx 4.5$ miles.

**16. B,** Mara will need $\frac{18 \times 24}{12} = \frac{432}{12} = 36$ ft of ribbon.

**17. B,** George ran $\frac{2,640}{1,760} = 1.5$ miles.

**18. C,** Kyle will need $\frac{(3)(448) + (2)(236)}{1,000} = \frac{1,816}{1,000} = 1.816$ kL of water.

**19. D,** They gathered $\frac{(5)(10) + (5)(1) + (5)(0.1)}{10} = \frac{55.5}{10} = 5.55$ cL of pond water.

**20. A,** The soccer team consumed $17 \div 1,000 = 0.017$ kL of water.

**21. B,** The tallest sapling is $121 \div 100 = 1.21$ m tall.

**22. A,** The difference in height is $\frac{57 - 33}{100} = \frac{24}{100} = 0.24$ m.

## UNIT 2 (continued)

**23. E,** The data from the following summer are missing, so there is not enough information.

**24. B,** One box of cereal contains $\frac{45 \times 8}{1,000} = \frac{360}{1,000} = 0.36$ kg of carbohydrates.

**25. A,** Jim consumed $250 \div 2 = 125$ mg of sodium. So, he consumed $125 \div 10 = 12.5$ cg.

**26. E,** Kyle crossed the finish line at 12:30 P.M. + 1:34 = 1:64 = 2:04 P.M.

**27. B,** Solve $5 = 7t$ for $t$ to find that $t \approx 0.714$ hour $\approx$ 42.8 minutes. Nick ran the course in about 43 minutes. Nick will finish at 11:35 A.M. + 0:43 = 11:78 = 12:18 P.M.

## LESSON 2, pp. 69–73

**1. D,** The area of the front wall with the door is (21)(44) = 924 ft. The area of the metal door is (12)(12) = 144 ft. The area that will need to be painted is 924 − 144 = 780 ft.

**2. E,** The perimeter of Parcel A is 340 + 250 + 340 + 250 = 1,180 ft.

**3. C,** Divide 340 by 2 to find 170 ft. The perimeter of Parcel C is 170 + 250 + 170 + 250 = 840 ft.

**4. C,** The deletion of Parcel C would not change the overall perimeter of the land.

**5. B,** The perimeter of the triangle is 8 + 8 + 8 = 24 ft.

**6. B,** Subtract $2 \times 8 = 16$ from 36 to get a difference of 20. Divide 20 by 2 to find $x = 10$ ft.

**7. D,** The perimeter of the shaded area is 2 • triangle side length + 2 • $x$ + width of rectangle = 2(8) + 2(10) + 8 = 44 ft.

**8. E,** The area of the bedroom floor is (12)(18) = 216 ft², and the area of the closet is (7)(6) = 42 ft². It will cost ($6)(216 + 42) = $1,548 to cover both.

**9. D,** It will cost ($6)(12)(18) = $1,296 to install flooring in the bedroom only.

**10. B,** The perimeter of the bedroom, without the closet door, is 12 + 18 + 6 + 18 = 54 ft. The perimeter of the closet, without the closet door, is 7 + 6 + 7 = 20 ft. Add 54 + 20 to get 74 ft of baseboard needed.

**11. D,** The area of a red strip is (7)(18) = 126 sq cm. The area of all of the red strips is (3)(126) = 378 sq cm.

**12. A,** The area of the dog pen is (0.5)(12)(12.5) = 75 sq m.

**13. B,** Subtract (2)(12) = 24 from 56 to get a difference of 32. Divide 32 by 2 to find a length of 16 ft.

**14. C,** The perimeter of one triangle is 16 + 12 + 20 = 48 ft.

**15. C,** The area is (12)(16) = 192 ft².

**16. A,** The total measurement of the two missing sides is 54 − (2)(16) = 22 cm. So, the width is 22 ÷ 2 = 11 cm.

**17. C,** The length of one side of one triangle is 15 ÷ 3 = 5 in.

**18. C,** The perimeter of one triangle is (5)(3) = 15 in.

**19. D,** The perimeter of the parallelogram is 15 + 5 + 15 + 5 = 40 in.

**20. C,** The new length will be 15 + 5 = 20 in., and the new perimeter will be 20 + 5 + 20 + 5 = 50 in.

**21. D,** The area of the rectangular piece of wood is (18)(32) = 576 in.².

**22. D,** The base of one triangle is 40 − 32 = 8 in. The area of one triangle is $\frac{1}{2}$ (8)(18) = 72 in.². The area of both triangles is 72 + 72 = 144 in.².

**23. C,** The area of the original piece of wood, a parallelogram, is (40)(18) = 720 in.².

**24. B,** Solve $A = \frac{1}{2} bh$ for $h$ when $A = 33.84$ and $b = 14.1$ to find that $h = 4.8$ cm.

**25. D,** The total area of the backyard is $\frac{1}{2}$(10)(40 + 16) = 280 ft².

**26. A,** The area of the lot that will be paved is 280 ÷ 2 = 140 ft².

**27. B,** The cost of pavers will be (140)($3) = $420.

**28. E,** They would spend ($420)(2) = $840 to pave their entire back yard.

## LESSON 3, pp. 74–78

**1. E,** Use $A = 3.14 \times r^2$ to find that $A = 3.14 \times 9^2 = 254.34$ in.².

**2. C,** If $d = 15$, then $r = 7.5$ cm; $A = 3.14 \times 7.5^2 = 176.625$, which rounds to 176.63 cm².

**3. B,** Use $C = 3.14 \times d$ to find that $C = 3.14 \times 25 = 78.5$ inches.

**4. C,** If $d = 18$, then $r = 9$ ft, and the area is $3.14 \times 9^2 = 254.34$ sq ft. The pavers will charge $1.59 \times 254.34 \approx $404.40.

**5. B,** If $r = 7$, then $d = 14$ ft; use $C = 3.14 \times d$ to find that $C = 3.14 \times 14 = 43.96$ ft, which rounds to 44 ft.

**6. D,** Use $A = 3.14 \times r^2$ to find that $A = 3.14 \times 4^2 = 50.24$, which rounds to 50.2 sq ft.

**7. E,** Use $A = 3.14 \times r^2$ to find that $A = 3.14 \times 7^2 = 153.86$ sq ft, which rounds to 154 sq ft.

**8. C,** First circle: $A = 3.14 \times 5.5^2 \approx 94.99$ sq cm. Second circle: $A = 3.14 \times 6.25^2 \approx 122.66$ sq cm. The second circle is about 122.66 − 94.99 = 27.67, which rounds to 27.7 sq cm larger.

**9. B,** Solve $C = 3.14(30)$ to find that the circumference is 94.2 in.

**10. D,** The radius is 132 ÷ 2. Use an area formula $A = \pi r^2$. So, the area of the plywood needed is $\pi \times \left(\frac{132}{2}\right)^2$.

**11. D,** The radius is 12 ÷ 2, or 6, ft. Solve $A = 3.14 \times 6^2$ to find an area of 113.04 ft².

**12. A,** Find 11 − 2 − 2 = 7, which is the diameter of the mirror without the frame. If $d = 7$, then $r = 3.5$ in.

**13. B,** Solve $A = 3.14 \times 3.5^2$ to find an area of 38.465 sq in., which rounds to 38.5 sq in.

**14. C,** Subtract the area of the mirror from the area of the entire frame and mirror: $(3.14 \times 5.5^2) - 38.465 = 56.52$, which rounds to 56.5 sq in.

**15. A,** The circumference is $3.14 \times 11 = 34.54$ in.

**16. C,** The tablecloth is $3.14 \times 4^2 = 50.24$ sq ft.

**17. B,** Subtract the area of the table from the area of the tablecloth: $50.24 - 3.14 \times 2.5^2 = 50.24 - 19.63 = 30.615$, which rounds to 30.62 sq ft.

**18. D,** The diameter will be 15 + 2 + 2 = 19 in., so the radius will be 9.5 in. The area is $3.14 \times 9.5^2 = 283.385$, which rounds to 283.4 sq in.

**19. E,** Subtract the area of the small circle from the area of the large circle: $(3.14 \times 4^2) - (3.14 \times 3^2) = 50.24 - 28.26 = 21.98$ sq in.

**20. E,** The area is $3.14 \times 0.5^2 = 0.785$, which rounds to 0.79 sq ft.

**21. A,** Divide the area of the circle by the area of the rectangle, and multiply by 100 to get $0.785 \div (2 \times 3) \approx 0.13$, and $0.13 \times 100 = 13\%$.

**22. B,** Divide the area of the large circle by the area of the small circle: $(3.14 \times 6^2) \div (3.14 \times 2^2) = 113.04 \div 12.56 = 9$.

**23. D,** Since $A = 3.14 \times r^2$, and $A = 144(3.14)$, solve $144(3.14) = 3.14r^2$ for $r$ to find that $r = 12$ ft. If $r = 12$, then $d = 24$, and $C = 3.14 \times 24 = 75.36$ ft.

**24. C,** Solve $19.625 = 3.14r^2$ for $r$ to find that $r = 2.5$. If $r = 2.5$, then $d = 5.0$ cm.

# UNIT 2 (continued)

**25. E,** The circumference is $3.14 \times 3.5 = 10.99$, so the circumference of both tires is $10.99 \times 2 = 21.98$ in.

**26. C,** Add the areas of the two circles: $(3.14 \times 7.5^2) + (3.14 \times 5^2) = 176.625 + 78.5 = 255.125$, which rounds to 255 sq m.

**27. D,** The diameter of the small section would increase to 10.45 m, which gives a radius of 5.225 m. Add the two areas: $(3.14 \times 7.5^2) + (3.14 \times 5.225^2) \approx 176.625 + 85.724 = 262.349$, which rounds to 262.35 square meters.

**28. A,** Solve $25.12 = 3.14d$ to find that $d = 8$, so $r = 4$ in.

**29. B,** Solve $C = 3.14d$ for $d$, where $C = 56.5$. The diameter is $d = 17.99 \approx 18.0$ cm.

## LESSON 4, pp. 79–83

**1. A,** Angles 1 and 5 are corresponding, so $m\angle 5 = 115°$; angles 5 and 6 are supplementary; $180° - 115° = 65°$.

**2. A,** The angles are supplementary; $180° - 100° = 80°$.

**3. C,** Vertical angles are congruent; 100°.

**4. B,** $90° - 25° = 65°$.

**5. D,** Angles 1 and 3 are vertical angles, so they are congruent.

**6. C,** Angles 2 and 4 are congruent because they are vertical. Angles 4 and 8 are congruent because they are corresponding.

**7. D,** $\angle LMN$ and $\angle LMP$ are complementary, so $m\angle LMP = 90° - 35° = 55°$. $\angle RMP$ is a right angle, so $m\angle RMP + m\angle LMP = 90° + 55° = 145°$.

**8. B,** $\angle 2$ and $\angle 4$ are vertical, so they are congruent; 55°.

**9. E,** $\angle 1$ and $\angle 3$ are vertical angles.

**10. D,** $\angle 3$ and $\angle 4$ are supplementary, so $180° - 52° = 128°$.

**11. D,** If two angles are congruent and complementary, then one angle has a measure of $90° \div 2 = 45°$. The supplement is $180° - 45° = 135°$.

**12. B,** The angles are supplementary, so $180° - 115° = 65°$.

**13. C,** The measures of $\angle ABD$, $\angle ABC$, and $\angle CBE$ have a sum of 180°; $180° - m\angle ABC = 180° - 90° = 90°$, and $90° \div 2 = 45°$.

**14. D,** $m\angle CBA + m\angle ABD = m\angle CBD$, so $m\angle CBD = 90° + 45° = 135°$.

**15. D,** If $b = $ the measure of $\angle B$, solve $b + (3b + 20) = 180$ for $b$ to find that $b = 40°$. Then $m\angle A = 180° - 40° = 140°$. To check, substitute 40 for $b$ in $3b + 20$ to get $120 + 20 = 140°$.

**16. D,** $\angle 2$ and $\angle 3$ are adjacent and supplementary because they share a side and have measures that add up to 180°.

**17. C,** $\angle 7$ and $\angle 8$ are supplementary; $180° - 60° = 120°$.

**18. C,** $\angle 6$ is an exterior angle and is on the other side of the transversal from $\angle 4$.

**19. E,** $\angle 3$ and $\angle 7$ are corresponding because they are in the same relative position.

**20. A,** $\overrightarrow{EF}$ and $\overrightarrow{EJ}$ form the angle, so it is called $\angle JEF$.

**21. E,** We do not have enough information.

**22. B,** $m\angle 1 + m\angle 2 = 90°$, so they are complementary.

**23. D,** $\angle 1$ and $\angle 2$ are supplementary, so $m\angle 2 = 180° - 35° = 145°$; $\angle 2$ and $\angle 4$ are vertical, so $m\angle 4 = 145°$; sum $= 145° + 145° = 290°$.

**24. D,** $\angle 1$ and $\angle 2$ are supplementary, and $\angle 3$ and $\angle 4$ are supplementary, so $180° + 180° = 360°$.

**25. D,** $\angle 4$ and $\angle 3$ are supplementary, so $180° - 30° = 150°$.

**26. C,** Each angle is 90°, so $90° + 90° = 180°$.

**27. A,** Figure A shows perpendicular lines, which form two adjacent 90° angles.

**28. D,** Figure D shows a right angle split into two angles.

**29. E,** Figure E shows one pair of vertical angles.

**30. C,** Figure C shows a straight angle, which is 180°, and two non-congruent angles that add to 180°.

## LESSON 5, pp. 84–88

**1. C,** $m\angle B = 55°$, $360° - 2(55°) = 250°$, $250° \div 2 = 125°$.

**2. A,** The base angles each have a measure of $180° - 115° = 65°$, so the missing angle of the triangle is $180° - 65° - 65° = 50°$.

**3. D,** Two angles are congruent, so two sides are congruent.

**4. C,** Solve $x + 2x + 90° = 180°$ to find that $x = 30°$ and the smaller angle measures 30°.

**5. C,** $m\angle SUV = 180° - 45° = 135°$, and $\angle R$ is congruent to $\angle SUV$.

**6. D,** $m\angle RST = m\angle RSU + m\angle UST$; $m\angle UST = 180° - 45° - 90° = 45°$, and $m\angle RSU = 180° - 135° = 45°$; $m\angle RSU + m\angle UST = 45° + 45° = 90°$.

**7. E,** Opposite angles are congruent; $360° - 35° - 35° = 290°$, and $290° \div 2 = 145°$.

**8. B,** $m\angle 1 = 70°$, because opposite angles of parallelograms are congruent.

**9. C,** Two side-by-side angles of a parallelogram have measures that sum to 180°, so $m\angle 2 = 180° - 70° = 110°$.

**10. A,** Opposite angles of parallelograms are congruent, so the missing angle is 110°.

**11. D,** The measure of the obtuse angle is $180° - 45° - 35° = 100°$.

**12. B,** All rectangles have four 90° angles.

**13. C,** The measure of the obtuse angle is $180° - 35° - 35° = 110°$.

**14. B,** A right triangle has one 90° angle, so $180° - 90° - 40° = 50°$.

**15. B,** A right angle has a measure of 90°, so half the measure would be 45°; $180° - 90° - 45° = 45°$.

**16. C,** $m\angle XYZ = 180° - 130° = 50°$, $m\angle X = 90°$, and $m\angle YZX = 180° - 50° - 90° = 40°$. The larger acute angle is 50°.

**17. D,** Subtract 40° from 90° to get 50°.

**18. A,** The angle has a measure of $180° - 90° - 60° = 30°$.

**19. E,** $\angle KLJ \cong \angle LHI$ because they are corresponding angles, so the angle measure is 60°.

**20. D,** The top and bottom sides are parallel, which are $\overline{LM}$ and $\overline{NO}$.

**21. D,** The sum of the measures of the interior angles of any quadrilateral is 360°.

**22. C,** Since $\angle R \cong \angle Q$ and $\angle T \cong \angle S$, $360° - 105° - 105° = 150°$, and $150° \div 2 = 75°$.

**23. B,** The sum of the measures of the interior angles of a triangle is 180°, which also is the measure of a straight angle.

**24. B,** An isosceles right triangle has one right angle and two congruent acute angles; $180° - 90° = 90°$, and $90° \div 2 = 45°$.

**25. A,** $\angle BDC \cong \angle BEF$ because they are corresponding angles.

**26. B,** $\angle FCD \cong \angle BFE$ because they are corresponding angles, so $m\angle BFE = 60°$.

**27. D,** The measure is 120° because $360° - 90° - 90° - 60° = 120°$.

## LESSON 6, pp. 89–93

**1. D,** Solve $a^2 + 5^2 = 10^2$ for $a$ to find that $a^2 = 75$, so $a \approx 8.66$ ft, which rounds to 8.7 ft.

**2. D,** Solve $15^2 + 30^2 = c^2$ for $c$ to find that $c^2 = 1{,}125$, so $c \approx 33.54$ ft, which rounds to 33.5 ft.

**3. B,** Solve $a^2 + 30^2 = 35^2$ for $a$ to find that $a^2 = 325$, so $a \approx 18.03$ ft, which rounds to 18 ft.

# UNIT 2 (continued)

**4. C,** Solve $a^2 + b^2 = c^2$ for $c$ to find $c^2 = 1,249$ and $c \approx 35.3$ ft. Subtract $35.3 - 33.5$. Since the values are approximate, the cable is now about 1.8 inches longer.

**5. D,** Solve $40^2 + 120^2 = c^2$ for $c$ to find that $c^2 = 16,000$, so $c \approx 126.49$ m, which rounds to 126 m.

**6. A,** Solve $20^2 + 120^2 = c^2$ for $c$ to find that $c^2 = 14,800$, so $c \approx 121.655$ m, which rounds to 122 m.

**7. E,** Plot the points on a grid; the distance between the points represents the hypotenuse; count the units for the triangle side lengths; solve $8^2 + 2^2 = c^2$ for $c$ to find that $c^2 = 68$, so $c \approx 8.246$, which rounds to 8.2.

**8. E,** Imagine that $\overline{AB}$ is the hypotenuse of a right triangle with vertices $A$, $B$, and $(3, -3)$. Solve $8^2 + 2^2 = c^2$ to find that $c \approx 8.25$.

**9. B,** Imagine that $\overline{BC}$ is the hypotenuse of a right triangle with vertices $B$, $C$, and $(3, -4)$. Solve $2^2 + 3^2 = c^2$ to find that $c \approx 3.61$.

**10. C,** The hypotenuse of the right triangle is $\overline{AC}$. The lengths of $a$ and $b$ are 1 and 10. Solve $1^2 + 10^2 = c^2$ to find that $c \approx 10.05$. The perimeter is $10.05 + 1 + 10 = 21.05$.

**11. A,** Solve $4^2 + 4^2 = c^2$ to find that $c^2 = 32$ and $c \approx 5.66$ in.

**12. B,** Solve $2^2 + 2^2 = c^2$ to find that $c^2 = 8$ and $c \approx 2.83$ in.

**13. D,** To get the variable alone on one side of the equation $16 + a^2 = 100$, Ella should have used the inverse operation and subtracted 16 from both sides of the equation. The correct next steps would be $a^2 = 84$ and $a \approx 9.17$.

**14. C,** Solve $16^2 + b^2 = 21^2$ to find that $b^2 = 185$ and $b \approx 13.6$ in.

**15. D,** Solve $75^2 + 63^2 = c^2$ to find that $c^2 = 9,594$ and $c \approx 97.9$ ft.

**16. C,** Write $x^2 + 48^2 = 50^2$ to represent the side lengths, with 50 as the hypotenuse length. Subtract $x^2$ to get $48^2 = 50^2 - x^2$. Take the square root of both sides to get $48 = \sqrt{50^2 - x^2}$.

**17. B,** Solve $6.5^2 + b^2 = 7.9^2$ to find that $b^2 = 20.16$ and $b \approx 4.49$ ft, which rounds to 4.5 ft.

**18. E,** Solve $5^2 + 8^2 = c^2$ to find that $c^2 = 89$ and $c \approx 9.43$ ft, which rounds to 9.4 ft.

**19. A,** Since $6.5^2 + 2.7^2 = 49.54$ and $7.04^2 = 49.5616$, the segments are very close to forming a right triangle.

**20. B,** Solve $12^2 + b^2 = 15^2$ to find that $b^2 = 81$ and $b = 9$ feet.

**21. A,** Solve $30^2 + 17^2 = c^2$ to find that $c^2 = 1,189$ and $c \approx 34.48$ yd, which rounds to 34.5 yd.

**22. C,** Solve $12^2 + b^2 = 40^2$ to find that $b^2 = 1,456$ and $b \approx 38.16$ yd, which rounds to 38.2 yd.

**23. D,** Since $15^2 + 33^2 = 1,314$ and $39^2 = 1,521$, then $a^2 + b^2 \neq c^2$. The segments cannot form a right triangle.

**24. E,** Solve $15^2 + 15^2 = c^2$ to find that $c^2 = 450$ and $c \approx 21.2$ inches.

**25. B,** The area is $\frac{1}{2} \times 15 \times 15 = 112.5$ square inches.

**26. B,** The perimeter would be $15 + 15 + 18 + 18 + 21.2$ (from question 24) $= 87.2$ inches.

## LESSON 7, pp. 94–98

**1. C,** $AC = 2 \times RT$, because $\overline{BC}$ is twice as long as $\overline{ST}$; $2 \times 1.2 = 2.4$ m

**2. D,** $\triangle XYZ$ appears to have the same side lengths and angle measures as $\triangle EFG$.

**3. A,** If the perimeter of Triangle 1 is 19 ft, then $EF$ is $19 - 6 - 9 = 4$ ft, and $\overline{EF} \cong \overline{XY}$, so $XY = 4$ ft.

**4. C,** Let $x = FG$. Solve $\frac{x}{5.4} = \frac{4}{2}$ to find that $x = FG = 10.8$ m.

**5. D,** Let $x = HG$. Solve $\frac{x}{5} = \frac{4}{2}$ to find that $x = 10$ m. The perimeter is $4 + 10 + 10.8 = 24.8$ m.

**6. B,** $M$ corresponds to $X$, and $N$ corresponds to $Y$, so $\overline{MN} \cong \overline{XY}$.

**7. A,** $\angle L \cong \angle F$ because they are corresponding parts.

**8. D,** $C$ corresponds to $M$, and $F$ corresponds to $N$, so $\overline{CF}$ corresponds to $\overline{ML}$.

**9. D,** The perimeter is $4.5 + 4.5 + 4.5 + 2 = 15.5$ cm.

**10. C,** $\angle A \cong \angle H$ because corresponding angles of similar figures are congruent.

**11. B,** The sides of Triangle 2 are half the length of the sides of Triangle 1. $AC = 1.5$ in., so the corresponding side is $1.5 \div 2 = 0.75$ in.

**12. D,** Use the Pythagorean theorem to find $AB$ by solving $(1.5)^2 + (1.25)^2 = AB^2$, to get $AB \approx 1.95$ in. The sides of Triangle 2 are half the length of the sides of Triangle 1, so the corresponding side is $1.95 \div 2 \approx 0.976$ in.

**13. C,** By the description above the figures, the two triangles are similar, so $\triangle TUV \sim \triangle XYZ$.

**14. D,** Corresponding sides of similar figures are proportional, and corresponding angles are congruent, so $\angle Z \cong \angle V$.

**15. C,** An obtuse triangle cannot have a right angle; triangles can be obtuse, right, or acute.

**16. B,** Subtract $180° - 75° - 45°$ to find $m\angle G = 60°$.

**17. C,** $\angle I \cong \angle C$ because corresponding angles of similar triangles are congruent.

**18. E,** The ratio of the corresponding sides is known and is 2:1. The length of side $\overline{BC}$, which corresponds to $\overline{GI}$, is unknown. So, the length of $\overline{GI}$ cannot be found.

**19. C,** $\overline{HI}$ corresponds to $\overline{CD}$, so $CD = 20$ m.

**20. B,** Subtract $180° - 75° - 75°$ to find $m\angle C = 30°$. Since $\angle C$ corresponds to $\angle H$, then $m\angle H = 30°$.

**21. C,** The base angles are congruent, so the triangles are isosceles, and the two longer sides are congruent. The perimeter is $20 + 20 + 5 = 45$ m.

**22. D,** Multiply 18 by $\frac{2}{3}$ to get $\frac{18}{1} \times \frac{2}{3} = \frac{6}{1} \times \frac{2}{1} = 12$ ft for the length of the smaller carpet.

**23. A,** The width of the smaller carpet is $\frac{9}{1} \times \frac{2}{3} = \frac{3}{1} \times \frac{2}{1} = 6$. The difference in areas is $(18 \times 9) - (12 \times 6) = 162 - 72 = 90$ ft².

## LESSON 8, pp. 99–103

**1. D,** Solve $\frac{1 \text{ in.}}{2.5 \text{ mi}} = \frac{5 \text{ in.}}{x}$ to find that $x = 12.5$ miles.

**2. B,** Use $\frac{\text{shadow}}{\text{tree}} = \frac{\text{shadow}}{\text{person}}$, or $\frac{1}{x} = \frac{2.5}{14}$.

**3. E,** Since she drove there and back, she drove $2 \times 2.5 = 5$ cm. Solve $\frac{1 \text{ cm}}{6 \text{ km}} = \frac{5 \text{ cm}}{x}$ to find that $x = 30$ km.

**4. D,** Solve $\frac{2 \text{ in.}}{4.8 \text{ mi}} = \frac{5.5 \text{ in.}}{x}$ to find that $x = 13.2$ miles.

**5. D,** Solve $\frac{6}{8.5} = \frac{4.25}{x}$ to find that $x \approx 6.02$ feet.

**6. E,** Solve $\frac{1 \text{ cm}}{20 \text{ km}} = \frac{1.5 \text{ cm}}{x}$ to find that $x = 30$ km.

**7. B,** Jack: $\frac{1 \text{ cm}}{20 \text{ km}} = \frac{2.5 \text{ cm}}{x}$; $x = 50$ km. Pedro: $\frac{1 \text{ cm}}{20 \text{ km}} = \frac{2 \text{ cm}}{x}$; $x = 40$ km. Jack drove 10 km more.

**8. A,** Carl drives a map distance of $2 + 2.25 + 2.25 + 2 = 8.5$ cm in all. Solve $\frac{1 \text{ cm}}{20 \text{ km}} = \frac{8.5 \text{ cm}}{x}$ to find that $x = 170$ km.

# UNIT 2 (continued)

**9. C,** Solve $\frac{12 \text{ in.}}{60 \text{ in.}} = \frac{1 \text{ in.}}{x}$ to find that $x = 5$ in., so 1 in. on the model represents 5 in. of the table.

**10. C,** Solve $\frac{1 \text{ in.}}{5 \text{ ft}} = \frac{x}{8 \text{ ft}}$ to find that $x = 1.6$ in.

**11. D,** Use proportions to find that the length is 20 ft and the width is 15 ft, so the area is 300 sq ft.

**12. E,** Solve $\frac{1 \text{ in.}}{5.5 \text{ mi}} = \frac{3.5 \text{ in.}}{x}$ to find that $x = 19.25$ miles.

**13. A,** Solve $\frac{1 \text{ in.}}{25 \text{ mi}} = \frac{x}{350 \text{ mi}}$ to find that $x = 14$ miles.

**14. D,** Solve $\frac{2 \text{ cm}}{6.5 \text{ km}} = \frac{3.25 \text{ cm}}{x}$ to find that $x \approx 10.6$ km.

**15. E,** Blue Harbor to Dodgeville through Westfield is a map distance of 6.25 cm, but through Crawford is a map distance of 6 cm. The difference is 0.25 cm. Solve $\frac{2 \text{ cm}}{6.5 \text{ km}} = \frac{0.25 \text{ cm}}{x}$ to find that $x \approx 0.8$ km.

**16. D,** Solve $\frac{0.5 \text{ in.}}{5 \text{ ft}} = \frac{1.5 \text{ in.}}{x}$ and $\frac{0.5 \text{ in.}}{5 \text{ ft}} = \frac{1 \text{ in.}}{x}$ to find that the dimensions are 15 ft and 10 ft.

**17. A,** Note that $\frac{7}{8} = 7 \div 8 = 0.875$. Solve $\frac{0.5 \text{ in.}}{5 \text{ ft}} = \frac{0.875 \text{ in.}}{x}$ to find that $x = 8.75$ ft, which also equals $8\frac{3}{4}$ ft.

**18. A,** Solve $\frac{0.5 \text{ in.}}{5 \text{ ft}} = \frac{0.75 \text{ in.}}{x}$ to find that the width is 7.5, or $7\frac{1}{2}$, ft. From question 17, the length is $8\frac{3}{4}$ ft.

**19. C,** Solve $\frac{2 \text{ in.}}{3.2 \text{ mi}} = \frac{x}{19.2 \text{ mi}}$ to find that $x = 12$ in.

**20. B,** Solve $\frac{48 \text{ km}}{4 \text{ cm}} = \frac{x}{1 \text{ cm}}$ to find that $x = 12$ km; the scale is 1 cm : 12 km.

**21. E,** Solve $\frac{22}{31.9} = \frac{55}{x}$ to find that $x = 79.75$ ft, which rounds to 79.8 ft.

**22. A,** Solve $\frac{4.2}{3.8} = \frac{x}{6.8}$ to find that $x \approx 7.5$ ft.

**23. C,** Solve $\frac{1}{32} = \frac{x}{108.8}$ to find that $x = 3.4$ in.

**24. E,** Use two proportions to find that the shorter side is 13.5 ft, and the longer side is 18.75 ft. The area is $13.5 \times 18.75 = 253.125 = 253\frac{1}{8}$ sq ft.

**25. B,** Solve $\frac{3 \text{ cm}}{18.6 \text{ km}} = \frac{10 \text{ cm}}{x}$ to find that Stacey drove 62 km. Use $d = rt$ to find that Stacey drove for about 0.689 hours, or about $0.689 \times 60 = 41.3$ minutes, which is about 40 minutes.

**26. D,** A square has four congruent sides. If the perimeter is 22 cm, then each side is $22 \div 4 = 5.5$ cm long. Solve $\frac{2 \text{ cm}}{5 \text{ yd}} = \frac{5.5 \text{ cm}}{x}$ to find that $x = 13.75$ yd.

**27. C,** Solve $\frac{3 \text{ in.}}{30 \text{ ft}} = \frac{1 \text{ in.}}{x}$ to find that $x = 10$ ft, so the scale is 1 in. : 10 ft.

**28. E,** Solve $\frac{0.5 \text{ in.}}{6.5 \text{ ft}} = \frac{1 \text{ in.}}{x}$ to find that both the length and width are 13 ft.

---

## LESSON 9, pp. 104–108

**1. D,** The volume is $3.14 \times 3^2 \times 8 = 3.14 \times 9 \times 8 = 226.08$, which rounds to 226 in.$^3$.

**2. E,** The volume of one hay bale is $(40)(20)(20) = 16,000$ in.$^3$. Multiply by 50 to find $50(16,000) = 800,000$ in.$^3$.

**3. B,** The area of the 2 bases is $2 \times \frac{1}{2}(8 \times 4) = 8 \times 4$. Each lateral face is a rectangle with length 9 in. Two faces are 5 in. wide and one face is 8 in. wide. So, the lateral area is $2(9 \times 5) + (9 \times 8)$. The total surface area is $(8 \times 4) + 2(9 \times 5) + (9 \times 8)$.

**4. B,** The volume of the prism in the diagram is $\frac{1}{2}(8)(4) \times 9 = 144$ in.$^3$. Solve $V = Bh$, or $144 = 24h$ to find that the height $h = 144 \div 24 = 6$ in.

**5. C,** The surface area is $2(8 \times 6) + 2(8 \times 10) + 2(6 \times 10) = 96 + 160 + 120 = 376$ in.$^2$.

**6. C,** The radius is $18 \div 2 = 9$. Solve $V = \pi r^2 h$, which is $9,156.24 = (3.14)(9^2)h$ to get $9,156.24 = 254.34h$ and the height $h = 36$ cm.

**7. B,** The volume of a cylinder is $V = \pi r^2 h$. Morgan divides the volume by the height to get $x$, so $\frac{V}{h} = \frac{\pi r^2 h}{h} = \pi r^2 = x$. Since $C = 2\pi r$, Morgan needs to find the radius of the cylinder. So, Morgan has $x = \pi r^2$ and should divide $x$ by $\pi$, or 3.14, to find $r^2$, and then take the square root to find $r$. To find the circumference, Morgan should multiply $r$ by $2\pi$, or 6.28.

**8. B,** The volume is $V = 3.14 \times 1.5^2 \times 10 = 70.65$ in.$^3$.

**9. B,** The volume is $V = 3.14 \times 5^2 \times 7 = 549.5$ m$^3$.

**10. A,** Rename 18 inches as 1.5 ft. The volume is $1.5 \times 1.5 \times 1.5 = 3.375$ ft$^3$, which rounds to 3 ft.

**11. A,** Rename 18 inches as 1.5 ft. The surface area of a cube is 6 times the area of the base: $6(1.5)(1.5) = 13.5$ ft$^2 \approx 14$ ft$^2$.

**12. B,** Solve $V = lwh$, or $97.5 = (2.5)(13)l$ to find the length $l = 3.0$ cm.

**13. C,** The surface area is $2\left(\frac{1}{2}\right)(10)(10) + 2(10)(50) + (14)(50) = 100 + 1,000 + 700 = 1,800$ cm$^2$.

**14. E,** The volume of a cube is $s^3$. The surface area of a cube is $6s^2$. For side length 8 ft, volume $= 8^3 = 512$ is greater than the surface area $= 6(8)^2 = 384$.

**15. C,** Solve $235.5 = 3.14 \times r^2 \times 3$ to get $235.5 = 9.42r^2$ and $r^2 = 25$, so $r = 5$. The diameter $d = 2(5) = 10$ in.

**16. D,** The area of cardboard needed = area of one base + lateral area. The lateral area is base circumference × cylinder height $= (31.4)(20) = 628$ cm$^2$. To find the radius of the base, solve $31.4 = 2\pi r$ to get $r = 5$ cm. The area of the base is $(3.14)(5^2) = 78.5$ cm$^2$. The area of cardboard needed is $628 + 78.5 = 706.5$ cm$^2$, which rounds to 707 cm$^2$.

**17. B,** The radius of the first cylinder is 5 cm (from question 16). Volume of first cylinder $= (3.14)(5^2)(20) = 1,570$ cm$^3$. Find the radius of the second cylinder by solving $1,570 = (3.14)(r^2)(15)$, so $r^2 \approx 33.33$ and $r$ is about 5.77 cm. The circumference is $2(3.14)(5.77)$, which is about 36.2 cm.

**18. D,** The volume of a triangular prism = volume of the rectangular prism $= (16)(5)(3) = 240$ in.$^3$. Find the base area of the triangular prism by solving $240 = B(12)$ to get $B = 20$ in.

**19. C,** Divide the volume of the full tank by 2: $V = (24)(10)(15) = 3,600$, and $3,600 \div 2 = 1,800$ in.$^3$.

**20. B,** The height of the water is $15 - 1 = 14$ in., and the volume is $V = (24)(10)(14) = 3,360$ in.$^3$.

**21. D,** The radius is $3 \div 2 = 1.5$ in. Solve $28.26 = 3.14(1.5)^2(h)$ to find that $h = 4$ in.

**22. A,** The volume of a cup would be $V = 3.14 \times 3.5^2 \times 10 = 384.65$ cm$^3$, which rounds to 384.7 cm$^3$.

**23. E,** The circumference of the base is $C = 2(3.14)(3.5) = 21.98$ cm. The lateral area is $(21.98)(10) = 219.8$, or about 220 cm$^2$.

**24. B,** The volume of the box is $V = (18)(10)(28) = 5,040$ in.$^3$. The volume of one can is $V = 3.14(2^2)(5) = 62.8$ in.$^3$. The space remaining is $5,040 - (40 \times 62.8) = 2,528$ in.$^3$.

**25. C,** The height of the soil is 24 − 3 = 21 in. The volume of the soil is $V = 3.14(9^2)(21) = 5,341.14$ in.$^3$, which rounds to 5,340 in.$^3$.

**26. C,** The volume is 4,050 = $(x)(2x)(25)$. Solve 4,050 = $50x^2$ to find $x^2 = 81$, so $x = 9$. Since $x$ is the base width, the length of the base is $2x = 2(9) = 18$ in.

**27. A,** The base is 9 in. by 18 in. (question 26). The surface area is 2(9 × 18) + 2(9 × 25) + 2(18 × 25) = 2(162) + 2(225) + 2(450) = 324 + 450 + 900 = 1,674 in.$^3$.

## LESSON 10, *pp. 109–113*

**1. B,** The volume is $V = \frac{4}{3} \times 3.14 \times 1.5^3 = \frac{4}{3} \times 3.14 \times 3.375 = 14.13$, which rounds to 14 in.$^3$.

**2. A,** Use $V = \frac{1}{3} \times 3.14 \times r^2 \times h$ to find that $V = \frac{1}{3} \times 3.14 \times 4^2 \times 12$.

**3. B,** The surface area of the cone is without the area of the base, so only $\pi rs$. Find 3.14 × 4 × 12.6 = 158.256, which rounds to 158 cm$^2$.

**4. B,** Use the radius to find the volume of the hemisphere, which is one-half the sphere's volume, or $\frac{1}{2} \times \frac{4}{3}\pi r^3 = \frac{2}{3}\pi r^3 = \frac{2}{3} \times$ 3.14 × 6$^3$ = 452.16 in.$^3$. Use the volume to solve for the height of the cone: 452.16 = $\frac{1}{3} \times 3.14 \times 6^2 \times h$. Multiply to get 452.16 = 37.68$h$, and then divide to find $h$ = 12 in.

**5. C,** The area of the base is 2$^2$ = 4, so the volume of the pyramid is $\frac{1}{3}$(4)(3) = 4 cm$^3$.

**6. B,** The volume of the pyramid (from question 6) is 4 cm$^3$, so twice the volume is 8 cm$^3$. Solve 8 = $\frac{1}{3} \times 3.14 \times r^2 \times 3$. Multiply to get 8 = 3.14$r^2$, and then divide to find $r^2$ = 2.5477. Take the square root to get $r ≈ 1.59615 ≈ 1.6$ cm.

**7. E,** The surface area of the pyramid is 6$^2$ + $\frac{1}{2}$(4 • 6)(5) = 36 + 60 = 96 square feet.

**8. B,** The volume is $\frac{4}{3}\pi r^3 = \frac{4}{3}(3.14)\left(\frac{15}{2}\right)^3 = 1,766.25$, which rounds to 1,770 cubic inches.

**9. A,** The base area is 36, so $V = \frac{1}{3} \cdot 36 \cdot 9 = 108$ cm$^3$.

**10. A,** Use the surface area, so solve 28.26 = 4 × 3.14 × $r^2$. Multiply to get 28.26 = 12.56$r^2$, and then divide to find $r^2$ = 2.25. Take the square root to get $r$ = 1.5 cm.

**11. B,** Use the volume, so solve 64 = $\frac{1}{3}$(8 • 8)$h$. Divide both sides by 64 to find 1 = $\frac{1}{3}h$ and $h$ = 3 ft.

**12. C,** Use the surface area to find the slant height, so solve 144 = (8 • 8) + $\frac{1}{2}$(4 • 8)$s$. Multiply to get 144 = 64 + 16$s$, and then subtract to find 16$s$ = 80 and the slant height $s$ = 5 ft.

**13. B,** The volume is $V = \frac{1}{3} \times 3.14 \times 3^2 \times 10 = 94.2$ cm$^3$, which rounds to 94 cm$^3$.

**14. B,** The cone is open at the base, so find the area of the curved surface of the cone. So, $\pi rs$ = 3.14(6 ÷ 2)(10.4) = 97.97 cm$^2$, which rounds to 98 cm$^2$.

**15. C,** Subtract the volume of the medium from the volume of the large. The large is $V = \frac{1}{3} \times 3.14 \times 5^2 \times 14 ≈ 366.33$, and the medium is $V = \frac{1}{3} \times 3.14 \times 4^2 \times 12 = 200.96$, so 366.33 − 200.96 = 165.37 cm$^3$, which rounds to 165 cm$^3$.

**16. C,** The volume is $V = \frac{1}{3} \times 6^2 \times 7 = 84$ cubic feet.

**17. D,** The volume is $V = \frac{1}{3} \times 10^2 \times 6 = 200$ cubic feet.

**18. A,** Pyramid C has volume 84 ft$^3$ and base edge length 10 ft. The volume is 84 = $\frac{1}{3}$(10$^2$)$h$. Multiply each side by 3 to get 252 = 100$h$. Then divide to find height $h$ = 2.52 ft, which rounds to 2.5 ft.

**19. B,** The surface area is 4 × 3.14 × 12$^2$ = 1,808.64 in.$^2$. The area of the circular hole is 3.14 × 3$^2$ = 28.26 in.$^2$. Subtract 1,808.64 − 28.26 to find 1,780.38 in.$^2$, which rounds to 1,780 in.$^2$.

**20. B,** The radius is 10 ÷ 2 = 5 in. The volume is $V = \frac{1}{3} \times 3.14 \times 5^2 \times 8 ≈ 209.33$, which rounds to 209 in.$^3$.

**21. C,** The area of the base is 12$^2$ = 144, so the volume of the pyramid is $\frac{1}{3}$(144)(15) = 720 ft$^3$.

**22. E,** Use the volume to solve for the cone's height: 550 = $\frac{1}{3} \times$ 3.14 × 5$^2$ × $h$. Multiply to get 550 = 26.16$h$, and then divide to find $h ≈ 21$ in.

**23. B,** Rename the side length as 2.5 ft, and the height as 3.25 ft. The volume is $V = \frac{1}{3} \times 2.5^2 \times 3.25 = 6.7708$ cubic feet, which rounds to 7 ft$^3$.

**24. C,** Solve 803.84 = $\frac{1}{3}$(3.14)(8$^2$)$h$ to find that $h$ = 12 ft.

**25. C,** Solve 563 = 3.14(8$^2$) + (3.14)(8)$s$ to get 563 = 200.96 + 25.12$s$. Subtract to find 362.04 = 25.12$s$, then divide to get $s$ = 14.41 feet. Convert: 0.41 ft = 12 × 0.41 ≈ 5 in. The slant height is about 14 feet 5 inches.

**26. D,** The volume of Greenhouse A is $V = \frac{1}{3}$(3.14)(8$^2$)(9) = 192 cubic feet.

**27. D,** The volume of Greenhouse B is 192 cubic feet (from question 27). Solve 192 = $\frac{1}{3}B$(16) to find the area of base $B$ = 36 square feet.

**28. D,** The amount of glass needed is the total area of the triangular faces, $\frac{1}{2}ps$. For Greenhouse A, the area is $\frac{1}{2}$(4 • 8)(9.5) = 152 ft$^2$. For Greenhouse B, the base area is 36 ft$^2$, so one side is 6 ft, and the area of the triangular faces is $\frac{1}{2}$(4 • 6)(7.2) = 86.4 ft$^2$. Greenhouse B uses less glass.

## LESSON 11, *pp. 114–118*

**1. D,** Add the areas of the two triangles and the square. Each triangle has an area of $\frac{1}{2} \times 3 \times 8 = 12$. The square has an area of 8 × 8 = 64, so 12 + 12 + 64 = 88 sq ft.

**2. E,** The two semicircles form one whole circle. Area of the circle = 3.14 × 2.5$^2$ = 19.625; area of rectangle = 8 × 5 = 40. So, 19.625 + 40 = 59.625, which rounds to 59.6 ft$^3$.

**3. C,** The volume of one cone = $\frac{1}{3} \times 3.14 \times 2^2 \times 3 = 12.56$; two cones: 12.56 × 2 = 25.12 m$^3$.

**4. C,** The volume of the cylinder = 3.14 × 2$^2$ × 6 = 75.36; add volume of two cones (from question 3) to get 75.36 + 25.12 = 100.48, which rounds to 100 cu m.

**5. E,** The top rectangle is 16 × 5 = 80 sq ft, the middle rectangle is 10 × 5 = 50 sq ft, and the bottom rectangle is 5 × 6 = 30 sq ft, so 80 + 50 + 30 = 160 sq ft.

**6. B,** The perimeter is 16 + 5 + 6 + 11 + 5 + 6 + 5 + 10 = 64 ft.

**7. E,** Multiply the area (from question 5) by 3 to get 160 × 3 = 480 ft$^3$.

**8. C,** Divide the figure into three horizontal rectangles. The top is 20 • 5 = 100 sq ft, the middle is (20 − 6) • (18 − 5 − 7) = 14 • 6 = 84 sq ft, and the bottom is 5 • 7 = 35 sq ft. The total area is 100 + 84 + 35 = 219 sq ft.

**9. B,** The perimeter is 18 + 20 + 5 + 6 + 6 + 9 + 7 + 5 = 76 ft.

**10. C,** The area of the whole pizza is $A$ = 3.14 × 8$^2$ = 200.96 sq in. If she ate three pieces, then $\frac{5}{8}$ are left, and $\frac{5}{8}$ = 5 ÷ 8 = 0.625, and 0.625 × 200.96 = 125.6 sq in.

## UNIT 2 (continued)

**11. D,** Add the volume of the cone and the volume of the cylinder. Cone: $V = \frac{1}{3} \times 3.14 \times 3^2 \times 4 = 37.68$ cm³. Cylinder: $V = 3.14 \times 3^2 \times 12 = 339.12$ cm³. The combined volume is $37.68 + 339.12 = 376.8$ cm³, which rounds to 377 cm³.

**12. B,** Divide the figure into a top rectangle and a bottom rectangle. The top is $40 \times 64 = 2,560$ in.², and the bottom is $40 \times 104 = 4,160$ in.², so the total is $2,560 + 4,160 = 6,720$ in.². Each tile has an area of 64 in.², so divide the area of the figure by 64. Adam needs $6,720 \div 64 = 105$ tiles.

**13. E,** The width of the stage is unknown, so the volume cannot be determined.

**14. B,** The diameter is 11 m, so the radius is 5.5 m, and $A = 3.14 \times 5.5^2 = 94.985$ m², which rounds to 95.0 m².

**15. C,** The area of the small circle is $A = 3.14 \times 4^2 = 50.24$ sq m, so the total area is about $95 + 95 + 50.24 = 240.24$ sq m, which rounds to 240 sq m.

**16. B,** Find the area of the semicircle by dividing the area of a circle with diameter 6 in half: $A = 3.14 \times 3^2 = 28.26$, and $28.26 \div 2 = 14.13$ cm². The area of the rectangle is $13 \times 6 = 78$ cm², so the total area is $14.13 + 78 = 92.13$ cm².

**17. C,** The circumference of the semicircle is $\frac{1}{2}(\pi d) = \frac{1}{2}(3.14)(6) = 9.42$ cm. The total perimeter is $13 + 6 + 13 + 9.42 = 41.42$ cm.

**18. E,** The new diameter would be 12 cm, so the new radius would be 6 cm. Semicircle: $A = \frac{1}{2} \times 3.14 \times 6^2 = 56.52$ cm²; Rectangle: $A = 12 \times 13 = 156$ cm², and $56.52 + 156 = 212.52$ cm².

**19. C,** The volume of the rectangular prism is $LWh$. The volume of the triangular prism is area of base × height $= \frac{1}{2}W(H - h) \cdot L$. Add the two volumes to get $LWh + \frac{1}{2}W(H - h)L$. Simplify to get $LWh + \frac{1}{2}WHL - \frac{1}{2}WhL$. Combine the two like terms and then factor out $\frac{1}{2}LW$ to get the total volume: $\frac{1}{2}WHL + (LWh - \frac{1}{2}WhL) = \frac{1}{2}WHL + \frac{1}{2}WhL = \frac{1}{2}LW(H + h)$.

**20. A,** The height of the triangle is $27 - 15.25 = 11.75$ ft.

**21. C,** The triangle is $\frac{1}{2} \times 14 \times 11.75 = 82.25$ ft², and the rectangle is $14 \times 15.25 = 213.5$ ft², so the total area is $82.25 + 213.5 = 295.75$ square feet.

**22. D,** Use the Pythagorean theorem ($a^2 + b^2 = c^2$) to calculate the hypotenuse of the right triangle. One leg is the height, 11.75; the other leg length is one-half the base of 14, or 7. So $c^2 = 11.75^2 + 7^2 = 187.0625$. Take the square root to find that $c \approx 13.68$ ft. Next, add to find the perimeter of the figure: $13.68 + 13.68 + 15.25 + 14 + 15.25 = 71.86$ ft, which rounds to 71.9 ft.

**23. A,** Add the perimeter of the triangle and the perimeter of the rectangle: $(14 + 14 + 14) + (14 + 15.25 + 14 + 15.25) = 100.5$ ft.

**24. B,** The volume is $\frac{1}{3} \times 3.14 \times 25^2 \times 8 = 5,233.33333$, which rounds to 5,233.3 cubic feet.

**25. D,** The volume of the cylinder-shaped section is $3.14 \times 25^2 \times 12 = 23,550$ ft³. Add the volume of the cone-shaped section to get $5,233.3 + 23,550 = 28,783.3$ ft³, which rounds to 28,783 ft³.

**26. A,** The surface area of the cylindrical wall is $2\pi rh$; the area of the floor is not included, as the question does not ask for it. The surface area of the conical ceiling is $\pi rs$; the area of the circular base is not relevant. Add the two surface areas to get $2\pi rh + \pi rs$, then factor out $\pi r$ to write $\pi r(2h + s)$.

## UNIT 2 REVIEW, pp. 119–123

**1. E,** $m\angle DBE + m\angle EBC = 90°$, so the angles are complementary.

**2. A,** $m\angle ABE = 150°$ because $m\angle ABD = 90°$, and $m\angle DBE = 60°$, so $90° + 60° = 150°$.

**3. A,** Solve $\frac{1 \text{ cm}}{15 \text{ km}} = \frac{8.5 \text{ cm}}{x}$ to find that $x = 127.5$ km.

**4. D,** The mass of two textbooks is 2 kg, which is equal to 2,000 g. So, 2,000 shoelaces would be needed.

**5. B,** Solve $d = rt$ for $t$ when $d = 850$ and $r = 500$ to find that $t = 1.7$ hours $= 1$ hour 42 minutes. The plane will arrive at Chicago at $(11:30 + 1:42) - 1$ hour (time change) $= 12:12$ P.M.

**6. B,** Use the formula $V = 3.14 \times r^2 \times h$. Solve $3,740 = 3.14 \times r^2 \times 17.5$ to find that $r^2 \approx 68.062$, so $r \approx 8.25$ cm. The diameter $d$ is about 16.5 cm.

**7. A,** Angles 2 and 4 are corresponding, so they have the same measure.

**8. D,** Solve $a^2 + b^2 = c^2$ for $c$ to find that $c^2 = 16,900$, so $c = 130$ yd. Wanda walked 130 yd instead of $50 + 120 = 170$ yd, so she walked $170 - 130 = 40$ fewer yards.

**9. B,** The two half-circles form one whole circle. $P = 3.14(4.5) + 5 + 5 = 24.13$ ft.

**10. C,** The circle is $A = 3.14 \times 2.25^2 \approx 15.896$ sq ft, and the rectangle is $A = 5 \times 4.5 = 22.5$ sq ft. The total area is $15.896 + 22.5 = 38.396$ sq ft, which rounds to 38.40 sq ft.

**11. B,** The sum of the interior angle measures of a triangle is 180°, so $m\angle G = 180° - 95° - 47° = 38°$.

**12. D,** Triangle $HFG$ has three different-sized angles, and one angle is greater than 90°, so the triangle is obtuse.

**13. A,** The volume of the pyramid is $\frac{1}{3}(14 \cdot 11)(18) = 924$ in.³. Since the volumes are equal, solve $924 = \frac{1}{3}(3.14)(9)h$ to get $924 = 84.78h$. Divide to find height $h \approx 10.8987$, which rounds to 10.9 in.

**14. C,** The volume of the prism is $(30)(20)(15) = 9,000$ cm³. The volume of the hemisphere is $\frac{1}{2}\left(\frac{4}{3} \times 3.14 \times 12^3\right) = 3,617.28$ cm³. So, the total volume is about $9,000 + 3,600 = 12,600$ cm³.

**15. C,** Solve $\frac{1 \text{ in.}}{3 \text{ ft}} = \frac{8 \text{ in.}}{x}$ to find that the length is 24 ft, and solve $\frac{1 \text{ in.}}{3 \text{ ft}} = \frac{7 \text{ in.}}{x}$ to find that the width is 21 ft.

**16. B,** Since $\angle C \cong \angle E$ because opposite angles are congruent, $360° - 125° - 125° = 110°$, and $110° \div 2 = 55°$.

**17. C,** The sum of the interior angle measures of a triangle is 180°, so the measure of the missing angle is $180° - 90° - 37° = 53°$.

**18. A,** The radius is $3.4 \div 2 = 1.7$. The volume of one cone is $V = \frac{1}{3}(3.14)(1.7^2)(4) \approx 12.09$ cubic inches. Divide 480 by 12.09 to get 39.7, or about 40 cones.

**19. C,** The volume of one pyramid is $\frac{1}{3}(12)(15) = 720$ cm³, so the volume of two pyramids is $720 \cdot 2 = 1,440$ cm³.

**20. D,** Use the Pythagorean theorem. $LN = 3$ units and $NM = 5$ units; $3^2 + 5^2 = c^2$, so $c^2 = 34$ and $c \approx 5.83$, which rounds to 6 units.

**21. D,** Solve $\frac{2 \text{ cm}}{29.8 \text{ km}} = \frac{2.8 \text{ cm}}{x}$ to find that $x = 41.72$ km.

**22. A,** The area of the circle is $3.14 \times 11^2 = 379.94$. Subtract the area of the grass from the area of the entire circle: $379.94 - 253.3 = 126.64$ square feet.

# UNIT 3 DATA ANALYSIS / PROBABILITY / STATISTICS

## LESSON 1, *pp. 126–130*

**1. C,** No flag can be repeated and order matters, so for 5 places, $6 \cdot 5 \cdot 4 \cdot 3 \cdot 2 = 720$.

**2. A,** Multiply all options for each choice: $2 \cdot 3 \cdot 6 \cdot 5$.

**3. C,** Add beads and charms for the total options in the third choice: $2 \cdot 3 \cdot (6 + 5) = 66$.

**4. C,** There is 1 way to get a pink band. There is 1 way to get a silver clasp. There are 6 beads taken 3 at a time, where order matters. Find $1 \cdot 1 \cdot (6 \cdot 5 \cdot 4) = 6 \cdot 5 \cdot 4$ different bracelets.

**5. B,** There is 1 way to get a gold band. There is 1 way to get a gold clasp. There are total of $6 + 5 = 11$ beads and charms. Take only 8 of the beads and charms to find $1 \cdot 1 \cdot (11 \cdot 10 \cdot 9 \cdot 8 \cdot 7 \cdot 6 \cdot 5 \cdot 4) = 11 \cdot 10 \cdot 9 \cdot 8 \cdot 7 \cdot 6 \cdot 5 \cdot 4$ different bracelets.

**6. D,** Each toss has 2 possible outcomes: $2 \cdot 2 \cdot 2 \cdot 2 = 2^4 = 16$, so the coin is tossed 4 times.

**7. B,** The learners are in 3 groups of 8. The new group takes one learner from each existing group, so there are 8 options for each choice: $8 \cdot 8 \cdot 8$.

**8. C,** Last year the permutations were $12 \cdot 11 \cdot 10 \cdot 9 \cdot 8 = 95{,}040$ possible arrangements of pictures. This year, he has 15 pictures, so $15 \cdot 14 \cdot 13 \cdot 12 \cdot 11 = 360{,}360$ arrangements. Subtract $360{,}360 - 95{,}040 = 265{,}320$ more.

**9. C,** She has 16 stones to choose from for her first pick, 15 stones to choose from for her second pick, and so there are 14 stones to choose from for her third pick.

**10. E,** The cards are being replaced, so each event of a card being chosen has 52 possible outcomes. You can use the fundamental counting principle to find $52 \cdot 52 \cdot 52 \cdot 52$.

**11. B,** Pick from a group to make smaller groups, so order doesn't matter: $\frac{9 \cdot 8 \cdot 7}{3 \cdot 2 \cdot 1} = 84$.

**12. B,** Each worker can be at a different house, so order matters: $9 \cdot 8 \cdot 7 = 504$ different ways.

**13. A,** There are conditions on each digit. The first digit cannot be a zero or the number would have only two digits. The second digit has no conditions, and the third digit must be odd: $4 \cdot 5 \cdot 2 = 40$ different 3-digit odd numbers.

**14. E,** A permutation does not repeat choices, so the numbers being multiplied must decrease by 1, so $13 \cdot 12 \cdot 11$.

**15. C,** There are 7 choices for the first dog to be groomed, 6 choices for the second dog, 5 choices for the third dog, and so 4 choices for the fourth dog.

**16. E,** There are 5 options for each of 10 questions: $5 \cdot 5 \cdot 5 \cdot 5 \cdot 5 \cdot 5 \cdot 5 \cdot 5 \cdot 5 \cdot 5 = 5^{10} = 9{,}765{,}625$.

**17. D,** There are $3(10) = 30$ questions. There are 5 options for each of 30 questions: $5^{30}$.

**18. A,** There are 10 multiple-choice questions so the student makes 10 selections or choices.

**19. B,** There are 10 multiple-choice questions and 5 options for each question. So, the test displays $10 \cdot 5 = 50$ options.

**20. A,** One test displays 50 options so 4 tests display $4 \cdot 50$ options.

**21. C,** Order matters because each elected office is different, so choose 4 of the members: $112 \cdot 111 \cdot 110 \cdot 109$.

**22. A,** Pick from a group to make smaller groups, where the order of the members in a committee doesn't matter: $\frac{112 \cdot 111 \cdot 110 \cdot 109}{4 \cdot 3 \cdot 2 \cdot 1}$.

**23. D,** Order matters because every one of the 112 members will get a unique number: $112!$.

**24. E,** Pick from a group to make smaller groups, where order of the cards in a hand doesn't matter: $\frac{52 \cdot 51 \cdot 50 \cdot 49 \cdot 48}{5 \cdot 4 \cdot 3 \cdot 2 \cdot 1}$.

**25. B,** Two choices (color and shape) have 6 options and 4 options, respectively, so use $6 \cdot 4$ to find all possible bead outcomes.

**26. D,** Since handshaking occurs between 2 people, find how many different pairs of people can be formed from 18 people. So, pick smaller groups from a larger group, where order of people shaking hands matters, by finding $\frac{18 \cdot 17}{2 \cdot 1}$.

**27. D,** Two choices (place and friend) each has 6 options, so multiply $6 \cdot 6 = 36$ different bike rides.

**28. C,** There are 5 snacks in a day, so there are 5 choices to be made. Each choice can be one of three snack types, so there are 3 options for each of 5 choices: $3 \cdot 3 \cdot 3 \cdot 3 \cdot 3 = 3^5 = 243$.

**29. C,** The scenario of finding different items involves three choices, with 7 options for each choice. So, use the fundamental counting principle: $7 \cdot 7 \cdot 7$. The other scenario involves one type of choice (a person) taken three times from a larger group of 7, without repeating which person is chosen each time. So use the formula $\frac{7 \cdot 6 \cdot 5}{3 \cdot 2 \cdot 1}$.

**30. D,** There are 5 options for noodles and 6 options for sauces. Because order doesn't matter when selecting two vegetables and no vegetable can be selected twice, use the formula for choosing smaller groups from a larger group: $\frac{12 \cdot 11}{2 \cdot 1}$. The number of different noodle bowls is $5 \cdot 6 \cdot \frac{12 \cdot 11}{2 \cdot 1}$.

**31. C,** There are 1,980 different noodle bowls (from question 30). So multiply $(1{,}980)(\$12.99)$ to get $\$25{,}720.20$.

## LESSON 2, *pp. 131–135*

**1. B,** By the fifth event, there are 4 striped + 2 black = 6 marbles left in the bag. The probability of selecting a black marble is 2:6 = 1:3.

**2. A,** The probability of spinning a 6 out of 8 choices is 1:8.

**3. C,** The probability of spinning a 4 or 8 out of 8 choices is $\frac{2}{8} = \frac{1}{4} = 0.25$.

**4. E,** Neither 4 nor 6 are odd, so Maude's experimental probability of spinning an odd number is $\frac{0}{2}$.

**5. A,** The experimental probability of receiving a clothing department complaint is $\frac{3}{6 + 4 + 2 + 3} = \frac{3}{15} = \frac{1}{5} = 20\%$

**6. D,** The experimental probability that the next complaint will concern electronics or housewares is $\frac{6 + 4}{6 + 4 + 2 + 3} = \frac{10}{15} = \frac{2}{3}$.

**7. C,** The probability it will not rain tomorrow is $100\% - 40\% = 60\% = \frac{60}{100} = \frac{3}{5}$.

**8. C,** The probability of spinning a white or striped wedge is 6:8 = 3:4.

**9. D,** The probability of spinning a red or striped wedge is $\frac{5}{8}$.

# UNIT 3 (continued)

**10. A,** There are no green wedges so the probability of landing on green is 0% and the event is impossible.

**11. B,** The probability of picking one of the 5 black marbles out of 12 total marbles is 5:12.

**12. C,** There are now 6 striped + 4 black = 10 total marbles. The probability of picking a striped marble is $\frac{6}{10} = 60\%$.

**13. B,** Out of three picks, she selected a striped marble twice. The experimental probability at this point of picking a striped marble is $\frac{2}{3}$.

**14. D,** There are 6 chances of spinning a red wedge or an odd number. The probability is 100%.

**15. C,** Marta has 1 chance of spinning a 4 out of 6 possibilities, for a probability of 1:6.

**16. D,** There are 5 chances of spinning a 6, 2, or white wedge. The probability is 5:6.

**17. D,** The factors of 12 are 1, 2, 3, 4, 6, and 12. The possible outcomes on the spinner for factors of 12 are 1, 2, 3, 4, and 6, which are 5 of the 6 wedges; the event is likely.

**18. A,** The probability of landing on a 2 out of the 6 sides is 1:6.

**19. E,** Chuck landed on 2 odd numbers in 2 rolls so the probability is 2:2 = 1:1.

**20. C,** The probability of rolling an even number is $\frac{3}{6} = 50\%$.

**21. B,** The probability that the next car will be blue or red is $(15 + 25) \div (32 + 15 + 25 + 18 + 10) = 40 \div 100 = 0.40$.

**22. B,** The probability of seeing a car of a different color is $\frac{10}{100} = 10\%$.

**23. C,** Black cars have been seen the most, so the probability of seeing a black car is greatest.

**24. D,** Julian has picked 2 red marbles out of 3 picks, so the experimental probability is $\frac{2}{3}$.

**25. E,** There is not enough information because Julian does not replace the marbles and we do not know how many black or red marbles there are.

## LESSON 3, pp. 136–140

**1. C,** The median height of the runners is $\frac{65 + 67}{2} = 66$ in.

**2. B,** The range of the runners' times is 17.2 – 11.8 = 5.4 seconds.

**3. C,** The median time in the race is $\frac{12.8 + 13.5}{2} = 13.15$ seconds.

**4. B,** The mean time of the runners is
$\frac{13.5 + 16.0 + 12.6 + 15.2 + 12.8 + 11.8 + 17.2 + 12.1}{8} = \frac{111.2}{8} = $ 13.9 seconds. The difference between Sarah's time and the mean time is 13.9 – 12.1 = 1.8 seconds.

**5. C,** The median is $\frac{22 + 24}{2} = 23$ milk shakes sold. The median is best to use when there is an extreme value (85).

**6. B,** Solve $\frac{5229 + 3598 + 6055 + 3110 + 3765 + x}{6} = 3{,}743.14$ for $x$ to find that $x = \$701.84$.

**7. A,** The range of the grades is 97 – 68 = 29.

**8. C,** The score 85 is the most frequent, so it is the mode.

**9. C,** Subtract Elena's grade, 75, from the median grade, 85, to find a difference of 10.

**10. A,** The median for January through June is ($11,820 + $18,560) ÷ 2 = $15,190.

**11. C,** The total sales for July through December is $24,450 + $22,110 + $23,450 + $19,300 + $15,340 + $16,980 = $121,630. The mean is $121,630 ÷ 6 ≈ $20,272.

**12. D,** The range of sales throughout the year is $26,890 – $7,200 = $19,690.

**13. E,** The range for the data set is 7.5 – 0.25 = 7.25.

**14. D,** The result 5.5 occurs the most frequently, so it is the mode of the data set.

**15. C,** The median number of hours is (5.5 + 6) ÷ 2 = 5.75.

**16. C,** The range for the number of hours is 7 – 2.25 = 4.75.

**17. B,** The median term is 1.5.

**18. D,** The mean of Wednesday's data is (5 + 5.5 + 0.5 + 5 + 1) ÷ 5 = 3.4. The mean of Sunday's data is (6 + 5.5 + 3 + 7 + 7.5) ÷ 5 = 5.8. The difference is 5.8 – 3.4 = 2.4.

**19. C,** The average number of points is (24 + 7 + 13 + 12 + 0 + 36) ÷ 6 ≈ 15.33.

**20. B,** The median is (12 + 14) ÷ 2 = 13.

**21. C,** The fastest runner is Ana because she had the lowest time.

**22. C,** The median time is 23:27.

**23. A,** The range of the data set is $6\frac{1}{2} - 0 = 6\frac{1}{2}$.

**24. E,** The range of the data set is 9 – 1 = 8.

**25. C,** The mean of the data is (3 + 9 + 3 + 4 + 7 + 5 + 6 + 3 + 1 + 5) ÷ 10 = 4.6.

**26. A,** The number 3 is the mode because it appears the most times.

## LESSON 4, pp. 141–145

**1. D,** She did not run 2 mi in Week 1. In Week 2, the only day she ran 2 mi is Day 3.

**2. E,** Subtract 1.35 – 1.3 = 0.05 mph.

**3. B,** Find the value 6.5 mph in the column Run Speeds and in the row AP.

**4. C,** Use a proportion $\frac{1.5 \text{ mi}}{60 \text{ min}} = \frac{1 \text{ mi}}{x \text{ min}}$ to find the number of minutes it takes to swim 1 mile. Use similar proportions, or divide 60 minutes by each rate. Swim: 60 ÷ 1.5 ≈ 40 min; Run: 60 ÷ 8 ≈ 7.5 min; 60 ÷ 18 ≈ 3.33 min. Add 40 + 7.5 + 3.33 = 50.83, which rounds to 51 minutes.

**5. D,** The run speeds are greater than the swim speeds. The difference between the run speed and swim speed at the alternating pace is 6.5 – 1.35 = 5.15 mph.

**6. C,** Hotel A food cost = (90)($12.95) = $1,165.50. Hotel B food cost = (90)($13.25) = $1,192.50. Subtract $1,192.50 – $1,165.50 = $27.00 more.

**7. D,** Total cost is $150 + (250)($13.40) = $3,500.00.

**8. B,** Brand 2 Calcium is 500 mg, which is the greatest quantity in the table.

**9. C,** The table for Iron has three values listed, so the matrix is 3 × 1 and the values are listed the same as the table.

**10. A,** Find the percent of calcium in the tablet to the daily dose: $\frac{250 \text{ mg}}{1000 \text{ mg}} = 0.25 = 25\%$.

**11. B,** Discount applies, because 100 shirts is more than the 50 shirts needed for a discount. Multiply (100)(5.50) = $550. Find discount by multiplying $550(0.30) = $165. Subtract to find discounted cost: $550 – $165 = $385 cost.

# UNIT 3 *(continued)*

**12. B,** Discount applies at first two stores only, because more than 50, but less than 75, shirts are purchased. *Look at Me*: (51) ($5.00) = $255.00. Discounted price is 1 − 25% = 1 − 0.25 = 0.75 of total cost. Multiply (0.75)($255.00) = $191.25, which is the least expensive at all stores.

**13. C,** Discount applies at all stores, because 100 or more shirts are purchased. *Got Shirt*: 20% discount = 80% of the price, so (0.80)($4.50) = $3.60 per T-shirt, the lowest price per shirt.

**14. E,** The discount at *Not Clear?* is the least amount of money at 5% of $4 = 0.05($4.00) = $0.20 per T-shirt.

**15. C,** Add the prices for each pack to get 3.99 + 12.99 + 4.50 + 5.25 + 2.99 = $29.72.

**16. C,** The unit rate is price of a pack divided by the number in a pack: $4.50 ÷ 8 ≈ $0.56 per bead.

**17. D,** Each item must be purchased 24 times. Find the factor of 24 for each number of items in a pack to see how many packs Arthur must buy. Then multiply that number by the cost per pack. Noisemakers: 24 ÷ 6 = 4 packs; 4($3.99) = $15.96. Eyeglasses: 24 ÷ 2 = 12 packs; 12($12.99) = $155.88. Beads: 24 ÷ 8 = 3 packs; 3($4.50) = $13.50. Hats: 24 ÷ 1 = 24 packs; 24($5.25) = $126.00. Poppers: 24 ÷ 5 = 4.8 packs, so round up to 5 packs; 5($2.99) = $14.95. Total cost = 15.96 + 155.88 + 13.50 + 126.00 + 14.95 = $326.29.

**18. D,** The greatest increase was $0.36, which occurred between Weeks 4 and 5.

**19. C,** The gas price of $3.89 is the lowest over the eight-week period.

**20. B,** Sales for January–April = $501.10 + $628.23 = $1,130.33. Subtract from Nov–Dec sales to find $1,512.81 − $1,130.33 = $382.48 more.

**21. C,** Expenses increased from $84.06 to $124.43. So, $\frac{124.43 - 84.06}{84.06} ≈ 0.48025 ≈ 48\%$ increase.

**22. D,** The two highest sales totals were from September to October and from November to December.

**23. D,** Ernesto has driven 636 + 91 = 727 miles altogether.

**24. D,** Ernesto will have to drive 165 + (148)(3) = 609 miles.

---

## LESSON 5, *pp. 146–150*

**1. A,** For data from a sample population to be considered valid, the sample must be made up of individuals chosen at random.

**2. D,** Observation; the police are not influencing whether or not drivers use Elm Street, and the video records the sample population of drivers on Elm during a period of time.

**3. B,** Observation; Inez and the volunteers are not influencing whether or not drivers use Elm Street, and they record information for only part of each day for a period of time.

**4. D,** A convenience sample; the police observed drivers that happened to pass while the cameras were set up.

**5. C,** Inez's study does not include data from the nighttime when she and her volunteers did not keep track of traffic.

**6. A,** The police study does not account for possible changes in traffic patterns that might occur during a weekend. Tracking traffic on Saturday and Sunday would provide a more accurate representation of traffic for all days of the week.

**7. D,** Using data from the police's study, 393 ÷ 1253 = 31%.

**8. E,** Inez and the volunteers did not collect data at nighttime, and it is possible that truck traffic was higher during that time.

**9. D,** Both questions reference television shows, so it is logical to assume that the ad agency wants to use responses from their sample to draw conclusions about the general television-viewing population.

**10. C,** The agency asked questions and collected responses from part of the population.

**11. D,** It is a convenience sample because the agency included people who were available, nearby, at the time the agency wanted to conduct the survey.

**12. B,** This question would provide more information not only about whether respondents do or do not view a certain way, but also in quantities that allow comparison.

**13. B,** The ad agency's poll is limited by the local convenience sample, which may not represent the viewing habits of the entire, national population.

**14. B,** At each draw all the names have an equal chance of being selected.

**15. D,** There are 40 names remaining in the bowl at the third drawing; $\frac{1}{40}$ = .025 = 25%.

**16. A,** The supervisor is not influencing any of the responses, and every individual in the population has an equal chance of being selected for the sample.

**17. B,** 5 ÷ 42 = .11904; about 12%.

**18. A,** This information is most likely already recorded within the school district, but if not, it is a manageable task to count all students participating in sports at one school.

**19. E,** Researchers need to know as much as possible about the population of the study to determine whether the sample selected provides useful data about the greater population.

**20. D,** A petition is a voluntary survey because petitioners who sign their names voluntarily agree to do so. A petition only requires signatures of a certain percentage of eligible voters, so it involves a sample and not an entire population.

**21. D,** Cameron basically conducted a one-question survey that asked, "Are you interested in my club?" He posted the survey in places he could easily get the attention of potential interested runners. The respondents constitute a volunteer sample of the running population.

**22. A,** His objective was to find enough interested people to form a club. This sample, drawn on convenience and voluntary participation, does not need to represent the whole population of runners.

**23. A,** Systematic sample survey; he used the system of alphabetical listing to select names for his sample.

**24. D,** Although Dylan's study records quantitative data, it also includes qualitative data by recording the reasons that each person supports or opposes the project.

---

## LESSON 6, *pp. 151–155*

**1. B,** The difference in rainfall between the two parks was greatest in April.

**2. A,** The winner jumped 20 ft, and Contestant A jumped 10 ft.

**3. C,** Bars B and E are the only bars with the same length, so Katie and Alana both jumped 15 ft.

**4. E,** The range of scores is 20 − 5 = 15.

## UNIT 3 (continued)

**5. E,** When percent correct is greater than 80, the number of hours is 8.

**6. B,** In general, as x (hours) increases, y (percent correct) also increases.

**7. D,** The sharpest decrease occurred from 1990 to 2000.

**8. B,** The bars for 1960 and 2000 are closest in height.

**9. C,** The heights of the bars increase and then decrease.

**10. D,** The dots on the scatter plot are highest at age 55, so the salaries are the greatest.

**11. B,** Up to a certain age, greater levels of experience generally lead to higher annual earnings by employees.

**12. A,** Gray squirrels show the only increase.

**13. E,** There is no bar for the opossum, so it was not seen in 2010.

**14. D,** The striped bar for deer is the same height as the solid bar for raccoons.

**15. B,** The lines are closest together in April, so the rainfall is about the same.

**16. D,** The widest gap between lines is in August, so the difference between rainfalls is greatest.

**17. D,** The line for Anchorage is highest during summer months (Jul, Aug) and the line for Hawaii is highest during winter months (Nov, Dec, Jan).

**18. B,** About 7,500 – 2,500 = 5,000 more women work in education.

**19. D,** The bar for computer specialist is twice as tall as the bar for retail.

**20. E,** There is not enough information, because this bar graph does not indicate future trends.

**21. C,** The points for June through August are above 17.

**22. D,** The approximate difference between the number of daylight hours in August and December is 17 – 6 = 11.

**23. B,** March has about 10 hours of daylight, which is twice the 5 hours in January.

---

## LESSON 7, pp. 156–160

**1. D,** The food section shows about 30%.

**2. E,** Slightly more than 50% of the employees drive to work, and the only answer choice greater than 50% is 60%.

**3. B,** A significant increase in the price of gasoline likely would result in more employees walking or taking a bus to work.

**4. E,** Poetry has a smaller section than crime.

**5. A,** Since they are the most popular categories of books in September, a librarian could make the best argument to order nonfiction and mystery titles.

**6. D,** The romance section represents about 15%. So, (30,000)(0.15) = 4,500 books.

**7. D,** Together, fuel oil and gas occupy 60% of the circle, which is greater than 50%.

**8. E,** The wood stove and electricity are both used by 15% of the people.

**9. D,** The percentage of the population that does not use gas is 100% – 25% = 75%.

**10. C,** The Democrat section occupies half of the graph.

**11. D,** The percentage of the population that voted Independent or Republican is slightly less than 50%, which is closest to 45%.

**12. C,** Together, walking and running occupy half of the circle.

**13. C,** Running occupies $\frac{1}{4}$ of the circle, and $1 - \frac{1}{4} = \frac{4}{4} - \frac{1}{4} = \frac{3}{4}$.

**14. E,** There is not enough information, because skiing is not on the graph.

**15. E,** Red maples occupy more than 30% and less than 50% of the circle, so about 40%.

**16. C,** Expect about 40% of 300 = (300)(0.40) = 120 red maple trees.

**17. E,** The percentage of students who speak more than two languages is 25% + 40% = 65%.

**18. D,** The percentage of students who do not speak Chinese is 20% + 15% = 35%.

**19. C,** English is included on each section of the graph.

**20. C,** Diego should spend (100)(0.25) = $25 on food.

**21. D,** Diego should put (2,200)(0.05) = $110 toward savings.

**22. E,** Food and rent are the two largest sections of the graph, so they make up the largest percentage.

---

## LESSON 8, pp. 161–165

**1. C,** The age group 36–45 has the most viewers between the hours of 7 P.M. and 11 P.M.

**2. C,** There were 3 + 5 = 8 bowlers who scored between 100 and 139.

**3. D,** The scores between 160 and 179 are the most frequent.

**4. E,** Because of how the intervals are organized, we cannot be sure of how many bowlers scored between 157 and 189.

**5. C,** The percentage of bowlers who scored between 160 and 179 is $\frac{9}{3 + 5 + 8 + 9 + 5} = \frac{9}{30} = 30\%$.

**6. C,** The number 3 is the mode because it has the most tally marks.

**7. C,** The median number of pets owned is 3.

**8. A,** The mean number of pets owned is [(1)(2) + (2)(5) + (3)(7) + (4)(2) + (5)(1)] ÷ 17 ≈ 2.7.

**9. B,** The mean would be [(1)(2) + (2)(5) + (3)(7) + (4)(2) + (5)(2)] ÷ 18 ≈ 2.83.

**10. C,** The show had high ratings among teens and adults under the age of 50.

**11. B,** The 18–34 age group produced the most viewers. To find the median, first order the ages from lowest to highest (18, 19, 20, 21 22, 23, 24, 25, 26, 27, 28, 29, 30, 31, 32, 33, 34). Since there are 17 data points, the median value will be at the 9th data point (26).

**12. B,** Weeks 9 through 16 show substantially higher visitation than the weeks before and after that period.

**13. E,** The histogram shows no measurable visitation for weeks 21–24.

**14. D,** Weeks 9 through 16 have more than twice the visitation of the next most active periods.

**15. C,** The numbers of rods in ranges on the histogram that measure outside the acceptable range are 3, 6, 4, and 2—a total of 15 rods. 15 ÷ 53 = .238 = 23.8%.

**16. C,** The two bars showing ranges below the acceptable 59.4 cm record 3 and 6 bars, totaling 9 that were too short.

**17. A,** To determine the mode, find the bar on the histogram that extends higher than any other bar. That tells you the mode, or the most frequent age group of viewers.

# UNIT 3 (continued)

**18. B,** Most of the viewers are between the ages of 10 and 19.

**19. C,** Each bar in the histogram shows a number of students who can type that many words; the numbers of students represented by each bar in the histogram add up to 20.

**20. C,** To determine the mode, find the bar on the histogram that extends higher than any other bar. That tells you the mode, or the most common amount of words typed per minute by students in the keyboarding class.

**21. D,** The numbers of students recorded for the three bars representing 28 or more words per minute are 5, 4, and 2—totaling 11 students.

**22. E,** Both histograms and frequency tables display data about numbers of occurrences; histograms do so using a visual display of bars.

**23. D,** Histograms can be used with any size data set. The scale can make a very short bar represent a very high value if needed.

**24. C,** "Number of Walkers" and "Number of Bus Riders" are data that can be shown as frequencies for each month.

---

## LESSON 9, pp. 166–170

**1. B,** Half of the data set has 16 points, and so the lower quartile point will be between the eighth and ninth data values, which are 6 and 7. The quartile point is halfway between the two values, or 6.5.

**2. C,** In the set of 40 data points, the median is halfway between the twentieth and twenty-first points, which are 7 and 8. The median is 7.5 hours.

**3. B,** The value with the most dots is 7 hours.

**4. D,** The maximum hours of sleep is 12 hours; the minimum is 4 hours. $12 - 4 = 8$.

**5. A,** Each dot represents a subject. There are 5 dots in the column representing 9 hours.

**6. C,** Students reported having 3 siblings with the most frequency.

**7. B,** There is an even number of values in the list, 16. The median falls halfway between the eighth and ninth values, which are 2 and 3.

**8. C,** First, find the mean. $[(2 \cdot 0) + (2 \cdot 1) + (4 \cdot 2) + (5 \cdot 3) + (2 \cdot 4) + (0 \cdot 5) + (1 \cdot 6)] \div 16 = 39 \div 16 = 2.4375 \approx 2.5$. The mode is 3, so the median is less than the mode.

**9. B,** In 30 tosses, the median is between the fifteenth and sixteenth occurrences in an ordered list; both values are 5, so the median is 5.0.

**10. A,** The total with the greatest number of dots associated with it is 6.

**11. B,** The minimum value in the plot is 2.

**12. C,** Two possible combinations produce a total of 3 (1 + 2 or 2 + 1); two possible combinations total 7 (4 + 3 and 3 + 4). Three possible combinations total 4 (1 + 3, 2 + 2, and 3 + 1) and 6 (2 + 4, 3 + 3, and 4 + 2). Four combinations total 5 (1 + 4, 2 + 3, 3 + 2, and 4 + 1). The number that can be rolled with the greatest number of possible combinations is 5, so that is the expected mode.

**13. B,** The median scores are denoted by the bars in the middle of the boxes. Comparison of the four boxes shows that Chef 2 has the highest median score.

**14. D,** The upper quartile corresponds to the top of each box. The chef with the box whose top is the highest is chef 4.

**15. C,** $61 \div 8 = 7.625$

**16. D,** For a mean of 8.0 after ten quizzes, the sum of scores must total 80. The scores of the first eight total 61, so the sum of the final two must be at least 19.

**17. E,** All of Thomas's test scores occur with equal frequency.

**18. A,** The range is the difference between the greatest and least values. $18 - 15 = 3$.

**19. B,** The values total to 195. $195 \div 12 = 16.25$; 16.25 years equals 16 years 3 months.

**20. C,** The mode is 16; the mean is 16.25. The mean is greater than the mode, so the distribution is positively skewed.

**21. A,** Regardless of whether you use the range or the difference between the lower and upper quartile to judge how well a thermostat maintains a relatively constant temperature, thermostat 1 is best.

**22. C,** The median is represented by the bar in the middle of the box; thermostat 3 has that closest to the set value of 68°.

**23. D,** The end points of the lines extending a box represent the minimum and maximum values; thermostat 4 has the largest range.

**24. A,** Thermostat 1 has the smallest range.

**25. C,** In 19 data points, the median value will be at the tenth data point. With the values ordered lowest to highest, the tenth value is 29.

**26. D,** The fewest words per minute is 18, and the most is 39. $39 - 18 = 21$.

---

## Unit 3 Review, pp. 171–175

**1. B,** The mean number of hours is
$$\frac{21.5 + 28 + 15.5 + 23 + 29 + 34 + 27 + 35}{8} = 26.625 \approx 26.6.$$

**2. A,** The median is $\frac{27 + 28}{2} = 27.5$, and the mean is 26.63, so the median is slightly greater than the mean.

**3. C,** The maximum number of hours is 35; the minimum number of hours is 15.5. $35 - 15.5 = 19.5$.

**4. C,** There are a total of six sides; three of the sides have even numbers. The probability of rolling an even number is 3:6, or 50%.

**5. B,** There are a total of six sides, of which two sides have either a 2 or a 4. The probability of rolling a 2 or a 4 is then $\frac{2}{6} = 0.333$, which is 33%.

**6. B,** For each of the six sides of one die, there are six possible results for the other die. The total number of possible combinations is $6 \cdot 6 = 36$ combinations. There is only one way of rolling a sum of a 2 (1 + 1); there are three ways of rolling a 4 (2 + 2, 1 +3, and 3 + 1). So, out of the 36 possible rolls, there are 4 that could produce totals of either 2 or 4. That gives a probability, expressed as a fraction, of $\frac{4}{36} = \frac{1}{9}$.

**7. B,** There are two years where the amount of bonuses increased—from 2003–2004 and from 2004–2005. The increase between 2004–2005 was greater, evidenced by the steeper line on the graph.

## UNIT 3 (continued)

**8. D;** The only year in which the data point is clearly below $5,000 is 2008, where the amount is between $4,000 and $5,000.

**9. D,** The probability of coming up with a three on the die is $\frac{1}{6}$. The probability of getting a head when flipping a coin is $\frac{1}{2}$.

The probability of all three events happening is $\frac{1}{6} \times \frac{1}{2} \times \frac{1}{2} = \frac{1}{24}$.

**10. B,** If the first box is green, the remaining three can be ordered green/red/red, red/green/red, and red/red/green. If the first box is red, the remaining three can be ordered green/green/red, green/red/green, and red/green/green—for a total of 6 unique arrangements.

**11. C,** When rolling three six-sided dice, there are $(6)^3 = 216$ possible results. Of those 216 possibilities, there are 6 where all three dice have the same number. That corresponds to a probability of $\frac{6}{216} = \frac{1}{36}$.

**12. D,** The probability of two heads coming up on a given toss is independent of previous tosses. There remain four possible results of the toss, with only one of the four being a pair of heads. So the probability is 25%.

**13. A,** There are eight sections on the wheel. Four are prize sections. The probability of winning a prize is four out of eight, or 0.5.

**14. C,** If the spin comes up "Cake," the player winds a cake; if the spin comes up "Your choice," the player can choose cake. The probability of winning a cake is two spins out of eight, or 25%.

**15. B,** The greatest daily amount raised is $7,600; the least is $4,400. $7,600 − $4,400 = $3,200.

**16. C,** The sum of the results for the five days is $28,500. Dividing that by five days gives the daily mean of $5,700.

**17. C,** The median is the middle figure in an odd-numbered data set. In this case, it is $5,400.

**18. B,** The goal for the seven-day total is $45,000; they raised $28,500 during the first five days. That leaves $16,500 to be raised in the final two days, which is a mean of $8,250 per day.

**19. C,** Players scored 4 a total of eight times, more frequently than any of the other scores.

**20. E,** There are $4 \times 3 \times 2 \times 1 = 24$ possible outcomes of dealing the four cards. There is only one way to get 1/2/3/4. The probability that 1/2/3/4 will be the result is, then, $\frac{1}{24}$.

**21. C,** At age 10 Kris is the tallest, then Mike, then Peter.

**22. B,** The sharpest rise of any of the lines occurs for Mike in the year following his 11th birthday.

**23. D,** This is a combination subgroup of a group and order does not matter. Compute $\frac{26 \times 25}{1 \times 2} = 325$. There are 325 possible combinations for the first round of matches.

**24. B,** Permutation without repetition. Compute 52!

**25. C,** The correct choice is a $3 \times 1$ matrix; the entries are (35 − 25), (35 − 45), (55 − 45).

## UNIT 4 Algebraic Concepts

### LESSON 1, pp. 178–182

**1. A,** If Gabe's sister's age = $x$, then Gabe's age = $3x$.

**2. D,** Kevin's yard is twice $g$ increased by 10, which is the same as $2g + 10$.

**3. C,** The number of employees that work in manufacturing is 500 less than $3s$, which is the same as $3s − 500$.

**4. A,** Let Michael's science score = $s$; then his math score $= 8 + \frac{1}{2}s = 8 + \frac{s}{2}$.

**5. E,** Julie left $\frac{1}{6}(48) + 2 = \frac{48}{6} + 2 = $10$ for a tip.

**6. D,** An adult ticket costs $2($12) − $4 = $24 − $4 = $20$.

**7. E,** The perimeter of the rectangle is represented by $2(2w − 3) + 2(w) = 4w − 6 + 2w = 6w − 6$.

**8. B,** Because area = length × width, and the length of this rectangle is $(2w − 3)$, the area of the rectangle is represented by $(2w − 3)w$, which is the same as $w(2w − 3)$.

**9. D,** Let $b$ = the number of boys; then there are $2b − 15$ girls.

**10. B,** Let $a$ = the number of adult tickets; then $\frac{1}{3}a + 56$ or $\frac{a}{3} + 56$ represents the number of children's tickets sold.

**11. B,** There are $3(12)(50)p − 1$ pencils left.

**12. E,** Let Wednesday = $x$ and Thursday = $y$; then Tuesday can be described by $4(x + y)$.

**13. B,** Let $y$ = the sophomore class; then the freshman class can be described by $\frac{3y}{4}$.

**14. C,** The perimeter can be represented by $(2b + 1) + b + (−4 + 3b)$. Simplify: $(2b + b + 3b) + (1 − 4) = 6b − 3$.

**15. A,** Let $x + y$ = the two grandchildren together; then Nick's grandfather can use the expression $5 + 2(x + y)$ or $2(x + y) + 5$ to describe his age.

**16. D,** The cyclist rode $3s − 20$ miles on Monday. Solve for $s = 30$; $3(30) − 20 = 70$.

**17. B,** Simplify $3(x + 2x)$ to $3(3x)$. Solve for $x = 4$; $3(3 \cdot 4) = 3(12) = 36$.

**18. C,** Leo is $2s − 21$. Solve for $s = 23$; $2(23) − 21 = 46 − 21 = 25$ years old.

**19. C,** Solving for $a = 207$ and $b = 134$, the theater takes in $15($207) + 25($134) = $3,105 + $3,350 = $6,455$.

**20. B,** There are $\frac{1}{2}m + 56$ female students. For $m = 347$, there are $\frac{1}{2}(374) + 56 = 243$ female students.

**21. E,** Let $m$ = the number of men's shoes sold; then $12 + 4m$ represents the number of women's shoes sold. However, the value of $m$ is unknown, so there is not enough information given.

**22. C,** The area of a triangle is one half the base times the height, or $A = \frac{1}{2}bh$. If $b$ = the base, then $3b − 3 = h$, so $A = \frac{1}{2}b(3b − 3)$.

**23. A,** Let $t = \frac{1}{2}B − 45$ and solve for $B$ to find $B = 2t + 90$.

**24. C,** The first number is represented by $\frac{1}{2}(x + y) = \frac{x + y}{2}$.

**25. D,** Sean swam $2(15) − 8 = 22$ laps.

**26. E,** The expression $4y − 8(3 − 2(−3))$ is equal to $4y − 8(3 − (−6)) = 4y − 8(9) = 4y − 72$.

**27. B,** There were $2(45) − 34 = 56$ students who scored above average.

**28. D,** Jada paid $\frac{1}{2}($84) − $5 = $37$ for gas.

**29. E,** Let $\frac{x}{y}$ represent the quotient of the second and third numbers; then $3\left(\frac{x}{y}\right)$.

**30. A,** The perimeter of the rectangle can be represented by $2w + 2\left(6 + \frac{2}{3}w\right) = 2w + 12 + \frac{4}{3}w = \frac{10}{3}w + 12$.

# UNIT 4 (continued)

## LESSON 2, pp. 183–187

**1. A,** $3x + 9 - 9 = 6 - 9$. Simplify: $3x = -3$. Next, divide both sides by 3: $\frac{3x}{3} = \frac{-3}{3}$. Simplify: $x = -1$.

**2. D,** Isolate the variable. Add 4 to both sides: $0.5x - 4 + 4 = 12 + 4$. Simplify: $0.5x = 16$. Divide both sides by 0.5: $\frac{0.5x}{0.5} = \frac{16}{0.5}$. Alternately, multiply both sides of the equation by 2: $2(0.5x) = 2(16)$. Simplify: $x = 32$.

**3. E,** Group variable terms and subtract $3y$ from both sides: $5y - 3y + 6 = 3y - 3y - 14$. Group like terms: $2y + 6 = -14$. Next, subtract 6 from both sides: $2y + 6 - 6 = -14 - 6$. Group like terms: $2y = -20$. Finally, divide both sides by 2: $\frac{2y}{2} = \frac{-20}{2}$. Simplify: $y = -10$.

**4. A,** To group variable terms on one side, subtract $\frac{1}{2}t$ from both: $\frac{1}{2}t + 8 - \frac{1}{2}t = \frac{5}{2}t - 10 - \frac{1}{2}t$. Group like terms and simplify fraction: $8 = 2t - 10$. Add 10 to both sides: $8 + 10 = 2t - 10 + 10$. Simplify $18 = 2t$. Divide both sides by 2: $t = 9$.

**5. C,** $1{,}200 + 0.08s = 2{,}800$. Solve for $s$. First, subtract 1,200 from both sides: $1{,}200 - 1{,}200 + 0.08s = 2{,}800 - 1{,}200$. Simplify: $0.08s = 1{,}600$. Divide both sides by 0.08: $\frac{0.08s}{0.08} = \frac{1600}{0.08}$; $s = 20{,}000$.

**6. B,** The perimeter is 84 feet; $4x + 8 = 84$. Solve for $x$. Subtract 8 from both sides: $4x + 8 - 8 = 84 - 8$. Simplify: $4x = 76$. Divide both sides by 4: $\frac{4x}{4} = \frac{76}{4}$. Simplify: $x = 19$. Since $x$ represents the width of the yard, add 4 feet to find the length. The yard is 23 feet long.

**7. C,** Isolation of the variable, subtraction, addition, and division are used to solve for $x$.

**8. C,** Solve the equation. Eliminate the parentheses: $3x = 8 - 0.25x - 3 + 0.75x$. Group like terms: $3x = 5 + 0.5x$. Subtract $0.5x$ from both sides: $3x - 0.5x = 5 + 0.5x - 0.5x$. Simplify: $2.5x = 5$. Divide both sides by 2.5: $\frac{2.5x}{2.5} = \frac{5}{2.5}$. Simplify: $x = 2$. Lucas' solution is incorrect.

**9. D,** Group variable terms on one side; add $3x$ to both: $-3x + 11 + 3x = x - 5 + 3x$. Simplify: $11 = 4x - 5$. Add 5 to both sides: $11 + 5 = 4x - 5 + 5$. Simplify: $16 = 4x$. Divide both sides by 4 to get $x = 4$.

**10. D,** Group variable terms; subtract $0.6y$ from both sides of the equation: $0.6y + 1.2 - 0.6y = 1.1y - 0.9 - 0.6y$. Simplify: $1.2 = 0.5y - 0.9$. Add 0.9 to both sides: $1.2 + 0.9 = 0.5y - 0.9 + 0.9$. Simplify: $2.1 = 0.5y$. Divide both sides of the equation by 0.5 to get $y = 4.2$.

**11. A,** Group variable terms. Subtract $\frac{3n}{2}$ from both sides: $\frac{n}{4} - \frac{3n}{2} - \frac{1}{2} = \frac{3n}{2} - \frac{3n}{2} + \frac{3}{4}$. Simplify: $\frac{n}{4} - \frac{3n}{2} - \frac{1}{2} = \frac{3}{4}$. Add $\frac{1}{2}$ to both sides: $\frac{n}{4} - \frac{3n}{2} - \frac{1}{2} + \frac{1}{2} = \frac{3}{4} + \frac{1}{2}$. Simplify: $\frac{n}{4} - \frac{3n}{2} = \frac{3}{4} + \frac{1}{2}$. Write each fraction with the common denominator of 4: $\frac{n}{4} - \frac{6n}{4} = \frac{3}{4} + \frac{2}{4}$. Simplify: $-\frac{5n}{4} = \frac{5}{4}$. Multiply both sides by $-\frac{4}{5}$: $\left(-\frac{4}{5}\right)\left(-\frac{5n}{4}\right) = \left(-\frac{4}{5}\right)\left(\frac{5}{4}\right)$. Simplify: $n = -1$.

**12. C,** Solve for $x$. Multiply each term inside the parentheses by 0.5: $0.5(4x) - 0.5(8) = 6$. Simplify: $2x - 4 = 6$. Add 4 to each side: $2x - 4 + 4 = 6 + 4$. Simplify: $2x = 10$. Divide both sides by 2: $\frac{2x}{2} = \frac{10}{2}$. So, $x = 5$.

**13. A,** Multiply each term inside the parentheses by 3: $3\left(\frac{2}{3}x\right) + 3(4) = -5$. Simplify: $2x + 12 = -5$. Subtract 12 from both sides: $2x + 12 - 12 = -5 - 12$. Simplify: $2x = -17$. Divide both sides by 2: $\frac{2x}{2} = \frac{-17}{2}$. So, $x = -8.5$.

**14. C,** Drew is 17; $17 = \frac{1}{2}t + 3$. Subtract 3 from both sides: $17 - 3 = \frac{1}{2}t + 3 - 3$. Simplify: $14 = \frac{1}{2}t$. Multiply both sides by 2: $2(14) = (2)\frac{1}{2}t$. Simplify: $28 = t$.

**15. B,** Expand the parentheses on both sides: $0.25(3x) - 0.25(8) = 2(0.5x) + 2(4)$. Simplify: $0.75x - 2 = x + 8$. Subtract $x$ from both sides: $0.75x - x - 2 = x - x + 8$. Simplify: $-0.25x - 2 = 8$. Add 2 to both sides: $-0.25x - 2 + 2 = 8 + 2$. Simplify: $-0.25x = 10$. Divide both sides by $-0.25$: $\frac{-0.25x}{-0.25} = \frac{10}{-0.25}$. So, $x = -40$.

**16. A,** Multiply each quantity inside the parentheses by $-2$: $10w + 8 - 2(3w) - 2(-4) = -12$. Simplify: $10w + 8 - 6w + 8 = -12$. Group like terms: $4w + 16 = -12$. Subtract 16 from each side: $4w + 16 - 16 = -12 - 16$. Simplify: $4w = -28$. Divide both sides by 4: $\frac{4w}{4} = \frac{-28}{4}$. So, $w = -7$.

**17. A,** Expand the parentheses: $12b - 2(b) - 2(-1) = 6 - 4b$. Simplify: $12b - 2b + 2 = 6 - 4b$. Next, group like terms: $10b + 2 = 6 - 4b$. Add $4b$ to both sides: $10b + 4b + 2 = 6 - 4b + 4b$. Simplify: $14b + 2 = 6$. Subtract 2 from both sides: $14b + 2 - 2 = 6 - 2$. Simplify: $14b = 4$. Divide both sides by 14 and simplify: $\frac{14b}{14} = \frac{4}{14} = \frac{2}{7}$.

**18. B,** If $(x - 4)$ is 5 more than $3(2x + 1)$, then $(x - 4) = 3(2x + 1) + 5$. Expand the parentheses: $x - 4 = 3(2x) + 3(1) + 5$. Simplify: $x - 4 = 6x + 3 + 5$. Group like terms: $x - 4 = 6x + 8$. Subtract $6x$ from both sides: $x - 6x - 4 = 6x - 6x + 8$. Simplify: $-5x - 4 = 8$. Add 4 to both sides: $-5x - 4 + 4 = 8 + 4$. Simplify: $-5x = 12$. Divide both sides by $-5$: $\frac{-5x}{-5} = \frac{12}{-5}$. So, $x = -2.4$.

**19. A,** Solve the equation. $-2(4x) - 2(-5) = 3x - 6 + x$; $-8x + 10 = 4x - 6$; $-8x - 4x + 10 = 4x - 4x - 6$; $-12x + 10 = -6$; $-12x + 10 - 10 = -6 - 10$; $-12x = -16$; $\frac{-12x}{-12} = \frac{-16}{-12} = \frac{4}{3}$. Ameila is correct. Brandon could not have arrived at an answer of 85 from an error grouping the constant terms on one side of the equation, so Brandon made an error expanding the parentheses.

**20. D,** Multiplying to resolve parentheses: $0.1q - 0.2(3q) - 0.2(4) = -0.3(5q) - 0.3(3)$. Simplify: $0.1q - 0.6q - 0.8 = -1.5q - 0.9$. Group like terms: $-0.5q - 0.8 = -1.5q - 0.9$. Add $1.5q$ to both sides: $-0.5q + .5q - 0.8 = -1.5q + 1.5q - 0.9$. Simplify: $q - 0.8 = -0.9$. Add 0.8 to both sides: $q - 0.8 + 0.8 = -0.9 + 0.1$. Simplify: $q = -0.1$.

**21. E,** The perimeter of the triangle is 9.5 feet; $9.5 = x + 2(3x - 4)$. Multiply terms within the parentheses by 2: $9.5 = x + 2(3x) - 2(4)$. Multiply: $9.5 = x + 6x - 8$. Combine like terms: $9.5 = 7x - 8$. Add 8 to both sides: $9.5 + 8 = 7x - 8 + 8$. Simplify: $17.5 = 7x$. Divide both sides by 7: $\frac{17.5}{7} = \frac{7x}{7}$. So, $x = 2.5$.

## UNIT 4 (continued)

**22. B,** The perimeter is 35 inches; $2x + 2(4x − 2) = 35$. Multiplying each term inside the parentheses by 2: $2x + 2(4x) − 2(2) = 35$. Simplify: $2x + 8x − 4 = 35$. Group like terms: $10x − 4 = 35$. Add 4 to both sides: $10x − 4 + 4 = 35 + 4$. Simplify: $10x = 39$. Divide both sides by 10: $x = 3.9$.

**23. E,** Solve for $y$. Isolation of the variable, multiplication, addition, and subtraction are used. Division is not required.

**24. E,** Expand the parentheses: $4\left(\frac{1}{8}t\right) + 4(2) = 2(t) − 2(8) − \frac{1}{2}t$. Simplify: $\frac{4}{8}t + 8 = 2t − 16 − \frac{1}{2}t$. Write fractions with like denominators: $\frac{1}{2}t + 8 = \frac{4}{2}t − 16 − \frac{1}{2}$. Group like terms: $\frac{1}{2}t + 8 = \frac{3}{2}t − 16$. Subtract $\frac{3}{2}t$ from both sides of the equation: $\frac{1}{2}t − \frac{3}{2}t + 8 = \frac{3}{2}t − \frac{3}{2}t − 16$. Simplify: $−\frac{2}{2}t + 8 = −16$. Simplify the fraction and subtract 8 from both sides: $−t + 8 − 8 = −16 − 8$. Simplify: $−t = −24$. Multiply both sides by −1: $t = 24$.

**25. A,** the length of the fencing is 65 feet; $2x + 2(x + 7.5) = 65$. Multiply terms inside the parentheses by 2: $2x + 2x + 2(7.5) = 65$. Simplify: $4x + 15 = 65$. Subtract 15 from both sides: $4x + 15 − 15 = 65 − 15$. Simplify: $4x = 50$. Divide both sides of the equation by 4: $\frac{4x}{4} = \frac{50}{4}$, so $x = 12.5$.

**26. C,** Keira previously found that $12x − 20 = −2x − 14$. To group together the variable terms, she must have added $2x$ to both sides of the equation: $12x + 2x − 20 = −2x + 2x − 14$. Simplify: $14x − 20 = −14$. Since $a$ represents the $x$-coefficient, $a$ is equal to 14.

**27. C,** Solve for $y$ and compare each step to Quentin's solution. Step 1: Expand the parentheses; $0.5(7y) − 0.5(6) = 2(2y) + 2(1)$. Simplify: $3.5y − 3 = 4y + 2$. Step 2: Subtract $4y$ from both sides: $3.5y − 4y − 3 = 4y − 4y + 2$. Simplify: $−0.5y − 3 = 2$. Step 3: Add 3 to both sides: $−0.5y − 3 + 3 = 2 + 3$. Simplify: $−0.5y = 5$. Quentin subtracted 3 from the right side of the equation in Step 3, so the error is in Step 3.

**28. C,** Expand the parentheses: $10 − 3(2b) − 3(4) = −4b − 2(5) − 2(−b)$. Simplify: $10 − 6b − 12 = −4b − 10 + 2b$. Group like terms: $−2 − 6b = −2b − 10$. Add $2b$ to both sides: $−6b + 2b − 2 = −2b + 2b − 10$. Simplify: $−4b −2 = −10$. Add 2 to both sides: $−4b − 2 + 2 = −10 + 2$. Simplify: $−4b = −8$. Divide both sides by −4: $\frac{−4b}{−4} = \frac{−8}{−4}$. So, $b = 2$.

**29. A,** Solve for $x$ and compare each step to Lisa's solution. Step 1: Expand the parentheses: $\frac{1}{4}(2x) + \frac{1}{4}(12) = 3\left(\frac{1}{4}x\right) − 3(2)$. Simplify: $\frac{2}{4} + 3 = \frac{3}{4}x − 6$. Lisa did not multiply −2 by 3, so the error is in Step 1.

**30. A,** Solve for $n$. Resolve parentheses: $0.2(6n) + 0.2(5) = 0.5(2n) − 0.5(8) + 3$. Simplify: $1.2n + 1 = n − 4 + 3$. Group like terms: $1.2n + 1 = n − 1$. Subtract $n$ from both sides: $1.2n − n + 1 = n − n − 1$. Simplify: $0.2n + 1 = −1$. Subtract 1 from both sides: $0.2n + 1 − 1 = −1 − 1$. Simplify: $0.2n = −2$. Divide both sides by 0.2: $\frac{0.2n}{0.2} = \frac{−2}{0.2}$. So, $n = −10$.

**31. C,** Solve for $w$. Resolve parentheses: $3w − 2(.05w) − 2(1) = −4(2) − 4(w)$. Simplify: $3w − w − 2 = − 8 − 4w$. Group like terms: $2w − 2 = − 8 − 4w$. Add $4w$ to both sides, so that $2w − 2 + 4w = − 8 − 4w + 4w$. Simplify: $6w − 2 = − 8$. Add 2 to both sides so that $6w = − 6$. Divide $6w$ by − 6 to find that $w = − 1$.

## LESSON 3, pp. 188–192

**1. C,** Solve $42 − 5 = 16 − 5 = 11$.

**2. C,** The pattern 2, 4, 16, 256, 65,536, … is the same as 2, $2^2$, $4^2$, $16^2$, $256^2$, …. so, the rule is to square the previous term.

**3. A,** The next term in the pattern is $65,536^2 = 4,294,967,296$.

**4. D,** A multiple of 5 will result in a whole number output, so $x = 25$.

**5. B,** Solve $1,035 = 230t$ for $t$ to find 4.5 hours.

**6. D,** When $x$ increases by 5, $y$ increases by $0.40, so $1.60 + $0.40 = $2.00.

**7. C,** Each term is divided by 2. The fifth term is $24 ÷ 2 = 12$, and the sixth term is $12 ÷ 2 = 6$.

**8. C,** To find $y$, multiply $x$ by 5 and then subtract 1, so $4(5) − 1 = 19$.

**9. A,** If $f(x) = 4$, then $4 = 2 − \frac{2}{3}x$. Solve for $x$ to find that $x = −3$.

**10. A,** To find $y$, multiply $x$ by 3 and then subtract 2 from the product, so $y = 3x − 2$.

**11. C,** Solve $y = 3x − 2$ with $x = 6$ to find that $y = 16$.

**12. C,** Multiply the previous term by 2 to get the next term.

**13. B,** Multiply 27 by 3 to get 81, and then multiply 81 by 3 to find that the sixth term is 243.

**14. B,** The pattern is $1^2$, $2^2$, $3^2$, $4^2$, … So, the next term is $5^2 = 25$.

**15. A,** Multiply the previous term by 2 to get the next term, so $−80 × 2 = −160$.

**16. D,** Let $x = 1$; then $50 − 12 = 49$.

**17. E,** Solve $1 = \frac{1}{2}x$ for $x$ to find that $x = 2$.

**18. B,** The pattern is subtract 2. Since $5 − 2 = 3$ and $3 − 2 = 1$, the fifth term will be 1.

**19. B,** The pattern is subtract 3, so the sixth term is $−15 − 3 = −18$, and the seventh term is $−18 − 3 = −21$.

**20. E,** For $f(x) < 1$, $x$ must be greater than 8. So, $x = 9$.

**21. C,** The next term of the sequence has 16 triangles.

**22. D,** Solve $250 = 1,000(r)(5)$ for $r$ to find that the rate is $0.05 = 5\%$.

**23. E,** Solve $220 = 55t$ for $t$ to find that $t = 4$.

**24. D,** Solve $I = 0.01(5,000)$ to find that there is an increase of 50 people.

**25. E,** Solve $80 = 9/5C + 32$ for C to find that $80°F ≈ 26.7°C$.

**26. D,** The equation $h = d − 0.2d^2$ works for each point.

**27. E,** Solve $h = 5 − 0.2(5)^2$ for $h$ to find that the height is 0 meters.

**28. D,** Kara's answer was $3(1^2) + 1 = 4$. Solve $4 = 3x^2 + 1$ for $x$ to find that another value of $x$ is −1.

**29. A,** For the function $y = x^3$, $1 = 1^3$.

**30. B,** The rule is multiply by −2, so the eighth term is −256.

**31. E,** Since $y = x^2$, we know that $x^2 = 4$, so $x = 2$.

**32. B,** As $d$ increases by 10, $p$ increases by 14.7. To get $p$, multiply by 1.47, and then add 14.7 to the product. So, $p = 1.47d + 14.7$ or $p = 14.7\frac{d}{10} + 14.7$.

## LESSON 4, pp. 193–197

**1. A,** Factor $x^2 + 5x − 6$ to find $(x + 6)(x − 1)$.

**2. D,** The product of $(x + 5)(x − 7)$ is $x^2 + 5x − 7x − 35 = x^2 − 2x − 35$.

**3. A,** The product of $(x − 3)(x − 3)$ is $x^2 − 3x − 3x + 9 = x^2 − 6x + 9$.

**4. D,** Factor $x^2 − 6x − 16$ to find $(x + 2)(x − 8)$.

**5. A,** The area of the rectangle is $(2x − 5)(−4x + 1) = −8x^2 + 22x − 5$.

**6. B,** The term $(x + 3)$ is multiplied by $(4x + 1)$ to find $4x^2 + 13x + 3$.

**7. E,** Divide both sides by 2, and then factor to find $(x + 6)(x + 3)$. Solve for $x$ to find that $x = −6$ and $x = −3$.

# UNIT 4 (continued)

**8. B,** Solve $w^2 - 12w + 32$ for $x$ to find that one of the widths is 4 m.
**9. C,** The product of $(x - 7)(x + 7)$ is $x^2 - 49$.
**10. C,** Use FOIL to find $(x + 5)(x - 4) = x^2 + x - 20$.
**11. D,** Use FOIL to find that the area is $(x - 4)(x - 4) = x^2 - 8x + 16$.
**12. B,** Let $x$ = the number of pencils per student; then there are $(x - 5)x$ pencils in all.
**13. A,** Factor $x^2 - 4x - 21$ to find $(x - 7)(x + 3)$.
**14. E,** Factor $x^2 + 8x - 20$ to find $(x - 2)(x + 10)$.
**15. A,** Factor $x^2 - 5x - 6$ to find $(x - 6)(x + 1)$.
**16. B,** Factor $x^2 - 7x - 30$ to find $(x - 10)(x + 3)$.
**17. A,** $A = lw$, so $(x + 2)(x - 5) = x^2 - 3x - 10$.
**18. D,** Factor $x^2 - 16$ to find that $x = 4$ and $x = -4$.
**19. B,** Factor $x^2 + 6x + 9$ to find $(x + 3)(x + 3)$. So, each side is represented by $(x + 3)$.
**20. B,** If the product of two consecutive integers is 42, then $x(x + 1) = 42$. So, the equation is $x^2 + x - 42 = 0$.
**21. D,** Use FOIL to find that $(x + 4)(x + 4) = x^2 + 8x + 16$. Set the expression equal to 49 to find that $x^2 + 8x - 33 = (x + 11)(x - 3)$. So, $x = -11$ and $x = 3$. Since length cannot be negative, use $x = 3$.
**22. D,** Solve $x^2 + (x + 1)^2 = 113$ for $x$ to find that $x = -8$ and $x = 7$. All answers are positive, so 7 and 8.
**23. D,** Solve $x(x + 2) = 35$ for $x$ to find that $x = -7$ and $x = 5$. Since the product is 35, the factors must be positive. So, 5 is the first integer and $x + 2 = 7$ is the second integer.
**24. C,** Use FOIL to find that $(x - 3)(x - 3) = 81$ is equivalent to $x^2 - 6x - 72 = (x + 6)(x - 12)$. So, $x = -6$ and $x = 12$.
**25. A,** Multiply $x$ by $2x$ and set the expression equal to 32 to find $2x^2 - 32 = 0$.
**26. D,** Factor $2x^2 - 32 = 0$ to find that $x = 4$ and $x = -4$. The lengths of the sides are 4 m and $2(4) = 8$ m.
**27. D,** Solve $0 = t^2 - 2t - 8$ for $t$ to find $t = -2$ and $t = 4$.
**28. B,** Solve $x(x + 1) = 110$ for $x$ to find $x = -11$ and $x = 10$. So, $x = -11$ and $-10$.
**29. A,** Use FOIL and then solve $x^2 + 7x - 8x - 56 - 4x + 4$ and simplify to $x^2 - 5x - 52$.
**30. D,** Solve $x(x + 2) = 10 + 5(x + x + 2)$ for $x$ to find the positive value of $x$ is 10, so $x + 2 = 12$.
**31. A,** Solve $x(2x + 4) = 160$ for $x$ to find $x = -10$ and $x = 8$.
**32. D,** Multiply $x$ by $x + 2$ and set the expression equal to 48 to find $x^2 + 2x - 48 = 0$.
**33. D,** Solve $x(x + 1) = 12$ for $x$ to find $x = -4$ and $x = 3$. The answer is $-3$ and $-4$.
**34. C,** Divide both sides of the equation by 2 to find $x^2 - 4x - 5$. Factor to find $x = -1$ and $x = 5$.
**35. A,** Solve $(x - 2)(x - 2) = 64$ for $x$ to find $x = -6$ and $x = 10$.
**36. A,** Solve $w(w + 8) = 84$ for $w$ to find $w = -14$ and $w = 6$.
**37. C,** The length is $w + 8 = 6 + 8 = 14$ ft.

## LESSON 5, pp.198–202

**1. E,** Solve the inequality $5x \leq 2x + 9$ for $x$ to find $x \leq 3$.
**2. D,** Solve $x + 5 > 4$ for $x$ to find $x > -1$.
**3. B,** The number line shows that $x$ is equal to or less than $-2$, so $x \leq -2$.
**4. D,** Let $5x$ = the product of a number and 5. The inequality is $5x + 3 \leq 13$.
**5. E,** The relationship can be represented by $80 \geq w(3w - 3)$.

**6. A,** Kara and Brett have a total of $\$15 + \$22 = \$37$ for tickets, so $37 < x$.
**7. D,** The taxicab charges $\$2 + \$0.50x$, where $x$ = number of miles. Solve $2 + 0.50x \leq 8$ for $x$ to find that Josie can ride 12 miles.
**8. B,** The relationship can be represented by $x + 12 \leq 5x + 3$.
**9. B,** Solve $8 - 3x > 2x - 2$ for $x$ to find that $x < 2$.
**10. D,** Solve $-5x > 30 - 3(x + 8)$ for $x$ to find that $-3 > x$.
**11. B,** $-3$ and all numbers greater are plotted.
**12. E,** Subtract 5 from both sides to find $x < 9$.
**13. C,** The expression $x \geq -3$ uses a closed circle because it is greater than or equal to.
**14. D,** Solve for $x$ to find $x \geq 3$.
**15. B,** Solve for $x$ to find $-\frac{1}{3} \geq x$ or $x \leq -\frac{1}{3}$.
**16. E,** The numbers $-1$ and lower are graphed, so $x \leq -1$.
**17. A,** Solve for $x$ to find $x \geq 11$.
**18. A,** The situation is represented by the inequality $4x + 3 > 5x - 2$.
**19. C,** Solve $12 + 0.10x \leq 25$ for $x$ to find $x \leq 130$.
**20. C,** Solve $3g > 9$ for $g$ to find $g > 3$, so the lowest price is $\$3.01$.
**21. D,** Solve $\dfrac{45 + 38 + 47 + x}{4} \geq 44$ for $x$ to find $x \geq 46$.
**22. E,** Solve $\dfrac{450 + 550 + x}{3} \geq 600$ for $x$ to find $x \geq 800$.
**23. B,** Solve $15 + 0.75x \leq 25$ for $x$ to find the possible number of games Cole can play.
**24. E,** Add 156 to both sides to find that $x < 190$.
**25. D,** Solve $1,500 + 0.03x \geq 3,000$ for $x$ to find $x \geq 50,000$.
**26. B,** Solve $\dfrac{14,000 + 9,000 + x}{3} > 10,000$ for $x$ to find $x > 7,000$.
**27, D,** Solve $1.60 + 0.95x \leq 4.50$ for $x$ to find $x \leq 3.05$. Then add the quantities of the differently priced packages to arrive at the answer.
**28. B,** Solve $b + 2.5b \leq 157.50$ for $b$ to find $b \leq 45$.
**29. A,** Solve $m + 3m - 400 < 2,000$ for $m$ to find $m < 600$.
**30. C,** Solve $\dfrac{0.266 + x}{2} > 0.300$ for $x$ to find $x \geq 0.334$.
**31. C,** Solve $2x - 30 \leq 100$ for $x$ to find $x \leq 65$.
**32. A,** Let $y$ = the second round. Then $y < 2(10)$, so $y < 20$.
**33. D,** Solve $3x < 45$ to find $x < 15$, so the greatest amount is $\$14.99$.
**34. D,** Solve $\dfrac{78 + 85 + 82 + 74 + x}{5} \geq 80$ for $x$ to find $x \geq 81$.

## LESSON 6, pp. 203–207

**1. A,** Point $C$ is 2 units right and 2 units up, so the coordinates are (2, 2).
**2. B,** Point $T$ is 4 units right and 4 units down, so the coordinates are (4, −4).
**3. B,** Point $S$ is 1 unit right and 0 units up or down, so the coordinates are (1, 0).
**4. A,** Point $P$ is 5 units left and 5 units down, so the coordinates are (−5, −5).
**5. E,** Point $D$ is 3 units right and 5 units down, so the coordinates are (3, −5).
**6. C,** Point $C$ is (−2, 3). The new $C$ is 2 units right and 3 units up, so the new coordinates are (2, 3).
**7. A,** Point $B$ is (−2, 5). The $y$-coordinate would decrease by 3, so (−2, 2).
**8. E,** Point $D$ is 2 units right and 5 units up, so (2, 5).
**9. B,** Six units down from point $C$ is (1, −5).

# UNIT 4 (continued)

**10. B,** To get from point $F$ to point $E$, move 3 units down and 2 units left. So, move 3 units down and 2 units left from point $C$ to find the point with coordinates $(-1, -2)$.

**11. A,** Point $D$ would be $(2 + 2, 5 - 5)$, so $(4, 0)$.

**12. E,** Point $C$ would be 1 unit below the $x$-axis instead of 1 unit above, so $(1, -1)$.

**13. C,** Quadrant 2 has negative $x$-coordinates and positive $y$-coordinates, so $(-2, 5)$.

**14. A,** Frank landed at $(4 + 2, -3 - 1) = (6, -4)$.

**15. C,** The $y$-coordinate would decrease by 3, so $(x, -6 - 3) = (x, -9)$.

**16. B,** The new location of point $M$ is $(-3 + 3, -2 + 2) = (0, 0)$.

**17. C,** Point $M$ is moved 4 units right and 4 units up.

**18. C,** Point $K$ will be 1 unit to the left of the $y$-axis instead of 1 unit to the right, so $(-1, 2)$.

**19. D,** Quadrant 3 has negative $x$- and $y$-coordinates, so $(-3, -2)$.

**20. A,** The bottom point should be 3 units down from the top, so $(-3, -5)$.

**21. E,** When the dots are connected, they form a right triangle.

**22. C,** The $x$-value increases by 2 while the $y$-value increases by 3.

**23. E,** For vector **a**, the $x$-value increases by 2 while the $y$-value increases by 3, regardless of the point at which the vector is shown to begin.

**24. B,** The $x$-value increases from the initial point, making the arrow point upward, and the $y$-value decreases making the arrow point to the left.

**25. A,** $\overline{WZ}$ would be 6 units long instead of 3 units. Six units to the right of $W$ is $(-2 + 6, 2) = (4, 2)$.

**26. B,** $\overline{ZY}$ would be 1 unit long instead of 3 units long. So, 1 unit down from $(1, 2)$ is $(1, 1)$.

## LESSON 7, pp. 208–212

**1. B,** Let $x = 3$ and $y = -1$; then solve $2(3) + (-1) = 5$ to find that $5 = 5$, so the solution is $(3, -1)$.

**2. C,** Let $x = 0$ and $y = 2$; then solve $0 + 2(2) = 4$ to find that $4 = 4$, so the solution is $(0, 2)$.

**3. A,** Let $x = 0$ and $y = 0$; then solve $2(0) - (0) = 0$ to find that $0 = 0$, so the solution is $(0, 0)$.

**4. D,** Solve $3 = 2x + 2$ for $x$ to find that $x = \frac{1}{2}$.

**5. C,** Use the distance formula. Solve $\sqrt{(-4 - 0)^2 + (3 - 0)^2}$ to find $\sqrt{16 + 9} = \sqrt{25} = 5$.

**6. D,** Let $x = -5$ and $y = 1$; then solve $-5 + 2(1) = -3$ to find that $-3 = -3$, so the solution is $(-5, 1)$.

**7. E,** Use the distance formula. Solve $\sqrt{(2 - 4)^2 + (5 - 3)^2}$ to find $\sqrt{4 + 4} = \sqrt{8} \approx 2.83$.

**8. C,** The distance from $(-5, 2)$ to $(-3, 1)$ is $\sqrt{(-5 + 3)^2 + (2 - 1)^2} \approx 2.236$. The distance from $(-3, 1)$ to $(-1, -4)$ is $\sqrt{(-3 + 1)^2 + (1 + 4)^2} \approx 5.385$. Marvin's total distance was $2.236 + 5.385 = 7.621 \approx 7.62$.

**9. C,** The ordered pair whose $x$-value is double the $y$-value is $(4, 2)$.

**10. A,** Solve $-2 = y + 1$ to find $y = -3$.

**11. B,** Solve $2x + 2(-3) = -8$ to find $x = -1$.

**12. E,** The graph would pass through $(2, -2)$ because $-2 = 4 - 3(2)$ is true.

**13. A,** Solve $3x - 1 = 5$ to find $x = 2$.

**14. C,** For the equation $x + 2y = 2$, when $x = 0$, $y = 1$. The only line that passes through $(0, 1)$ is choice 3.

**15. D,** Use the distance formula to find $\sqrt{(4 - 0)^2 + (7 - 0)^2} = \sqrt{65} \approx 8.1$.

**16. E,** The $x$- and $y$-values of each point have a sum of 5, so $x + y = 5$.

**17. A,** Use the distance formula to find $\sqrt{(-3 + 2)^2 + (-8 + 5)^2} = \sqrt{10} \approx 3.2$.

**18. A,** The graph would pass through $(1, -3)$ because $-3 = -2(1) - 1$ is true.

**19. C,** Use the distance formula to find $\sqrt{(-2 - (-4))^2 + (-4 - 3)^2} = \sqrt{53} \approx 7.3$.

**20. C,** Use the distance formula to find $\sqrt{(4 - (-4))^2 + (4 - 3)^2} = \sqrt{65} \approx 8.1$.

**21. D,** Find $QR$: $\sqrt{(4 - (-2))^2 + (4 - (-4))^2} = \sqrt{100} = 10$. Add all sides to find the perimeter is $7.3 + 8.1 + 10 = 25.4$.

**22. E,** Ordered pair $(4, 5)$ is a solution because $2(4) - 3 = 5$ is true.

**23. C,** The line contains the points $(0, 1)$ and $(1, 3)$. Both points satisfy the equation $2x + 1 = y$.

**24. E,** Point $(3, -2)$ is found on the graph because $-2 = 4 - 2(3)$ is true.

**25. C,** Use the distance formula to find $\sqrt{(4 - 2)^2 + (0 - 4)^2} = \sqrt{20} \approx 4.5$.

**26. D,** Count 7 units up, so the police officer drove 7 miles straight north.

**27. B,** Use the distance formula to find $\sqrt{(5 - (-5))^2 + (5 - 1)^2} = \sqrt{116} \approx 10.8$.

## LESSON 8, pp. 213–217

**1. C,** Use the slope formula to find $\frac{4 - 3}{1 - (-1)} = \frac{1}{2}$.

**2. A,** Pick two points and use the slope formula to find $\frac{5 - 2}{1 - 0} = \frac{3}{1} = 3$.

**3. D,** Use $y = mx + b$ with $m = 3$. Use any point's coordinates to solve for $b$, such as $2 = 3(0) + b$, so $b = 2$. The formula is $y = 3x + 2$.

**4. C,** The slope is $\frac{\text{rise}}{\text{run}} = \frac{2}{32} = \frac{1}{16}$.

**5. C,** If $f(x) = 2$, then, using form $y = mx + b$, $f(x) = 2 + 0x$, so the slope is 0.

**6. E,** When the equation is written in slope-intercept form, $y = -2x + 4$, we can see the slope is $-2$. Since choice E has a slope of $-2$, they are parallel.

**7. A,** The slope of line $H$ is $\frac{3 - 4}{2 - (-4)} = \frac{-1}{6}$.

**8. A,** The slope of the line is $\frac{-4 - (-2)}{-3 - (-1)} = \frac{-2}{-2} = 1$.

**9. B,** The only graph with a slope of $\frac{1}{2}$ is $y = \frac{1}{2}x - 3$.

**10. C,** Solve $-2 = -1(4) + b$ to find that $b = 2$. So, the equation is $y = -x + 2$. If $x = 2$, then $y = 0$, so $(0, 2)$.

**11. A,** The $y$-intercept is $y = 3$, and the slope is $\frac{\text{rise}}{\text{run}} = \frac{4}{-10} = -\frac{2}{5}$. So, the equation is $y = -\frac{2}{5}x + 3$.

**12. E,** Solve $1 = 2(4) + b$ to find $b = -7$, so $y = 2x - 7$.

**13. D,** The initial fee $= b = \$20$, and the slope is $\frac{\text{rise}}{\text{run}} = \frac{30}{1} = 30$. So, $y = 30x + 20$.

**14. B,** The slope of line $T$ is $\frac{-2 - (-3)}{3 - (-3)} = \frac{1}{6}$.

**15. D,** Solve $\frac{1}{3} = \frac{4}{x}$ for $x$ to find that $x = 12$. The span is $12 \times 2 = 24$.

**16. A,** Solve $-3 = -\frac{1}{2}(-4) + b$ to find $b = -5$. So, $y = -\frac{1}{2}x - 5$.

**17. D,** Line $D$ is the only line that rises from left to right.

# UNIT 4 (continued)

**18. E,** Line $E$ is the only horizontal line.
**19. C,** Line $C$ is the only vertical line.
**20. A,** Lines $A$ and $B$ are always an equal distance apart.
**21. A,** The slope would be $\frac{-3-3}{-2-(-4)} = -\frac{6}{2} = -3$.
**22. E,** The only line with slope $\frac{1-(-3)}{3-(-2)} = \frac{4}{5}$ is $y = \frac{4}{5}x - 2$.
**23. D,** The slope would be $\frac{1-3}{3-(-4)} = -\frac{2}{7}$.

---

## LESSON 9, pp. 218–222

**1. D,** The $x$-value of the minimum is given by
$x = \frac{-b}{2a} = \frac{-1}{2\left(\frac{1}{3}\right)} = -\frac{3}{2} = -1.5$.
Substituting $x$ into the equation gives $\frac{1}{3}(-1.5)^2 + (-1.5) - 4 = .75 + -5.5$, so $y = -4.75$. So the coordinate of the minimum is $(-1.50, -4.75)$.
**2. B,** The curve crosses the $x$-axis when $y = 0$. Substituting, and factoring the expression in $x$ gives $(x + 4)(x - 2) = 0$. There are two solutions: $x = -4$ and $x = 2$.
**3. A,** The curve crosses the $y$-axis when $x = 0$. The $y$-intercept is the constant term from the equation, so the answer is $-8$.
**4. B,** The minimum occurs at $x = \frac{-b}{2a}$, where $b = 2$ and $a = 1$ (the value for $x^2$). That gives an $x$-value of $\frac{-2}{2} = -1$.
**5. C,** The curve crosses the $y$-axis when $x = 0$. For this equation, substituting $x = 0$ into the equation of $y = 2x^2 - 5x + 3$ gives $y = 3$.
**6. D,** The coefficient of the $x^2$ term ($a$) is less than 0, which means the function goes through a maximum. The $x$-value of the maximum is given by $\frac{-b}{2a}$. In this case, that is equal to $\frac{-12}{2(-3)} = 2$.
**7. E,** Quadratic equations with negative values of $a$ feature maxima and turn a curve upside-down. Curves $D$ and $E$ are upside-down (negative) curves with maxima.
**8. A,** Quadratic equations with $b = 0$ feature maxima or minima centered on the $y$-axis. Curve $B$ is the only curve centered on the $y$-axis.
**9. D,** Curves achieve maxima or minima at $x$-values equal to $\frac{-b}{2a}$. If $\frac{b}{2a}$ is negative, the $x$-value of the maximum or minimum must be positive on the graph. Curves $C$ and $E$ both fall entirely within a positive $x$ range.
**10. D,** The value of $c$ is the $y$-value at which curves cross the $y$-axis. The curves that cross the $y$-axis at $y = 0$ are Curves $B$ and $D$.
**11. C,** The given point at $y = -2$ has an $x$-value of $+1$, which is three units to the right of the $x$-value of the maximum. A corresponding point on the curve is three units to the left of the maximum, so that $x = -2 - 3 = -5$.
**12. B,** The curve crosses the $y$-axis at $y = c$, where $c$ is the constant term in the equation. In the case of the right-hand curve, that is $y = -24$.
**13. A,** Setting the left-hand expression equal to zero and multiplying through by 2 gives: $x^2 + 8x + 8 = 0$. The equation does not factor, so use the quadratic formula $\left(\frac{-b \pm \sqrt{b^2 - 4ac}}{2a}\right)$. The resulting $x$-values are:

$x = \frac{-8 \pm \sqrt{64 - 32}}{2} = \frac{-8 \pm \sqrt{32}}{2} = \frac{-8 \pm 5.656}{2} = \frac{-4 \pm 2.828}{1}$,
which leads to rounded values of $-6.8$ and $-1.2$.
**14. A,** The equation has the form $y = ax^2 + bx + c$. Because the ball is thrown horizontally, its maximum value occurs at $x = 0$, which means $b = 0$. At $x = 6$, the value of $y$ equals 0. Only choice A satisfies that condition.
**15. D,** Quadratic functions with $a < 0$ are characterized by curves that go through maxima and turn a curve upside down. Of the curves shown, $B$, $D$, and $E$ feature maxima.
**16. B,** If $b = 0$, the location of the maximum or minimum of the curve $\left(x = \frac{-b}{2a}\right)$ also must be zero, regardless of the values of $a$ or $c$. Curves $B$ and $D$ have maxima at $x = 0$.
**17. A,** The maximum or minimum of a quadratic function occurs at $x = \frac{-b}{2a}$. If $\left(\frac{b}{2a}\right)$ is positive, then the maximum or minimum occurs at a *negative* value of $x$. Only Curve $A$ falls entirely within a negative $x$ range.
**18. B,** Negative values of $a$ imply curves that go through maxima. Of the three curves with $a < 0$, curve $B$ is the steepest, so its equation features the most negative value of $a$.
**19. D,** The curve crosses the $x$-axis when $y = 0$. Setting the expression in $x$ to zero and dividing both sides by $-2$ gives: $x^2 + 2x - 3 = 0$. This factors into $(x + 3)(x - 1) = 0$. The solutions to this are $x = -3$ and $x = +1$.
**20. D,** The curve crosses the $y$-axis when $x = 0$, so that $y = -2(0)^2 - 4(0) + 6$. When simplified, only the final term ($c$) remains, so $y = +6$.
**21. D,** The $x$-value of the minimum is given by $\frac{-b}{2a}$ or, in this case, $\frac{-(-4)}{2(-2)} = -1$
**22. B,** The portion of the curve that is given includes the point $(-2, 3)$, four units to the left of the $x$-value where the minimum occurs. The other point where $y = 3$ (the same distance to the right of the minimum) is $x = 6$.
**23. D,** When the second player catches the ball, $y = 6$. Substitute 6 into the equation: $6 = -\frac{1}{144}x^2 + x + 6$. Next, subtract 6 from both sides and multiply through by $-144$: $x^2 - 144x = 0$. The two solutions correspond to the point the ball is thrown ($x = 0$) and the point the ball is caught ($x = 144$). The distance between the two players is 144 ft.
**24. B,** The maximum of the curve is located at $x = \frac{-b}{2a}$, where $b = 1$ and $a = \frac{-1}{144}$. Substituting, the $x$-value of the maximum is 72 feet.
**25. C,** Substituting $x = 72$ into the equation $y = -\frac{1}{144}72^2 + 72 + 6$ gives a $y$-value of 42 feet.
**26. C,** The equation representing the path is a quadratic function, so symmetry with respect to the maximum still exists regardless of how far the ball is thrown. A change that causes a 20-foot increase in the distances in which it achieves its peak height will result in a total distance increasing by 40 feet.

---

## LESSON 10, pp. 223–227

**1. C,** The rate of change is given by the $x$-coefficient. In answer choice C, the $x$-coefficient is 1, greater than the rate of change shown in the graph and less than the rate of change shown in the table.

# UNIT 4 *(continued)*

**2. C,** The function represented in the graph crosses the $y$-intercept at $y = 2$. In answer choice C, the $y$-intercept is 2, which is the same as the $y$-intercept shown in the graph.

**3. B,** The rate of change is 2 and the graph crosses the $y$-axis at $(0, 2)$, so the $y$-intercept is 2. For the ordered pairs, the ratio of vertical change to horizontal change is $\frac{6-2}{0-(-2)} = \frac{4}{2} = 2$. So, the rate of change of the two functions is the same. The ordered pair $(0, 6)$ shows that the $y$-intercept of the function is 6. So, the $y$-intercepts of the two functions are different.

**4. D,** Looking at the graph, when $x = -2$, $f(x) = -2$. Evaluate each function for $x = -2$ and compare. For answer choice D, $f(x) = 6x + 10 = 6(-2) + 10 = -12 + 10 = -2$.

**5. A,** In the table, $y = 0$ at the $x$-intercepts, so the $x$-intercepts are $-2$ and 2. Next, set $f(x) = 0$ for each function and solve for $x$. For answer choice A, $f(x) = \frac{1}{2}x^2 - 2$, so if $f(x) = 0$, $\frac{1}{2}x^2 - 2 = 0$ and $\frac{1}{2}x^2 = 2$. Multiply each side of the equation by 2: $x^2 = 4$, and $x = \pm 2$. So, $f(x) = \frac{1}{2}x^2 - 2$ has the same $x$-intercepts as the function represented in the table. Alternately, evaluate each function for $-2$ and 2. If the value of the function is 0, then the function has the same $x$-intercepts as the function represented in the table.

**6. A,** From the graph, the maximum value of the function is 4. The greatest value shown of function represented in the table $(y = 4, x = 0)$ is 4. Since the value of the function decreases symmetrically as the $x$-value changes, this is the maximum value of the function.

**7. B,** At $x = 0$, the quadratic function has a value of $-1$. At $x = 3$, the quadratic function has a value of 3.5. So, the average rate of change of the function is $\frac{3.5}{3} \approx 1.17$. The rate of change of a linear function is given by the $x$-coefficient. Only answer choice B has an $x$-coefficient less than 1.17.

**8. A,** The rate of change of $f(x) = -2x - 3$ is $-2$. Find the average rate of change of the quadratic equation over each interval and compare to $-2$. In answer choice A, $f(-4) = 7$ and $f(0) = -1$. So, the average rate of change is $\frac{7 - 1(-1)}{-4 - 0} = \frac{8}{-4} = -2$.

**9. D,** The $y$-intercept of the function represented in the graph is 1. The $y$-intercept is the constant term, or $b$. In answer choice D, the $y$-intercept is 1.

**10. C,** The $x$-intercept represented in the graph is 2. The $x$-intercept of a function represented algebraically is the value of $x$ for which the value of the function is 0. Set $f(x) = 0$ for each answer choice and solve for $x$. For answer choice C, $0 = -x + 2$, so $x = 2$.

**11. D,** The maximum value of the function represented in the graph is 5. In answer choice C, the $y$-intercept is 3, so the maximum is 3. In answer choice D, the $y$-intercept is 5, so the maximum is 5.

**12. B,** The speed of Car 1 is represented by the slope of the graph. Choose two points on the graph and find the ratio of the vertical change to the horizontal change. The graph passes through $(3, 150)$ and $(0, 0)$, so slope $m = \frac{150 - 0}{3 - 0} = 50$. Car 1 is traveling at a speed of 50 miles per hour. The speed of Car 2 is represented by the $x$-coefficient of the equation, which is 45.

**13. D,** The slope represented in the graph is 0.4. The rate of change of the function $f(x) = 0.75x - 1$ is 0.75. The rate represented in the graph is less than the rate of the function listed.

**14. B,** The slope represented in the graph is 0.4. Since the slope of the graph of the function $g(x)$ is twice that, its slope is 0.8. The $y$-intercept of the function represented in the graph is $-2$. So, the $y$-intercept of the graph of $g(x)$ is also $-2$. Therefore, $g(x) = 0.8x - 2$. Test each ordered pair to see if it makes the equation true. For answer choice B, $g(-5) = 0.8(-5) - 2 = -4 - 2 = -6$.

**15. E,** The function represented in the graph has a minimum value of $-3$, when $x = -2$. The function represented by the equation $f(x) = x^2 - 2$ has a minimum value of $-2$. Therefore, Alaina is incorrect.

**16. C,** The average rate of change of the function represented in the graph is 2. $f(x) = 2x - 6$ has the same rate of change as the average rate of change of the quadratic function over the given interval.

**17. D,** The rate of change represented in the table is $-0.5$. Therefore, the given function has the same rate of change as the function represented in the table.

**18. B,** The rate of change of the function represented in the table is $-0.5$ and the $y$-intercept is 4, so the function can be represented algebraically as $f(x) = -0.5x + 4$. When the function has a value of 0, $0 = -0.5x + 4$, so $0.5x = 4$ and $x = 8$. Therefore, the $x$-intercept of the function represented in the table is 8. For the function $f(x) = x + 8$, when the function has a value of 0, $0 = x + 8$ and $x = -8$. For the function $f(x) = 0.25x - 2$, when the function has a value of 0, $0 = 0.25x - 2$ and $x = 8$. So, the function $f(x) = 0.25x - 2$ has the same $x$-intercept as the function represented in the table and Sam is correct.

**19. B,** The function represented in the table has a rate of change of $-0.5$. Since the rate of change is negative, the function is decreasing. The rate of change of the function represented by the ordered pairs is $1 - (-1) -2 - (-6) = 24 = 0.5$. So, the function represented by the ordered pairs is increasing at the same rate that the function represented in the table is decreasing.

---

## UNIT 4 REVIEW, *pp. 228–232*

**1. A,** The painter earns \$20 per hour, and the assistant earns \$15 per hour. If $h$ = number of hours, then $20h + 15(h + 5) = 355$.

**2. D,** Since $x^2 = 36$, $\sqrt{x} = 6$ or $-6$. So, $2(6 + 5) = 22$.

**3. C,** Sara's new balance is $\$1,244 + \$287 - \$50 = \$1,481$.

**4. B,** Subtract 0.5 from the previous term 0 to find the next term is $-0.5$.

**5. B,** Let $w$ = the number of women. The number of men is represented by $5 + \frac{1}{2w}$ or $\frac{1}{2w} + 5$.

**6. E,** Solve $3x + 0.15 = 1.29$ for $x$ to find $x = 0.38$.

**7. E,** The slope of the line is $\frac{2 - 0}{3 - 0} = \frac{2}{3}$.

**8. C,** Use point $(0, 0)$ to find $b = 0$. The equation of line $Z$ is $y = \frac{2}{3}x$.

**9. B,** Let $x$ = the number of people under 25. The number of people over 25 is represented by $2x - 56$.

**10. D,** Solve $5 + 1.25x \le 65$ for $x$ to find that $x \le 48$.

**11. C,** Move the decimal point eight places to the left to find $1.496 \times 108$.

**12. B,** The point will be 3 units to the right of the $y$-axis, so the new coordinates are $(3, -2)$.

# UNIT 4 *(continued)*

**13. C,** The slope of $\overline{JL}$ is $\frac{-4+4}{-1+5} = \frac{0}{4} = 0$.

**14. D,** Solve $x(x+1) = 19 + (x + x + 1)$ for $x$ to find that $x = -4$ or $x = 5$. The integers are negative, so the two integers are −4, and −4 + 1 = −3. The greater integer is −3.

**15. B,** The distance between the two points is
$\sqrt{(-5-0)^2 + (4-1)^2} = \sqrt{34} = 5.83$.

**16. A,** The change in Emmit's account is −$64 × 3 = −$192.

**17. D,** Solve $y = \frac{3}{4}(2)$ to find that the missing number is $\frac{3}{2}$.

**18. D,** His new position is 786 − 137 + 542 = +1,191 feet.

**19. A,** The number line shows that $x$ is greater than or equal to 1, so $x \geq 1$.

**20. D,** The cost for one child is $\frac{1}{2}(230) - 30 = \$85$. The cost for three children is $85 × 3 = $255.

**21. B,** Since side $JK$ is 4 units long, the opposite side must be 4 units long as well. So, the coordinates for the fourth point will be (2, 1).

**22. C,** Use points (−2, 1) and (2, 5) to find $m = 1$ and $b = 3$. The equation of this line would be $y = x + 3$.

**23. B,** Solve $0 = -16t^2 - 48t + 160$ for $t$ to find that $t = -5$ or $t = 2$. So, the ball takes 2 seconds to reach the ground.

**24. A,** Let $c$ = the calf's weight; then the mother's weight is $4c + 200$.

**25. C,** Keenan paid $8x + 4.16 = \$73.36$ for the lights. Solve for $x$ to find that each light cost $8.65.

# Index

Note: Page numbers in **boldface** indicate definitions or main discussion and examples. Page ranges indicate practice.

random sample, **146**, 147–150
sample population, **146**, 147–150
**Data analysis**
bar and line graphs, **151**, 152–155
box plots, **166**
circle graphs, **156**, 157–160
correlation, **151**, 152–155
dot plot, **166**
frequency tables, **161**, 162
histograms, **161**, 163, 164
lower and upper quartiles, **166**, 167–170
matrices, **141**, 142–145
maximum and minimum values, **166**, 167–170
mean, **136**, 137–140
measure of variation, **136**, 137–140
measures of central tendency, **136**, 137–140
median, **136**, 137–140
mode, **136**, 137–140
normal and skewed distribution, **166**, 167–170
probability, **131**, 132–135
range, **136**, 137–140
scatter plot, **151**, 152, 153
tables, **141**, 142–145, **161**, 162–165
**Data table, 125**, **141**, 142–145, **161**, 162–165
**Decimal point**
in percents, **37**
placement of, 2, 32
in scientific notation, 52
**Decimals, 1**, 2
comparing and ordering, 2–6
converting to and from fractions and percents, **37**, 38–41
fractions written as, 2
mixed numbers written as, **37**
multiplying by ten, 2
operations with, **32**, 33–36
place value, **22**, 23–27
probability expressed as, 131
rounding, 2
values on circle graphs as, 156
**Denominator, 1**
in fractions, **8**
slope of a line, 213
**Dependent events, 131**
**Diameter, 63**, 74, 75–78
**Difference, 1**, **4**
**Dilations, 177**, 203
**Dimension, 125**, **141**
**Direction of vectors**, 203
**Distance between two points**
finding with Pythagorean theorem, **89**, 90–93
formula for, 208
using scale drawings, 99–103

**Distance measurements**, 32
**Distribution of values, 166**, 167–170
**Distributive property, 177**, 178
**Dividend, 1**, **4**
**Division**
with decimals, 32–36
with exponents, **52**, 53–56
of fractions, 22–26
of integers, 17–21
inverse operation of, 183
order of terms, 178
of whole numbers, **4**, 5–8
**Divisor, 1**, **4**, 32
**Dot plots, 125**, **166**, 167–170

# E

**Edge**, 109
**Endpoint, 79**
**Equal sign** (=), 2
**Equations, 177**
functions written as, 188
graphing, **208**, 209–212, **218**, 219–222
one-variable, **183**, 184–187
quadratic, **193**, 194–197
slope-intercept form of a line, **177**, 213, 214–217
solving, **183**, 184–187
two-variables linear, **208**, 209–212
**Equilateral triangles, 63**, 84
**Equivalent equations**, 183
**Equivalent units of measure**, 64
**Estimating answers**, 12, 74
**Evaluating algebraic expressions, 178**, 179–182
**Events, 125**
dependent and independent, **131**, 132–135
impossible and certain, **131**, 132–135
likely and unlikely, **131**, 132–135
**Expanded form, 52**
**Experimental probability, 125**, **131**, 132–135
**Experimental study, 146**, 147–150
**Exponents, 1**, **52**
negative and positive, 52
operations with, **52**, 53–56
in scientific notation, **52**, 53–56
zero and one as, 52
**Expression.** *See* Algebraic expressions
**Exterior angles, 63**, **79**, 80–83

# F

**Faces**
of cone, 109
of pyramids, 109
**Factoring, 193**, 194–197
**Factors, 177**, **193**, 194–197
**Fluid ounces, 64**
**FOIL method, 193**, 194–197
**Foot, 64**
**Formulas**
for area of a circle, 74
for area of a triangle, 104
for area of polygons, 69
for base of triangular prism, 104
for circumference, 74
for distance between two points, 208
for maximum or minimum of quadratic equations, 218
for percent, 42
Pythagorean theorem, 63, 89
for slope of a line, 213
for subgroup of *x*, 126
for surface area of a cylinder, 104
for volume of a cylinder, 104
**Fractions, 1**
converting to and from decimals and percents, **37**, 38–41
improper, 22
mixed numbers, 22, **37**
operations with, **22**, 23–26
probability expressed as, 131–135
ratios written as, 27
reducing, 22
values on circle graphs as, 156
written as decimals, 2
**Frequency, 125**, **161**, 162–165, 166
**Frequency tables, 125**, **161**, 162
**Functions, 177**, **188**, 189–192
comparing, **223**, 224–227
**Fundamental counting principle, 125**, **126**, 127–130

# G

**Gallon, 64**
**Geometric patterns, 188**, 190, 191
**Geometry**
angles, **63**, **79**, 80–83
circles, 74, 75–78
composite figures, **114**, 115–118
congruent figures, **94**, 95–98
indirect measurement, **63**, 64–67, **89**, 90–93
lines, **63**, **79**, 80–83

# Q

Quadrants (of coordinate grid), **177**, **203**

Quadratic equations, **177**
  factoring, **193**, 194–197
  graphing, **218**, 219–222

Quadratic expressions, factoring, **193**, 194–197

Quadrilaterals, **63**, **84**, 85–88

Qualitative data, **125**, **146**, 147–150

Quantitative data, **125**, **146**, 147–150

Quart, 64

Quotient, **1**, 4

# R

Radius
  of circle, **63**, **74**, 75–78
  of sphere, 109

Random sample, **125**, **146**, 147–150

Range, **125**, **136**, 137–140

Rate, **1**, **27**, 28–31, **42**, 43–46

Rates of change, **223**, 224–227

Ratios, **1**, **27**, 28–31, **125**, 132–135
  probability expressed as, 131
  in scale drawings, 99–103
  slope of a line as, **223**, 224–227
  unit rates, **27**
  written as fractions, 27

Rays, **63**, **79**, 80–83

Reciprocal, 22, 52

Rectangles
  area of, **69**, 70–73
  perimeter of, 32
  sides and angles of, 84, 85–88

Rectangular prism, 104, 105–108

Reducing fractions, 22

Reflections, **177**, 203

Regrouping, 4

Relations, **188**, 192

Rhombus, **63**, 84–88

Right triangles, **63**, 84
  Pythagorean theorem, **89**, 90–93
  sides and angles of, **89**, 90–93

Rise, **177**, 213, 223

Root of a number, **47**, 48–51

Root signs, 47

Rounding numbers, 2

Rows of matrices, 141

Rule for mathematical patterns, 188

Run, **177**, 213, 223

# S

Sample population, **125**, **146**, 147–150

Scale, **99**, 151

Scale drawings, **99**, 100–103

Scalene triangles, **63**, **84**

Scatter plot, **125**, **151**, 152, 153

Scientific notation, **1**, **52**, 53–56

Sequence of numbers, **188**, 189–192

Sides
  classifying triangles by, **84**
  of congruent and similar figures, 94
  of quadrilaterals, **84**
  of right triangles, **89**, 90–93
  of trapezoids, **84**

Similar figures, **63**, **94**, 95–98

Simplifying algebraic expressions, **178**, 179–182

Skewed distribution, **125**, **166**, 167–170

Slant height, **109**, 110–113

Slope-intercept form of a line, **177**, 213, 214–217

Slope of a line, **177**, **213**, 214–217
  comparing functions, **223**, 224–227
  finding from two points, **213**
  of parallel lines, **213**
  positive, negative, and zero, 213
  rate of change, **223**, 224–227
  of a vertical line, 213

Solid figures, **63**
  composite figures, **114**, 115–118
  cones, 104, **109**, 110–113, **114**
  cubes, 104
  cylinders, **104**, 105–108
  lateral area, **104**, 105–108
  prisms, **104**, 105–108
  pyramids, **63**, 104, **109**, 110–113
  rectangular prism, 104, 105–108
  slant height, **109**
  spheres, **109**, 110–113
  square pyramids, **63**, 109, 110–113
  surface area, **104**, 105–108, **109**, 110–113, **114**, 115–118
  triangular prism, **104**, 105–108
  volume of, **63**, **104**, 105–108, 109, 110–113, **114**, 115–118

Solution, **177**

Solution sets for inequalities, **198**, 199–202

Solving
  equations, **183**, 184–187
  inequalities, **198**, 199–202
  linear equations, **183**, 184–187
  for missing length of similar figures, 94
  multiple-step word problems, 69
  proportions, 27–31
  quadratic equations, **193**, 194–197
  word problems, 12–17

# S (continued)

Spheres, **109**, 110–113

Square of a number, **1**, **47**, 48–51, 208

Square pyramids, **63**, 109, 110–113

Square roots, **1**, **47**, 48–51

Squares, 85–88
  area, **69**, 70–73
  sides and angles of, **84**, 85–88

Square units, 69, 74

Standard deviation, 236

Statistics, **236**. *See also* data; data analysis

Subtraction
  with decimals, 32–36
  with exponents, **52**, 53–56
  of fractions, 22–26
  of integers, 17–21
  inverse operation of, 183
  order of terms, 178
  of whole number, **4**, 5–8

Sum, **1**, **4**, 5–8

Supplementary angles, **63**, **79**, 80–83

Surface area, **63**
  of composite figures, **114**, 115–118
  of a hemisphere, **109**, 110–113
  of prisms and cylinders, **104**, 105–108

Survey, 146, 147–150, **236**

Symbols
  angles, 79
  congruence, 94
  equal sign, 2
  functions, 188
  greater than or equal to symbol, 198
  greater than symbol, 2, 198
  inequalities, 198
  "is congruent to," 94
  "is similar to," 94
  less than or equal to symbol, 198
  less than symbol, 2, 198
  lines and rays, 79
  measure of an angle, 79
  multiplication, 178
  percents, **37**
  similarity, 94

Symmetry, 218

Systems of measure
  metric system, **64**, 65–69
  U. S. customary system, **64**, 65–69

# T

Tables, **161**, 162–165
  comparing functions, **223**, 224–227
  frequency tables, **161**, 162
  title and column headings, 161

Tens place, 2

Tenths, 2

**INDEX**

# X

*x*-axis, 203, 213
*x*-intercept
    of a function, 223
    of a line, **177**, 213, 214–217
*x*-value, 188, 203

# Y

**Yard**, 64
*y*-axis, 203, 213
*y*-intercept, **177**
    of a function, 223
    of a line, 208, 213, 214–217
    of quadratic equations, 218–222
*y*-value, 188, 203

# Z

**Zero**
    as a placeholder, 2, 4
    as a power of a number, 52
    as slope of line, 213